茅以升

MAOYISHENG
QUANJI

全集

[第2卷]

桥梁工程（下）

◎ 北京茅以升科技教育基金会 主编

U0359320

天津出版传媒集团

天津教育出版社

TIANJIN EDUCATION PRESS

图书在版编目（ＣＩＰ）数据

桥梁工程.下 / 北京茅以升科技教育基金会主编.
-- 天津 ：天津教育出版社，2015.12
　（茅以升全集；2）
　ISBN 978-7-5309-7818-4

　Ⅰ．①桥… Ⅱ．①北… Ⅲ．①桥梁工程—文集 Ⅳ.
①U44-53

中国版本图书馆CIP数据核字（2015）第191708号

茅以升全集 第2卷　桥梁工程（下）

出 版 人	胡振泰
主　　编	北京茅以升科技教育基金会
选题策划	田　昕
责任编辑	孙丽业
装帧设计	郭亚非

出版发行	**天津出版传媒集团** 天津教育出版社 天津市和平区西康路35号　邮政编码　300051 http://www.tjeph.com.cn
经　　销	新华书店
印　　刷	北京雅昌艺术印刷有限公司
版　　次	2015年12月第1版
印　　次	2015年12月第1次印刷
规　　格	32开（880毫米×1230毫米）
字　　数	460千字
印　　张	24
印　　数	2000
定　　价	55.00元

目

CONTENTS

录

茅以升
全集
❷

学位论文

ZhongGuo GuQiao JiShu

中国古桥技术

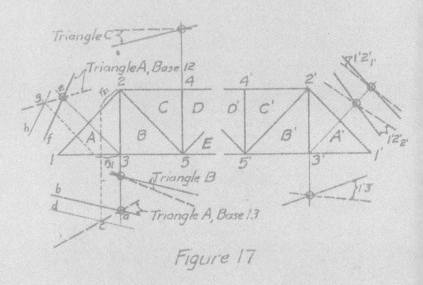

Figure 17

中国石拱桥

　　石拱桥的桥洞成弧形，就像虹。古代神话里说，雨后彩虹是"人间天上的桥"，通过彩虹就能上天。我国的诗人爱把拱桥比作虹，说拱桥是"卧虹""飞虹"，把水上拱桥形容为"长桥卧波"。

　　石拱桥在世界桥梁史上出现得比较早。这种桥不但形式优美，而且结构坚固，能几十年几百年甚至上千年雄跨在江河之上，在交通方面发挥作用。

　　我国的石拱桥有悠久的历史。《水经注》里提到的"旅人桥"，大约建成于公元 282 年，可能是有记载的最早的石拱桥了。我国的石拱桥几乎到处都有。这些桥大小不一，形式多样，有许多是惊人的杰作。其中最著名的当推河北省赵县的赵州桥，还有北京附近的卢沟桥。

　　赵州桥横跨在洨河上，是世界上最伟大的古代石拱桥，也是建成后一直使用到现在的最古的石桥。这座桥修建于公元 605 年左右,到现在已经一千三百多年了,还保持着原来的雄姿。到解放的时候,桥身有些残损了,在人民政府的领导下,经过彻底整修,这座古桥又恢复了青春。

赵州桥非常雄伟,全长五十点八二米,两端宽约九点六米,中部略窄,宽约九米。桥的设计完全合乎科学原理,施工技术更是巧妙绝伦。唐朝的张嘉贞说它"制造奇特,人不知其所以为"。这座桥的特点是:(一)全桥只有一个大拱,长达三十七点零二米,在当时可算是世界上最长的石拱。桥洞不是普通半圆形,而是像一张弓,因而大拱上面的道路没有陡坡,便于车马上下。(二)大拱的两肩上,各有两个小拱。这个创造性的设计,不但节约了石料,减轻了桥身的重量,而且在河水暴涨的时候,还可以增加桥洞的过水量,减少洪水对桥身的冲击。同时,拱上加拱,桥身也更美观。(三)大拱由二十八道拱圈拼成,就像这么多同样形状的弓合拢在一起,做成一个弧形的桥洞。每道拱圈都能独立支撑上面的重量,一道坏了,其他各道不致受到影响。(四)全桥结构匀称,和四周景色配合得十分和谐;就连桥上的石栏石板也雕刻得古朴美观。唐朝的张𬸦说,远望这座桥就像"初月出云,长虹饮涧"。赵州桥高度的技术水平和不朽的艺术价值充分显示了我国劳动人民的智慧和力量。桥的主要设计者李春就是一位杰出的工匠,在桥头的碑文里还刻着他的名字。

永定河上的卢沟桥,在北京附近,修建于公元1189~1192年间。桥长二百六十五米,由十一个半圆形的石拱组成,每个石拱长度不一,自十六米到二十一点六米。桥宽约八米,路面平坦,几乎与河面平行。每两个石拱之间有石砌墩,把十一个石拱连成一个整体。由于各拱相联,所以这种桥叫做联拱石桥。永定河发水时,来势很猛,以前两岸河堤常被冲毁,但这座桥却从没出过事,足见它的坚固。桥面用石板铺砌,两旁有石栏石柱。每个柱头上都雕刻着不同姿态的狮子。这些石刻狮子,有的母子相抱,有的交头接耳,有的像倾听水声,千态万状,惟妙惟肖。

早在13世纪,卢沟桥就闻名世界。那时候有个意大利人马可·波罗来过中国,他的游记里,十分推崇这座桥,说它"是世界上独一无二的",并且特别欣赏桥栏柱上刻的狮子,说它们"共同构成美丽的奇观"。在国内,这座桥也是历来为人们所称赞的。它地处入都要道,而且建筑优美,"卢沟晓月"很早就成为北京的胜景之一。

卢沟桥在我国人民反抗帝国主义侵略战争的历史上,也是值得纪念的。在那里,1937年日本帝国主义发动了对我国的侵略战争。全国人民在中国共产党领导下英勇抗战,终于彻底打败了日本帝国主义。

　　为什么我国的石拱桥会有这样光辉的成就呢？首先，在于我国劳动人民的勤劳和智慧。他们制作石料的工艺极其精巧，能把石料切成整块大石碑，又能把石块雕刻成各种形象。在建筑技术上有很多创造，在起重吊装方面更有意想不到的办法。如福建漳州的江东桥，修建于八百年前，那座桥有的石梁一块就有二百吨重，究竟怎样安装上去的，至今还不完全知道。其次，我国石拱桥的设计施工有优良传统，建成的桥，用料省，结构巧，强度高。再其次，我国富有建筑用的各种石料，便于就地取材，这也为修造石桥提供了有利条件。

　　两千年来，我国修建了无数杰出的石拱桥。解放后，全国大规模兴建起各种形式的公路桥与铁路桥，其中就有不少石拱桥。1961 年，云南省建成了一座世界最长的独拱石桥，名叫"长虹大桥"，长达一百一十二点五米。在传统的石拱桥的基础上，我们还造了大量的钢筋混凝土拱桥，其中"双曲拱桥"是我国劳动人民的新创造。我国桥梁事业的飞跃发展，表明了我国劳动人民的勤劳勇敢和卓越才能。

原载 1962 年 3 月 4 日《人民日报》

洛阳桥与江东桥

    洛阳桥与江东桥都是福建的古桥，一在泉州，一在漳州。这两地的古桥真是多，据《古今图书集成》所引这两州的《府志》，其有名称、地址及事迹可考的，泉州本府有桥六十四座，所辖各县有桥一百三十二座；漳州本府有桥五十八座，所辖各县有桥一百三十九座。不但泉漳两州如此，福建其他各地的桥也多，比如，建宁本府就有桥一百九十四座，其中有三座曾为13世纪的意大利人马可·波罗在他的游记中称道过。福建的桥，不但数目多，而且有不少宏伟的结构，凡是到过福建的人，都会感到"闽中桥梁甲天下"（《闽部疏》）之说，确非过誉，同时也会承认"泉州桥梁甲闽中"。如果在这里，泉州是"冠军"，那么，漳州应当是"亚军"。洛阳桥是泉州桥的代表作，江东桥是漳州桥的代表作。

    且看《泉州府志》是如何记载这些巨大桥梁的。万安桥

（即洛阳桥），在洛阳江，宋蔡襄造，"长三百六十余丈，广一丈五尺左右"；安平桥（即五里桥），在安海港，宋绍兴间僧祖派造，"长八百有十一丈，广一丈六尺"；石笋桥，在笋江，宋绍兴间僧文会造，"长八十余丈"；顺济桥，在笋江下游，宋嘉定间造，"长一百五十余丈"。以上是著名的泉州四大名桥。此外，还有：苏埭桥，宋绍兴间僧守徽建，"桥凡四，共长二千四百余丈"；玉澜桥，宋绍兴间僧仁惠修，"跨海千余丈"；普利大通桥，宋绍兴间造，"长二百丈"；北平桥，宋绍兴间造，"长百丈有奇"；龙津桥，宋庆元间造，"长六十八丈"；獭窟屿桥，在大海中，宋开禧间僧道询建，"六百六十间，直渡海门，凡五里许"；金鸡桥，嘉定间僧守静建，"长一百丈有奇"；通济桥，宋时建，"长一百八十九丈"；宏济桥，宋时建，"长一千三百丈有奇"；下辇桥，元至正僧法助建，"凡六百二十间"。这些记载，有的显然失实，不能置信，也许是把水中堤道混作桥梁之故，但就现存的各桥来说，经过勘测，其原来面目，确属惊人。

我国各省地方志，留下许多珍贵史料，实是一笔丰富遗产。可惜的是，对于工程技术，语焉不详，一座桥只记它有多长多宽，有多少孔，至于何种结构，如何施工，一般就都不提了。幸亏福建各桥几乎全部都是石头造成的，坚固耐久，虽时隔千年而规模犹在，把今天眼前的桥和《府志》里的记载，对照来看，还不难得其真相。这样一对照，就可发现关于泉漳两地的桥梁有许多值得深入研究的问题。

第一，泉州的古老大桥，除去洛阳桥外，几乎全都是南宋时期兴建的，特别是绍兴年间（公元 1131~1162 年）的更多，这是什么原因呢？泉州得名，由于当地有泉山，"上有石乳泉，清洁甘美"，这对滨海居民是非常可贵的，然而也可见那时人口之少。后来西汉时代的统治者为了镇压革命，还"迁其民于江淮间"，人就更少了。到了东晋，北来的人口大增，流经泉州的晋江，据说就是由于"晋南渡时衣冠士族避于此，沿江而居"而得名的。再后来，到了宋室南渡，北方人民更是大批地移居江南，从浙江到福建的人，沿着海边平原而定居在泉漳等地的，就更多了。泉州自南朝陈武帝永定二年（公元 558 年）起，就开始了海外交通，到了唐代更成为对外贸易的广州、泉州、扬州、交州的四大港之一。在南宋绍兴年间，经济更为繁荣，交通上的桥梁需要，当然日益迫切，而且又有物力、人力上的可能条件，于是在泉州出现了桥梁勃兴时代，这对当时全国的桥梁来讲，也是异常突出的。元代以后，泉州的经济便由稳定而趋于衰落，在明清两代，对于大桥就只有修理，而没有什么新建的，至多不过造些小桥。

第二，泉州的桥梁，由僧人修建的特别多，在泉州本府的六十四座桥梁中，由僧人修建的就有二十座，而且所有的大桥，他们都有份，这又是什么原因呢？过去把修桥造路当作是做"功德"，和尚募缘修庙造桥，原不足奇，有时他们中也出现有工程师，如修理赵州桥的僧怀丙，但像泉州有这样多的僧人，负责建造这样多的桥，却是别处所无有的。在这些僧人中，有的技术特别高，在《府志》中都留下了他们的名字，如义波、宗善、祖派、道询等都是杰出的工程师。元朝的僧人法助，一个人就主持造了七座桥，此外，间或也有个别道士，参加工事。我很怀疑，《府志》中的所谓"僧人"是否都是佛教徒。在西安碑林中，有唐代的《大秦景教流行中国碑》，其文中所谓"僧"，实际都是天主教徒，而且他们也有汉名，看不出与佛教徒的区别。

第三，泉州大桥的长，确是长得惊人。这是由于地处海滨，有的桥不是过江而是在江口跨海的。比如安平桥，从晋江县安海市，跨海而与南安县的水头镇相接，宋绍兴八年（公元1138年），僧人祖派等倡造石桥，绍兴二十一年续建，经一年时间而完成。这座桥俗名五里桥，不是距城五里，而是五里长之故。据实测，这桥的现存长度两千零七十米，桥宽三到三点八米，桥墩三百一十四座，全用花岗石造成。对这座桥，当地流传着"天下无桥长此桥"的话。确实，直到今天，除去郑州黄河铁路桥比它较长外还没有比它更长的桥，它确是为泉州赢得了福建桥梁"冠军"的一个"健将"。当然，桥长并不意味着它一定是艰巨的，然而泉州这些大桥都是用石头在波涛险恶的江海里造成的,工程浩大是可想而知的。那时没有机械,全凭人工操作,一块石头重达几吨,几十吨,甚至几百吨,是如何从山上搬到桥上的呢?

上面关于泉州的这些问题,也适用于漳州。漳州得名,由于漳水,水以漳名,"取其清浊相杂,而有文章"。漳字从"泣"从"早",以前写匾额,恐书法"犯忌","不得已为篆书"(《闽部疏》)。因此,在漳州,关于桥梁故事的神奇传说,也不亚于泉州。

现在简单介绍一下关于洛阳桥与江东桥的故事。

洛阳桥位于晋江、惠安两县夹界的洛阳江入海处。洛阳江的得名由于唐宣宗未即位时(公元847年以前)避居泉州,"微行览山水胜概,有类吾洛阳之语"(《漳州府志》),洛阳桥,宋时名万安渡石桥,宋蔡襄的《万安桥记》云"始造于皇祐五年(公元1053年)四月庚寅,以嘉祐四年(公元1059年)十二月辛未讫功,累址于渊,酾水为四十七道(桥孔),梁空以行,其长三千六百尺,广丈有五尺"。造桥时先于水中抛石,铺满桥址,形成水下"海堤",然后在上面筑墩,同时,"种蛎于础以为固"(《宋史·蔡襄传》),利用浅海里"蛎房"的繁殖,把石基胶固,使成整体。就因首创了这个筑墩方法,洛阳桥成为后来泉漳各桥的先行者和带动者。关于这桥的神话很多,最著名的是,蔡襄造此桥,"限以涛势,不能案址(建基),

乃檄江神,得一醋字,公云,二十一日酉时为之"(《泉南杂志》)。蔡襄是状元,状元造桥,流为佳话。戏剧中有《洛阳桥》灯彩戏,其主题是形容桥成后,"三百六十行过桥"时人民的欢乐情景。可惜这出戏早就停演了,能否编一出《新洛阳桥》,来描写我国从资本主义过渡到社会主义的桥梁,显示出新的三百六十行的全国人民的欢欣鼓舞呢? 洛阳桥完成后,先后经过修理和重建十六次,但大修不过四次,一为飓风损坏,二为大水冲毁,三为地震和大水倾毁,四为连年地震、大水和飓风所毁,于1761年修复,即现存的石梁桥。明末时,郑成功曾据此桥,以抗清兵。(《读史方舆纪要》)1932年在桥上添建了钢筋混凝土的公路桥面,失去了本来面目。

江东桥在漳州东四十里柳营江上,根据《漳州府志》,其地"为郡之寅方,因名虎渡",而桥亦名虎渡桥。原系板桥,"垒石为址,酾为十五道",后来宋嘉熙二年(公元1238年)改建,"以石为梁","四年而桥告成,长二百余丈"。这桥的最大特色是最大的桥孔上,只用三根巨大石梁跨过,每根石梁"长八十尺,广博皆六尺有奇"(宋黄朴《虎渡桥记》)。当地人称这石梁为"一根扁担厚",估计最大的一根梁的重量约为两百吨,其开采琢制,固已不易,而如何运来江边,架到墩上,至今还是个不解之谜。这样巨大的石梁,可算是漳州在福建桥梁中赢得"亚军"的一个"健将"。参加江东桥工程的也有"佛者廷浚,与其徒净音、德屋等"(《虎渡桥记》)。关于这座桥,也有神话:"江南桥梁,虎渡第一,昔欲为桥,有虎负子渡江,息于中流,探之有石如阜,循其脉沉石绝江,隐然若梁,乃因垒址为桥,故名虎渡。"(《读史方舆纪要》)江东桥曾经屡坏屡修,现存的是清康熙四十八年(公元1709年)重修的。1933年在老墩上筑新墩,新墩上架钢筋混凝土桁梁,支持公路桥面,使原来结构完全改观了。

泉州洛阳桥、漳州江东桥等等的"甲天下"的闽中桥梁,都是福建人民的光荣,中国人民的骄傲!

原载1962年4月15日《人民日报》

洛阳桥

桥名谈往

万物皆有名，有的还要有专名，就像人有名字一样。既然是名，当然就要起得好。孔夫子不是早就说过吗，"必也正名乎""名不正则言不顺"。为了名要"正"，就费了不知多少人的心血。比如人名，在《永乐大典》或《四库全书》中，就有不少关于它的汇考、艺文、纪事、杂录等等的记载。至于对物题名，也是非同小可，既要意义恰当，又要工整大方。《红楼梦》"大观园试才题对额"一回中，贾政说："这匾对倒是一件难事，论理该请贵妃赐题才是……偌大景致，若干亭榭，无字标题，任是花柳山水，也断不能生色。"一座花园里的题名是件难事（要论理，要生色），那么，一座桥梁的题名，也就不简单了。

我国近代桥梁，受了西方影响，题名时，总是从地理观点出发的。只要能指出它的所在地，使人一望而知，这个名就算"正"了。铁路公路上，更是用里程标记作名字，如同某某路上的"345,678 公里桥"，那才真是确切不移的。然而我国古时桥名，不是这样。它总要有些文学气息，使人见了，不由得发生情感，念念不忘。或是记事抒情，引起深思遐想；或有诗情画意，为之心旷神怡。这样，通过慎重题名，一座桥的历史、作用或影响，就立刻表现出来，因而容易流传。桥的"身价"也因此而抬高。桥出了名，它的名字还会跟着多起来，除了正名，还有俗名、别名等等，就像人名，除了学名，还有别名、小字等等。有的是在民间自然而然地逐渐形成的，有的却是文人学士，要借此而为自己题名的。总之，桥成就要题名，成为风气，也是我国古代文化的一个特色。

桥的题名，字不在多，如同人名一样，一般都是两个字，有时只有一个字。就只这一两个字，而能显示出桥的特征，正是我国文字的妙用。这是由于我国历史上的典故多，和文学里的成语丰富的缘故。文史里的财富，大为桥名增光。然而桥多了，关于它的典故和成语也反过来为文史服务。比如，《史记》里"信如尾生"一词来自桥的典故，《阿房宫赋》里"长桥卧波"一词来自桥的成语。桥的名字题得好，它对文史就可有贡献了。桥名的重要，有如此者。

现在来介绍一些桥名，借以窥知我国桥梁文学的丰富，它也许是世界无双的。先谈单名。较著名的有鹊桥，神话传说，"七夕织女当渡河，使鹊为桥"（《风俗记》）；蓝桥，在陕西蓝田县蓝溪上，"传其地有仙窟，即唐裴航遇云英处"（《清一统志》）；枫桥，在苏州，唐张继有《枫桥夜泊》诗，又唐杜牧诗"长洲茂苑草萧萧，暮烟秋雨过枫桥"；断桥，在杭州西湖，唐张祜诗"断桥荒藓合"，明莫仲玙有《断桥残雪》词；虹桥，宋张择端画的《清明上河图》里有河南开封的虹桥等等。但单名之桥往往不限一地，其中以材料或形状为名的则更多，如石桥，梁简文帝即有《石桥》诗"写虹便欲饮，图星逼似真"；铁桥，明吴兆元有《渡铁桥》诗"宝筏群生渡，金绳八道开"；板桥，即木桥，唐温庭筠诗"鸡声茅店月，人迹板桥霜"；浮桥，古称"河桥"，晋杜预传，"预曰'造舟为梁'，则河桥之谓也"，唐太宗有《赋得浮桥》诗等等。又有泛指桥的所在或形状而有可能是专名的，如山桥，梁简文帝诗"卧石藤为缆，山桥树作梁"；江桥，唐杜甫诗"山县早休市，江桥春聚船"；野桥，唐刘长卿诗"野桥经雨断，涧水向田分"；市桥，杜甫诗"市桥官柳细，江路野梅香"；方桥，唐韩愈诗"君欲问方桥，方桥如此作"；斜桥，宋欧阳修诗"波光柳色碧溟濛，曲渚斜桥画舸通"；画桥，宋范与求诗"画桥依约垂杨外，映带残阳一抹红"；朱桥，唐郑谷诗"朱桥直抵金门路，粉堞高连玉垒云"；双桥，唐李白诗"两水夹明镜，双桥落彩虹"等等。有的桥名很自然，如天桥，山西太原保德州及云南大理都有，大理的"下断上连,石梁跨之,两岸激水溅珠,宛如梅绽,人呼为不谢梅"(《云南通志》);花桥,福建宁德县,湖北长阳县及广西桂林都有,桂林的有"花桥烟雨"之称;柳桥,在杭州西湖,宋周邦彦词"水涨鱼天拍柳桥";竹桥,杜甫有《观造竹桥》诗等等。有的比较特殊,如草桥,在北京右安门外;席桥,在山东东平县,"相传宋真宗东封泰山,车驾经行,以席铺借"(《山东通志》);瓜桥,浙江富阳县,"世传孙钟设瓜于此桥"(《浙江通志》);鸭桥,在陕西陕城,见《初学记》;金桥,在山西上党,唐潘炎有《金桥赋》,序云"常有童谣云,圣人执节渡金桥"等等。更有奇特的如暗桥,在安徽建平县,"旧传伍员奔吴,避于山中,追者至此,云气护之,员及桥而天暗"(《江南通志》);鬼桥,《初学记》"上方有鬼桥";赤桥,在山西太原晋水北渠上,"宋太宗凿卧龙山,血出成河,因更今名"(《山西通志》)等等。

桥名用两个字是最普遍而又标准化的,单名的桥已经不少,双名的更是多得多。试思每桥皆有名,在我们古老的大国,该有多少桥名啊!然而在这成千上万的单名和双名中,重复的究竟不多。如果把这所有的桥名都搜集起来,编成一部《中国桥名录》,该是够洋洋大观的了。

现在再来举一些双名的例,说明桥名的丰富多彩。根据反映内容,一部《桥名录》可分为六章。

第一章是"表扬"，首先是表扬桥成德政的，如安济桥，即赵州桥，在河北赵县南洨河上，一名大石桥，制造奇特，"隋匠李春之迹也"，见唐张嘉贞《赵郡南石桥铭》；万安桥，即洛阳桥，在福建泉州，为渡海用，"去舟而徒，易危以安，民莫不利"，见宋蔡襄《万安渡石桥记》；灭渡桥，在江苏吴县，桥成"南北往来者踊跃称庆，名灭渡，志平横暴也"，见元张亨《灭渡桥记》；安平桥，在福建晋江，建成于宋绍兴二十二年，全长两千零七十米，俗名五里桥，旧有"天下无桥长此桥"的传说；弘仁桥，在北京左安门东，明李贤《敕建弘仁桥碑记》云"为建石桥，以便往来……名之曰弘仁，盖弘者廓而大之，而仁则不忍人之政也"；万福桥，在湖南湘乡县，上部桥墩似楹柱，为石桥中罕见。其次是表扬造桥人物和轶事的，如济美桥，在安徽天长，是"父作子述"之桥，见明陈继儒《济美桥记跋》；玉带桥，在江苏苏州，"今呼为小长桥，相传为唐王仲舒建，捐宝带助费，故名"（《苏州府志》）；绩麻桥，在湖北孝感县，"世传居民女绩麻所建"（《湖广通志》）；夫妇桥，即四川灌县竹索桥，清何先德造，未完，其妻续成之；葛镜桥，在贵州平越，明万历间葛镜建，"屡为水决，三建乃成，靡金巨万，悉罄家资"（《贵州通志》）；大夫桥，在浙江会稽，唐颜真卿《张志和碑铭》云"门墙流水，十年无桥，观察使陈少游为建造，行者谓之大夫桥"；葫芦桥，在浙江余杭，相传东汉时隐士张俨"植葫芦以贸,积钱造桥,故名"(《浙江通志》)。

第二章是"记事",记载有关桥上的流传故事,如万里桥,在四川成都南门外,"昔孔明于此饯费祎聘吴,日万里之行,始于此矣"(见《寰宇记》),又据《四川通志》云"唐史载,明皇狩蜀过此,问桥名,左右对以万里,明皇叹曰,开元末僧一行谓,更二十年国有难,朕当远游至万里之外,此是也。因驻跸于成都焉",唐陆肱有《万里桥赋》,宋吕大防有《万里桥》诗,杜甫诗"万里桥西宅,百花潭北庄",唐张籍诗"万里桥边多酒家,游人爱向谁家宿";宋苏轼诗"我欲归寻万里桥,水花风叶暮萧萧",宋陆游诗"雕鞍送客双流驿,银烛看花万里桥";驷马桥,即升仙桥,在四川成都城北,《四川通志》云"司马相如尝题柱云,大丈夫不乘驷马车,不复过此桥",唐岑参有《升仙桥》诗"及乘驷马车,却从桥上归",宋京镗有《驷马桥记》云"兹建桥以驷马名,自是长卿之遗踪亦不泯矣";兰亭桥,在浙江绍兴,"晋王右军修禊处,桥下细石浅漱,水声昼夜不绝"(《浙江通志》);百口桥,在江苏苏州,"汉顾训五世同居,聚族百口,故因其居名桥"(《江南通志》);洗耳桥,在河南汝州,"即许由洗耳处"(《河南通志》);虎渡桥,在福建漳州,亦名江东桥,"江南桥梁,虎渡第一,昔欲修桥,有虎负子渡江,息于中流……乃因垒址为桥"(《读史方舆纪要》);宵市桥,在江苏扬州,即小市桥,"相传隋炀帝时于此开夜市"(《江南通志》）。

第三章是"抒情"，通过桥名，来表达思想感情，如销魂桥，即灞桥，在陕西西安，"东汉人送客至此桥，折柳赠别"（《三辅黄图》），因"取江淹别赋句，又呼为销魂桥"（《陕西通志》），唐王之涣诗"杨柳东风树，青青夹御河，近来攀折苦，应为别离多"，宋柳永词"参差烟树灞陵桥，风物尽前朝，衰杨古柳，几经攀折，憔悴楚宫腰"，明葛一龙词"桥上飞花桥下水，断肠人是过桥人"；思乡桥，在河北丰润，"宋徽宗北辕过桥，驻马四顾，泫然曰，吾过此向大漠，安得似此水还乡矣……人乃谓思乡桥也"（《畿辅通志》）；至喜桥，在四川广安，"昔欧阳修自吴入蜀，喜路险至此始平"（《四川通志》）；情尽桥，在四川简阳，唐雍陶《题情尽桥》诗有序云"阳安送客至情尽桥，问其故，左右曰送迎之地止此"，诗云"从来只有情难尽，何事名为情尽桥，自此改名为折柳，任他离恨一条条"；忘恩桥，在陕西西安，"中官初入选，进东华门，门内有桥曰皇恩桥……俗呼曰忘恩桥，以中官既富贵，必仇所生，盖耻之也"（《客窗偶谈》）。

第四章是"写景"，美化桥身及四周景物，如垂虹桥，在江苏吴江，桥身环如半月，长若垂虹，宋王安石《垂虹桥》诗云"颇夸九州物，壮丽无此敌"；春波桥，在浙江绍兴，贺知章诗云"'离别家乡岁月多，近来人事半消磨，唯有门前鉴湖水，春风不改旧时波',故取此名桥"(《会稽志》);海棠桥,在湖北黄州,"宋时桥侧海棠丛开,秦观尝醉卧于此,明日题其柱"(《湖广通志》);胭脂桥,在江西饶州,"鄱阳王萧俨宅宫人,尝遗胭脂水流出,故名"(《江西通志》);月样桥,在山东青州,"采石凝结如天成"(《山东通志》);绿杨桥,在湖北蕲水,"因东坡醉卧桥上,有'解鞍欹枕绿杨桥'之句,遂名"(《湖广通志》),按坡公《西江月》词自序云"春夜行蕲水中……至一溪桥上,解鞍曲肱,醉卧少林,及觉已晓……书此语桥柱上",即是桥也。

第五章是"纪念",多指名人故事,但未必与造桥有关,如留衣桥,在江西南城,原名万寿桥,"清咸丰间有知府治郡有惠政,卸职时郡人饯行,留衣于此,遂改名留衣桥"(《南城县志》);惠政桥,在江苏邵伯,旧谓谢安以政惠民,故名;甘棠桥,在湖北崇阳,宋张乖崖为令时劝民种桑,兴修水利,乡民为立甘棠桥,以志不忘;王公桥,在江西饶州,"宋知州王十朋,徙知夔州,民走诸司乞留不得,至断其桥,十朋乃以车从间道去,众葺断桥,因名"(《江西通志》);莱公桥,在江西峡江,因寇莱公谪潮时经此而得名;斩蛟桥,在江苏宜兴,"宜兴长桥元丰元年火,四年邑宰褚理复修,立榜曰欣济,东坡过之,为书曰晋周侯斩蛟之桥"(《游宦纪闻》);国士桥,在山西赵城,"昔豫让为智伯报仇欲杀赵襄子,伏于其下……后人改名曰国士桥"(《山西通志》)。

第六章是"神话",把神仙和桥梁联系起来,如乌鹊桥,即鹊桥,《白帖》云"乌鹊填河成桥,而渡织女",杜甫有诗云"江光隐现鼋鼍窟,石势参差乌鹊桥",唐宋之问诗"鸳鸯机上疏萤度,乌鹊桥边一雁飞";圣女桥,在陕西白水,传说为三神女一夜成之;白鹤桥,在江苏句容,"汉永元间茅氏兄弟三人,乘鹤至此,有白鹤桥,大茅君驾白鹤会群仙处"(《句容县志》);集仙桥,在江西安福,"相传居人夜闻桥上仙乐缭绕,且往视之,唯见书吕洞宾字于桥柱"(《江西通志》);乘鱼桥,在江苏苏州,"昔琴高乘鲤升仙之地"(《江南通志》);照影桥,在湖北石首,"相传有仙人于此照影"(《湖广通志》)。

以上是单字和双字的桥名录。三字、四字或更多字的桥名,当然也有,但为数极少。三字桥名中著名的有二十四桥,在江苏扬州。不过这种用数目字当专名的桥,并无他例,有的只是以数为序而已,如杜甫诗"不识南塘路,今知第五桥"。四字或以上的如旧浣花桥、杨柳河桥、德阳王桥(系德阳王创建,以上均在成都府)、新学前桥、建富木桥(以上均在南昌府)、新饭店石桥(四川温江县)等等,有的是地名或人名关系,其余都是把单名或双名的桥加以解释,并非完整的专名。因此,中国桥名,基本上只有单名和双名两种,在《桥名录》中把这两种搜集齐全,所余就无几了。

但是，我国古桥并非个个都有专名。有的本来并无名称，后来有人随便叫它一下，逐渐也就成了名字，如大桥、小桥、新桥、旧桥、长桥、短桥、南桥、北桥等等。这些俗名，时间一久，就成为正名了，如福州的小桥，因在万寿桥的大桥附近而得名，就此成为专名。有的桥，名气非常之大，但实际上并无此桥，如陈桥，在河南开封，宋赵匡胤"黄袍加身"处，实系陈桥驿，"在京师陈桥、封丘二门之间，唐为上元驿，朱全忠纵火，欲害李克用之所，艺祖启运立极之地也"（《玉照新志》）。但更多的桥，却是不声不响地在那里服务，而它们的名字是早就湮没不彰了。可见，桥和桥名，都有幸与不幸，有的是有名无桥，有的是有桥无名。然而，虽是无名，难道就不是"英雄"！

原载 1962 年 7 月 22 日《光明日报》

名桥谈往

古往今来，芸芸大众，得名者极少，其能流芳百世的就更少。桥也是一样。自有历史以来，就有人造的桥。最早有记载的是夏禹用"鼋鼍以为桥梁"（《拾遗记》），后来在渭河上，先是"造舟为梁"（《诗经》中《大明篇》），逐渐地就"以木为梁""以石为梁"（《初学记》），于是桥梁日多，布满全国。四千来历代所建桥梁，据说有几百万座之多。由于我国文化昌盛，这许多桥梁，一般都有名字，就像人有名字一样（参见1962年7月22日《光明日报》撰作《桥名谈往》）。然而，虽然个个有名字，真正"出名"的却不多，人是如此，桥也是如此。不过，桥不像人，从未有过"遗臭万年"的。尽管桥上会有遗臭的事，但桥的本身总是流芳的。流芳有长短和远近的不同，决定于桥本身的技术和艺术，桥在历史上的作用，桥上的故事传说和有关桥的文艺、神话、戏剧等等。这几方面当然是互有影响的,在一方面出了名,其他方面也会跟着附和。然而各方面未必相称。小桥可以享大名,而大桥未必尽人皆知,甚至简直无名,桥的有名无名,要看它在群众中的"威望"。现在以此为准,来谈谈我国传统的各地名桥。所谓传统的桥就是我国固有的各种形式的桥,并非从西方输入的近代形式的桥。

技术上的名桥

我们常常自谦,说是科学技术落后,比不上世界上的先进国家。这是近百年来受了帝国主义压迫的结果。但是,回顾过去数千年的历史,我国不但文化悠久,光辉灿烂,而且就是在科学技术上,也曾盛极一时,桥梁就是一例。我国有许多桥梁,其技术在当时是大大超过世界水平的。这有实物为证。首先要提到的是赵州桥,这是全世界桥梁史上的一座最突出的桥。它的技术是大大超过时代的。它是在一千三百五十多年前(隋代)由"总工程师"李春造成的一座石拱桥,直到现在,还可使用。

其次应当提出的是福建泉州洛阳桥。这是一座石梁桥,修建于南宋皇祐、嘉祐年间(公元 1053~1059 年),长三百六十丈,有四十七孔。洛阳江入海处水流湍急,波涛汹涌,建桥当然不易，而且当时福建沿海各河上，除有少数浮桥外，几无一处有石桥，洛阳桥的建成，实是划时代的巨大贡献。

也应当提到广东潮州的湘子桥，它所跨越的韩江，就是唐代大文学家韩愈驱逐鳄鱼的所在，那时就名为"恶溪"，可见水深流急，造桥之不易了。这座桥全长五百一十八米，分为三段，东段十二孔，长二百八十四米，西段七孔，长一百三十七米，中段一大孔，长九十七米。东西两段，皆石墩石梁，中段是浮桥，由十八只木船组成。这桥的特点就在中段，那里的木船，可以解缆移动，让出河道以通航。这就是近代的所谓"开合桥"，合时通车，开时走船，对于水陆交通，是两不妨碍的。然而这样一座结构巧妙的桥梁，却是建成于南宋乾道年间（公元 1169～1173 年），距今已将八百年了。

万年桥，在江西南城县，是国内罕见的极长的联拱石桥，计石拱二十三孔，全长四百余米长。所谓"联拱"就是把许多拱联成一线，形成一个整体，每一拱上的载重，由全部各拱共同负担，因而是个很经济的设计。这座桥在宋代初建时为浮桥，到明代崇祯时（公元 1634 年）更建为石桥。

西津桥，在甘肃兰州，俗名卧桥或握桥，在阿干河上，是伸臂式的木结构桥，其木梁由两岸伸向河心，节节挑出，在河心处，于两边挑梁上铺板，接通全桥。传说这桥建自唐代，经历代重修，现存的是公元 1904 年重建的。

珠浦桥，在四川灌县，位于都江堰口，横跨岷江，是用竹缆将桥面吊起的悬桥，共长三百三十米，分十孔，最长跨度六十一米，竹缆锚碇于两岸的桥台中。

以上六座桥，代表六种类型，即拱桥、梁桥、开合桥、联拱桥、伸臂桥和悬桥。从今天看来，所有近代桥梁的主要类型，"粲然具备矣"。当然，在每一类型中还有其他名桥，比如拱桥类有建于元代的江苏吴江垂虹桥；梁桥类有福建泉州的五里桥，有"天下无桥长此桥"的传说，福建漳州的江东，最大一根石梁重至二百吨，均建于南宋时代；联拱桥类有建于清初的安徽歙县的太平桥；悬桥类有建于明代的贵州盘江桥等等。这许多名桥的技术有一个共同特点，就是把桥造得坚固耐久。

艺术上的名桥

桥不在水上，就在山谷，而山与水又往往相邻，构成图画，"山水"成为风景的代名词，桥在这样的天然图画中，如果本身不美，岂不大杀风景。桥的美首先表现在形体，亦即桥身的构造，要它在所处环境中，显得既不可少，又不嫌多，"秾纤得衷，修短合度"。其次在艺术布置上处理得当，绝不画蛇添足。一条重要法则是技术和艺术的统一，不因此害彼。上述的几座名桥,特别是赵州桥,就都能达到这种境界。特别在艺术上驰名的,还有很多,这里举几个例。

宝带桥在江苏苏州,是座联拱石桥,全长约三百一十七米,分五十三孔,其中三孔联拱特别高,以通大型舟楫,两旁各拱路面,逐渐下降,形成弓形弧线。建于唐代(约公元806年),重修于宋(约公元1232年)。全桥风格壮丽,堪称"长虹卧波,鳌背连云"。这座桥的工程浩大,构造复杂,而又结构轻盈,奇巧多姿,成为江南名胜。

玉带桥在北京颐和园,建于清代(约公元1770年),桥拱作蛋尖形,特别高耸,桥面形成双向反曲线,据说是美国纽约狱门桥设计的张本。这是座小桥,庄严而又玲珑,大为湖山生色。

程阳桥在广西三江,长达一千余米,是座伸臂式桥,用大木节节伸出,跨度二十余米。每一桥墩上建有宝塔式楼阁四层,约五米见方,高十余米。各墩楼阁之间,用长廊联系,上有屋盖,为行人遮阳蔽雨。这桥的构造奇特,结合桥梁与建筑为一体,形成一座水上的游廊。

鱼沼飞梁在山西太原的晋祠内,是个游览胜地。这是座在鱼沼上建成的十字形的飞梁,就像两条路的十字交叉一样。飞梁的中心是个六米见方的广场,东西向和南北向的两头各有挑出的翼桥,长六米,形成十八米长的两桥交叉。这桥的构造曲折,整齐秀雅,富丽堂皇。

五亭桥在江苏扬州瘦西湖,也是个十字交叉的飞梁桥,在中心广场和东南西北的四个翼桥上,各有一亭,桥下正侧面共有十五个桥孔,月满时每孔各衔一月,波光荡漾,蔚为奇观。

历史上的名桥

桥是交通要道的咽喉,军事上在所必争,历史上记载的与桥有关的战役,真是太多了,往往一桥得失影响到整个战争局面。在和平建设上,有的桥也起过重大历史作用。现举历史上的几个著名的例子。

泸定桥,即大渡河铁索桥,是公元1935年我红军长征强渡大渡河的所在。这座桥建成于清代(公元1706年),计长一百零三米,宽约三米,桥面木板铺在九根铁链上,铁链锚碇于两岸桥台。

卢沟桥,在北京广安门外永定河上,是1937年日本帝国主义对我国发动侵略战争的爆发地,也是我国人民解放战争中永远值得纪念的一座桥。这是座联拱石桥,共长二百六十五米,由十一孔石拱组成,建成于金代(公元1192年)。13世纪时,意大利人马可·波罗在他的游记中提到这座桥。经过他的宣传，卢沟桥早就闻名世界。

阴平桥在甘肃文县，从文县至四川武县的阴平道，即三国时魏将邓艾袭蜀之路。姜维闻有魏师，请在阳安关口阴平桥头防御。这座桥于清代（公元 1729 年）重建，是一个有名的石拱桥。

孟盟桥在山西蒲州，春秋时秦将孟明伐晋，"济河焚舟，盟师必克"，晋师不敢出，遂霸西戎，故以名桥。这里所谓"舟"，就是浮桥。

在桥梁史上，有的桥是先行者，成为后来建桥的楷模。晋杜预以孟津渡险，建河桥于富平津，当时反对者多，预曰，造舟为梁，则河桥之谓也，及桥成，晋武帝司马炎向他祝酒说，非君此桥不立也。后来，"杜预造桥"故事，成为一种鼓舞力量。福建自洛阳桥兴建成功，泉漳两地相继修成"十大名桥"，为桥梁技术开辟了新纪元，致有"闽中桥梁甲天下"之誉。洛阳桥又是明代抗倭的一个要塞，明末时，郑成功更据此桥抗清，取得胜利。

有的历史上的名桥，实际并非桥，比如，宋代赵匡胤制造的"陈桥兵变，黄袍加身"的陈桥，就不是桥而是个"驿"名，唐时名上元驿，朱全忠曾在此放火，谋害李克用。

故事中的名桥

历史上有许多有名的故事,在这些故事里所牵涉到的桥也往往成为名桥。

有的桥是为纪念名人的,如惠政桥、斩蛟桥、甘棠桥、王公桥、留衣桥等。

有些桥的故事流传甚广,但其确址难考。如汉张良游下邳,遇圯上老人命取履,圯就是桥,这桥当然在下邳了,但河南归德府永城县有赞城桥,"一名圯桥,即张良进履处"(见《河南通志》)。

文艺中的名桥

桥是地上标志,又是克服困难把需要变成可能的人工产物,因而桥的所在和有关故事,最能动人,成为文艺上的极好题材。在文学中诗、词、歌、赋里以桥命题的固然多不胜数,到了近代文学里,它为群众服务的作用,就更显得重要了。同样,在绘画、雕塑等等的艺术作品中,桥也是重要对象。现就文艺遗产中举几个例。

灞桥在陕西西安,"东汉人送客至此桥,折柳赠别"(《三辅黄图》）。"灞陵有桥,来迎去送,至此黯然,故人呼为销魂桥"（《开元遗事》）。唐王之涣诗:"杨柳东风树,青青夹御河,近来攀折苦,应为别离多。"宋柳永词:"参差烟树灞陵桥,风物尽前朝,衰杨枯柳,几经攀折,憔悴楚宫腰。"

枫桥在江苏苏州,因唐张继《枫桥夜泊》诗而名闻中外,其中,"江枫渔火对愁眠"句,有人谓是指"江桥"和"枫桥"两个桥。又唐杜牧有诗:"长洲茂苑草萧萧,暮烟秋雨过枫桥。"其实枫桥只是一个较小的石拱桥。

在古代绘画中,桥虽多,但知其名的很少。可以提出的是宋张择端画的《清明上河图》中的河南开封的虹桥。名画中的桥,多半是拱桥,但这幅画中的名桥却是个木结构的拱形伸臂桥。它的结构非常奇巧,堪称举世无双。

神话中的名桥

由于桥是从此岸跨到彼岸,凌空飞渡,不管下界风波,这就引起人们的美丽幻想,特别是爱把桥比作"人间彩虹",把彩虹当作是人间到天上的一条通路。既然上天,神仙就少不了了。

鹊桥,是神话中牛郎织女在银河上的相会处。《淮南子》云:"乌鹊填河成桥,而渡织女。"《风俗记》说:"七夕织女当渡河,使鹊为桥。"神仙本来是会腾云驾雾的,然而在银河上还需要桥。人们把桥的作用抬高到天上去了!

蓝桥,在陕西蓝田县蓝溪上,"传其地在仙窟,即唐裴航遇云英处"(《清一统志》)。

照影桥,在湖北石首,"相传有仙人于此照影"(《湖广通志》)。

此外,各地桥以"升仙"为名的特别多,也是人们在封建统治下不堪压迫向往出头的一种反映。

戏剧里的名桥

出名的人物故事,总会搬上戏剧舞台,桥当然不例外。京剧里演出的名桥故事就不少,最著名的是《长坂坡》,即长坂桥,见《三国演义》。《三国志·张飞传》载:"曹公入荆州,先生奔江南,使飞将二十骑拒后,飞据水断桥,瞋目横矛……敌皆无敢近者,遂得免。"还有《金雁桥》也是三国故事戏。关于恋爱戏,有《鹊桥相会》《断桥相会》《虹桥赠珠》《草桥惊梦》等等。直接宣扬造桥故事的有《洛阳桥》灯彩戏,描述建桥如何艰巨以及桥成后三百六十行过桥时人民的欢乐情景。

今天造桥的传统

上述的这些名桥中,有四座已经在我国纪念邮票中发表了,就是赵县安济桥(即赵州桥)、苏州宝带桥、灌县珠浦桥和三江程阳桥。此外值得纪念的还有很多,特别是泸定桥、卢沟桥、洛阳桥和湘子桥。有很多古桥的传统,已经成为民族遗产中的财富,有的更发展为今天造桥的传统,如云南南盘江上的公路石拱桥,跨度达一百一十四米,成为世界上最大的石拱桥,就是继承了赵州桥的传统而发展成功的。这种古为今用的发展前景,将是不可限量的。往时名桥虽多,然而"俱往矣",数宏规巨构,"还看今朝"!

原载 1962 年 9 月 29 日《文汇报》

桥　话

　　人的一生,不知要走过多少桥,在桥上跨过多少山与水,欣赏过多少桥的山光水色,领略过多少桥的画意诗情。无论在政治、经济、科学、文艺等各方面,都可以看到各式各样的桥梁作用。为了要发挥这个作用,古今中外在这"桥"上所费的工夫,可就够多了。大至修成一座桥,小至仅仅为它说说话。大有大用,小有小用,这就是这个《桥话》的缘起。诗话讲诗,史话讲史,一般都无系统,也不预订章节。有用就写,有话就长。桥话也是这样。

最早的桥

　　首先要说清楚:什么是桥?如果说,能使人过河,从此岸到彼岸的东西就是桥,那么,船也是桥了;能使人越岭,从这山到对山的东西就是桥，那么，直升飞机也是桥了。船和飞机当然都不是桥，因为桥是固定的，而人在桥上是要走动的。可是，拦河筑坝，坝是固定的，而人又能在坝上走，从此岸走到彼岸，难道坝也是桥吗？不是的，因为桥下还要能过水，要有桥孔。那么，在浅水河里，每隔一步，放下一堆大石块，排成一线，直达对岸，上面走人，下面过水，而石块位置又是固定的，这该是一座桥了（这在古时叫做"鼋鼍以为桥梁"，见《拾遗记》，近代叫做"汀步桥"）。然而严格说来，这还不是桥，因为桥面是要连续的，不连续，不成路。但是，过河越谷的水管渠道，虽然具备了上述的桥的条件，而仍然不是桥，这又是何故呢？因为它上面不能行车。这样说来，矿山里运煤的架空栈道，从山顶到平地，上面行车，岂非也是桥吗，然而又不是，因为这种栈道太陡，上面不能走人。说来说去，桥总要是条路，它才能行车走人，不过它不是造在地上而是架在空中的，因而下面就能过水行船。

其次，怎样叫早？是自然界历史上的早呢，还是人类历史上的早。是世界各国的早呢，还是仅仅本国的早。所谓早是要有历史记载为根据呢，还是可凭推理来臆断。早是指较大的桥呢，还是包括很小的在内的，比如深山旷野中的一条小溪河上，横跨着一根不太长的石块，算不算呢？也就是说，是指有名的桥呢，还是无名的桥。这样一推敲，就很难落笔了。姑且定个范围,那就是:世界上最初出现的人造的桥,但只指桥的类型而非某一座桥。

在人类历史以前,就有三种桥。一是河边大树,为风吹倒,恰巧横跨河上,形成现代所谓"梁桥",梁就是跨越的横杆。二是两山间有瀑布,中为石脊所阻,水穿石隙成孔,渐渐扩大,孔上石层,磨成圆形,形成现代所谓"拱桥",拱就是弯曲的梁。三是一群猴子过河,一个先上树,第二个上去抱着它,第三个又去抱第二个,如此一个个上去连成一长串,为地上猴子甩过河,让尾巴上的猴子,抱住对岸一棵树,这就成为一串"猿桥",形式上就是现代所谓"悬桥"。梁桥、拱桥和悬桥是桥的三种基本类型,所有千变万化的各种形式,都由此脱胎而来。

因此,世界上最初出现的人造的桥就离不开这三种基本形式。在最小的溪河上,就是单孔的木梁。在浅水而较大的河上,就是以堆石为墩的多孔木梁。在水深而面不太宽的河上,就是单孔的石拱,在水深流急而面又宽的大河上,就是只过人而不行车的悬桥。

应当附带提一下,我国最早的桥在文字上叫做"梁",而非"桥"。《诗经》"亲迎于渭,造舟为梁"。这里的梁,就是浮桥,是用船编成的,上面可以行车。这样说来,在历史记载上,我国最早的桥,就是浮桥,在这以前的"杠""榷""礿"

"圯"等等，都不能算是桥。

古桥今用

古代建筑，只要能保存到今天，总有用。也许是能像古时一样地用它，如同四川都江堰；也许不能完全像古时那样地来用它，如同北京故宫；也许它本身还有用，但现在却完全不需要了，如同万里长城。更多的是，它虽还有小用，但已不起作用，如果还有历史价值，那就只有展览之用了。古桥也是这样，各种用法都有，不过专为展览用的却很少。要么就是完全被荒废了，要么就是经过加固，而被充分大用起来。值得提出的是，有一些古桥，并未经过改变，"原封不动"，却能满足今天的需要，担负起繁忙的运输任务。这是中国桥梁技术的一个特点。不用说，这种古桥当然是用石头造起来的。

在抗日战争时期，大量物资撤退到后方，所经公路，技术标准都不是很高的，路线上常有未经加固的古桥。但是，撤退的重车，却能安然通过，起初还限制行车速度，后来就连速度也放宽了。古桥是凭经验造起来的，当然没有什么技术设计。奇怪的是，如果用今天的设计准则，去验算这些古桥的强度，就会发现，它们好像是不能胜任这种重车的负担的。

然而事实上，它们是竟然胜任了，这是什么缘故呢？

原来我国古桥的构造，最重视"整体作用"，就是把全桥当作整体，不使任何部分形成孤立体。这样，桥内就有自行调整的作用，以强济弱，减少"集中负荷"的影响。比如拱桥，在拱圈与路面之间有填土，而桥墩是从拱圈脚砌高到路面的。拱圈脚、填土和路面都紧压在墩墙上，因而路面上的重车就不仅为下面的拱圈所承载，同时还为两旁墩墙的"被动压力"所平衡。但在现时一般拱桥设计中，这种被动压力是不计的，因而在验算时，这类古桥的强度就显得不足了。提高墩墙就是为了整体作用，其他类似的例子还很多。这都说明，古代的修桥大师，由于实践经验，是很能掌握桥梁作用的运动规律的，尽管不能用科学语言来表达它。正因为这样，我国古桥比起外国古桥来，如罗马、希腊、埃及、波斯的古桥，都显得格外均匀和谐，恰如其分，不像它们的那样笨重。北京颐和园的十七孔桥和玉带桥都能说明这一点。

古桥保存到今天，当然不是未经损坏的。除去风雨侵蚀，车马践踏外，还会遇到意外灾害，如洪水、暴风、地震等等，也许原来施工上的弱点，日后暴露出来，这都需要修理。而修理对于建桥大师，正是调查研究的好机会。他们从桥的损坏情况，结合历年来外加影响，就能发现问题所在，因而利用修理机会，予以解决。每经一次修理，技术提高一步。数千年来的修桥经验，是我国特有的宝贵民族遗产。

赵州桥，建成于一千三百多年前，从那时起，一直用到今天，可算是古桥今用的最突出的例子。更可贵的是，它今天还是原来的老样子，并未经过大改变。西班牙的塔霍河上，有一座石拱桥，建成于罗马特拉杨大帝时，距今已达一千八百多年，现仍存在，但其中有六百年是毁坏得完全不能使用的，其服务年限之长，仍然不及赵州桥。在古桥今用这件事上，我国是足以自豪的。

桥的运动

桥是个固定建筑物，一经造成，便屹立大地，可以千载不移，把它当作地面标志，应当是再准确不过的。《史记·苏秦列传》里有段故事："信如尾生，与女子期于梁下，女子不来，水至不去，抱柱而死。"就因为桥下相会，地点是没有错的，桥是不会动的。但是这里所谓不动，是指大动而言，至于小动、微动，它却是和万物一般，是继续不断，分秒不停的。

车在桥上过，它的重量就使桥身变形，从平直的桥身变为弯曲的桥身，就同人坐在板凳上，把板凳坐弯一样。板凳的腿，因为板的压迫，也要变形，如果这腿是有弹簧的，就可看出，这腿是被压短了。桥身的两头是桥墩，桥上不断行车，桥墩就像板凳腿一样,也要被压短而变形。把板凳放在泥土上,坐上人,板凳腿就在泥土上留下痕迹,表现泥土有变形。桥墩也同样使下面的基础变形。桥身的变形表示桥上的重量传递给桥墩了,桥墩的变形表示桥身上的重量传递给基础了,基础的变形表示桥墩上的重量传递给桥下的土地了。通过桥身、桥墩和基础的变形,一切桥上的重量就都逐层传递,最后到达桥下的土地中,形成桥上的重量终为地下的抵抗所平衡。物体所以能变形,由于内部分子的位置有变动,也就是由于分子的运动。因而一座桥所以能在有车的重量下保持平衡,就是因为它内部的分子有运动的缘故。

车在桥上是要走动的,而且走动的速度可以很高,使桥梁全部发生震动。桥上不但有车,而且还受气候变化的侵袭;在狂风暴雨中,桥是要摆动或扭动的;就是在暖冷不均、温度有升降时,桥也要伸缩,形成蠕动。桥墩在水中,经常受水流的压迫和风浪的打击,就有摇动、转动和滑动的倾向而影响它在地基中的移动。此外,遇到地震,全桥还会受到水平方向和由下而上的推动。所有以上的种种的动,都是桥的种种变形,在不同的外因作用下而产生的。这些变形,加上桥上重量和桥本身重量所引起的变形,构成全桥各部的总变形。任何一点的总变形,就是那里的分子运动的综合表现。因此,一座桥不论是在有重车疾驰、狂风猛扑、巨浪急冲或气温骤变的时候,或是在风平浪静、无车无人而只是受本身重量和流水过桥的影响的时候,它的所有的一切作用都可很简单地归结为一个作用,就是分子运动的作用。

桥是固定建筑物,所谓固定就是不在空间有走动,不像车船能行走。但是,天地间没有固定的东西。至多只能说,桥总是在动的平衡状态中的,就是桥的一切负担都是为桥下的土地所平衡的,这是总平衡。拆开来看,桥身是处于桥上车重和两头桥墩之间的平衡状态的,桥墩是处于桥身和基础之间的平衡状态的,基础是处于桥墩和土地之间的平衡状态的。再进一步来分析,桥身、桥墩和基础的内部的任何一点,也无不在它四周的作用和反作用的影响下而处于平衡状态的。平衡就是矛盾的统一。矛盾是时刻变化的,因而平衡也不可能是稳定的,更不可能是静止的。就是在桥上的一切动的作用都停止的时候,在桥上只有本身重量起作用的时候,桥的平衡也不是稳定的,因为桥和土地的变形,由于气候及其他关系,总是在不断的变化中的。桥的平衡只能是瞬息现象,它仍然是桥的运动的一种特殊状态。

恩格斯说:"运动是物质的存在形式。"可见,桥的运动是桥的存在形式,一切桥梁作用都是物质的运动作用。

桥梁作用

桥梁是这样一种建筑物,它或者跨过惊涛骇浪的汹涌河流,或者在悬崖断壁间横越深渊险谷,但在克服困难、改造了大自然以后,它便利了两岸的往来,又不阻挡山间水上的原有交通。

桥是为了与人方便而把困难留给自己的。人们正当在路上走得痛快时,忽然看到前面大河挡路,而河上正好有一座桥,这时该暗自庆幸,果然路是走对了。

造桥是不简单的。它要像条纽带,把两头的路,连成一体,不因山水阻隔而影响路上交通。不但行车走人,不受重量或速度的限制,而且凡是能在路上通过的东西,都要能一样地在桥上通过。如果能把桥造得像路一样,也就是说,造得有桥恍同无桥,这造桥的本领就够高了。桥虽然也是路,但它不是躺在地上而是悬在空中的,这一悬,就悬出问题来了。所有桥上的一切重量、风压、震动等等的荷载都要通过桥下的空间,而传到水下的土石地基,从桥上路面到水下地基,高低悬殊,当中有什么"阶梯"好让上面荷载,层层下降,安然入地呢? 这就是桥梁结构:横的桥身,名为"上部结构",竖的桥墩,名为"下部结构"。造桥本领就表现在这上下结构上。

　　桥的上下结构是有矛盾的。要把桥造得同路一样牢固，上部结构就要很坚强，然而它下面是空的，它只能靠下部结构的桥墩做支柱，桥墩结实了，还要数目多，它才能短小精悍，空中坐得稳。但是，桥墩多了，两墩之间的距离就小了，这不但阻遏水流，而且妨碍航运。从船上人看来，最好水上无桥，如果必须造桥，也要造得有桥恍同无桥，好让他的船顺利通过。桥上陆路要墩多，桥下水路要墩少，这矛盾如何统一呢？很幸运，在桥梁设计中，有一条经济法则，如果满足这个法则，就可统一那个矛盾。这个法则就是上下部结构的正确比例关系。

　　桥的上下部结构是用多种材料造成的。材料的选择及如何剪裁配合，都是设计的任务。在这里有两个重要条件，一是要使上层建筑适应下面的地基基础，有什么样的基础，就决定什么样的上层建筑，上层建筑又反过来要为巩固基础而服务；一是要把各种不同性质、不同尺寸的材料，很好地结合起来，使全座桥梁形成一个整体，没有任何一个孤立"单干"的部分。纵然上部结构和下部结构各有不同的自由活动，也要步调一致，发挥集体力量。桥的"敌人"是既多且狠的：重车的急驶、狂风的侵袭、水流的冲击、地基的沉陷等等而外，还有意外的地震、爆破、洪水等灾害。桥就是靠着它的整体作用来和这些敌人不断斗争的。

桥的上下部结构要为陆路水路交通同等服务,而困难往往在水路。水是有涨落的,水涨船高,桥就要跟着高,这一高就当然远离陆路的地面了。地面上的交通如何能走上这高桥呢? 这里需要一个"过渡",一头落地,一头上桥,好让高低差别,逐渐克服,以免急转直上。这种过渡,名为"引桥",用来使地面上的路,引上正桥。引桥虽非正桥,却往往比正桥更长更难修。

可见,一座桥梁要在水陆交通之间,起桥梁作用,就要先在它自己内部很好地发挥各种应有的桥梁作用。整体的桥梁作用是个别桥梁作用的综合表现。

原载 1963 年 2 月 6 日、9 日,3 月 20 日,4 月 27 日《人民日报》

二十四桥

中国修桥，向来有个"收尾工程"是别国所无的，那便是要为这桥题个好听的名字，借以记事或抒情，而不是简单地用个地名就够了。题名本来是好事，但有时也会因此而引起误会。诗文中常有一个桥字和上面两三个字连在一起，看去好像是有这么一座专名的桥，但又像是不过是桥的说明而已。唐代杜牧诗中的"二十四桥"就是一例。这到底是一座桥的专名呢，还是说有二十四座这么多的桥。千年来不少人为这问题作了考证，直到现在，报上还不时有争论。不知什么原因，二十四桥这个词很能引人入胜，我也为它引出一些话来。

杜牧的原诗是《寄扬州韩绰判官》："青山隐隐水迢迢，秋尽江南草未凋，二十四桥明月夜，玉人何处教吹箫。"用很少的字，表达出对扬州的繁华消逝，人去楼空之感。二十四桥和未凋的草一样，还依然存在，但江南秋尽，那桥上的吹箫玉人，却不知何处去了，这在明月之夜更是显得凄凉。这里的玉人"何处"，应作玉人"何在"解，来形容往事的不堪回首。往事越繁华，越显出今日的萧瑟。这首诗中，只有二十四桥这几个字指出往事的遗迹，因而这几个字就要能充分反映出昔日的盛况。如果这只是一座名叫二十四的桥，它如何能体现全扬州的繁华呢？扬州的桥很多，而且从隋朝起，就分布在全城。《一统志》云："扬州二十四桥，在府城，隋置，并以城门坊市为名。"杜牧的《樊川集》云"扬州胜地也，每重城向夕，倡楼之上，常有绛纱灯万数，辉耀罗列空中。九里三十步街中，珠翠填咽，邈若仙境"。可见杜牧是很醉心于这九里三十步街中的繁华的。二十四桥既以城门坊市为名，就当然是罗布于这九里三十步街中了。因此，杜牧诗中的二十四桥定然不只是一座桥。

正因为这个缘故，沈括的《梦溪笔谈》中，才为这二十四桥，考证出它们的名字，并且指出，在那时（约公元1064年），二十四桥中存在的仅有八座了。后来当然就全部损坏了。到了明朝，就有程文德的诗："倒看飞鸟穿林外，俯听吹箫出境中，二十四桥都不见，更从何处见离宫？"（见《扬州府志》）

二十四桥逐渐减少，就出现一种传说，认为这仅仅是一座桥的专名。如清代李斗的《扬州画舫录》云"二十四桥即吴家砖桥,一名红药桥,在熙春台后";清代吴倚《扬州鼓吹词序》云"出西郭二里许有小桥,朱栏碧甃,题曰烟花夜月,相传为二十四桥旧址,盖本一桥,会集二十四美人于此,故名";清代梁章钜《浪迹丛谈》云"或谓二十四桥只是一桥,即在孟玉生山人所居宅旁"等等,其实都是没有什么确证的。

比较有力的说法,是引南宋姜夔的《扬州慢》:"二十四桥仍在,波心荡,冷月无声。念桥边红药,年年知为谁生。"所谓"仍在""桥边",似乎应当是指一座桥,否则如是很多桥,难道每座桥边,都有红药吗?但是,紧接这几句词的下面,还有相关的几句:"杜郎俊赏,算而今,重到须惊。纵豆蔻词工,青楼梦好,难赋深情。"可见,"二十四桥"这句和"豆蔻""青楼"两句一样,都是从杜牧的诗句来的,既然在"豆蔻""青楼"两句的上面,有一个"纵"字,这个"纵"字就应当贯串到"二十四桥"这句,这才使姜夔从慨叹"玉人何处"而不由得"念"到桥边红药;玉人不在,红药何用,无怪杜牧的"明月"也就变成"冷月"了。显然,这首词里所指的二十四桥,就是杜牧诗里的二十四桥,而不会是另指一座桥;尽管沈括所见的六座桥,经过金人南下,到姜夔时,也许只剩下一座了。

原载 1963 年 2 月 6 日《文汇报》

五 桥 颂

——泸定桥、卢沟桥、安平桥、安济桥、永通桥

　　桥梁是代表文化的一种物质建设。我国文化悠久，全国各地都有大量的古桥，其中屹立到今天的，数不胜数。有的桥，在历史上起过重大作用；有的桥在艺术上独具一格，显露出各民族的无穷智慧；有的桥在科学技术、创造发明上做出巨大贡献。它们随着时代前进所取得的日新月异的成就，是我国文化发展的一个重要标志。这样的民族遗产、人民财富，在历史、艺术、科学上的价值，是无法估计的。然而，在过去数千年的封建压迫、反动统治下，有珍贵文物价值的桥梁，是从未得到当时政府的重视和保护的。因而很多重要桥梁在自然侵蚀和人为破坏下，遭到了严重的甚至毁灭性的摧残，造成了无可补偿的损失。见过这样情景的人，无不感到痛心疾首，总希望有一天，所有幸逃浩劫的这类桥梁，终会得到抢救。这个愿望，在解放后不久，果然实现。我们党的伟

大正确,社会主义制度的优越性,使我国古代桥梁也获得了新的生命。全国桥梁工作者为此感到鼓舞,感到兴奋!

在国务院公布的第一批全国重点文物保护单位的名单中,列有五座桥:四川省泸定县的泸定桥,北京市丰台区的卢沟桥,福建省晋江县的安平桥和河北省赵县的安济桥、永通桥。它们都是具有重大的历史、艺术、科学价值的文物,都属于国家重点保护的范围。

从技术方面看,这五座桥代表着四种形式:泸定桥是"悬桥",卢沟桥是"连续桥",安平桥是"梁桥",安济桥、永通桥是"拱桥"。这四种形式的桥,再加两种"伸臂桥"和"开合桥",构成近代桥梁中的六种基本形式。在这六种形式中,我们国家这次公布保护的就占了三分之二。从建筑材料看,这五座桥中只泸定桥是铁索桥,其余四座桥都是石桥。在古桥中,永久性桥梁只有铁石两种,我们所保护的也就都有了。从建成年代看,这五座桥中,年代最近的泸定桥,建成于公元1706年,距今二百五十多年;最古的是安济桥,建成于公元605年左右,距今已一千三百五十多年了,成为我国现存的最古老的桥梁。可见这五座受到国家保护的桥梁都各有其代表性,而且它们的价值都不限于一方面。

泸定桥

　　泸定桥,在四川省甘孜藏族自治州泸定县,是 1935 年红军长征途中强夺铁索桥战役的纪念地。桥在大渡河上,是从西面的雪山往东面的二郎山的必经之路。当我红军强渡大渡河时,正逢初夏,河水暴涨,奔流湍急。两岸悬崖峭壁,高耸入云。我大军沿河前进,就在这下临深渊的河岸上行军,而且日夜不停。在将到桥头的前夕,大雨倾盆,天黑地滑,更是步履维艰,稍有不慎,就会堕入惊涛骇浪中。在到达桥头时,敌人早将桥上木板烧毁,只剩下几根摇晃的铁索,高悬在倾泻奔腾的洪流上。对岸桥头正在县城西门下,敌人在城上凭险据守,弹如雨下。我英勇红军,踏上铁索,奋勇前进。每人带着一块木板,边走边铺。有的木板为敌人弹中起火。有二十二位英雄,就踩着燃烧的木板,冒着弹雨,攻到对岸,于是全桥为我军占领,胜利完成了强夺大渡河的战役,在我人民革命的长征历史上,留下了"大渡桥横铁索寒""三军过后尽开颜"(毛主席《长征》诗)的光辉诗篇。

　　泸定桥是一座铁索悬桥,东西长三十一丈一尺(合一百零三点七米),宽九尺(合三米)。桥身用铁链九根,系于两岸,悬挂空中,铁链外径九厘米,上铺木板,形成桥面,以通行人。桥两旁各有铁链两根，横贯东西，用作扶栏。两岸各有桥台一座，用条石砌成，形同长方碉堡，高二十米。桥台内有落井，宽两米，长五米，深六米，内有生铁铸成的铁桩八根，直径二十厘米，竖立井内，四面用灰浆块石胶固。铁桩后面有四米长的锚桩一根，直径为两米，横卧贴紧，上绕过河的九根铁链，扶栏铁链则系于铁桩上。这样，全部十三根铁链就都牢固地锚碇于两岸桥台了。然而，铁链桥身，不论上下左右，都不是固定的，摇曳空中，容易摆动，风雨交加的时候，更是晃荡得厉害，人行其上，随桥起伏，不能自主，俯视深渊，目眩心悸，故自古有"绳桥惊险"之叹。过桥如此，造桥的困难就可想而知了。

《四川通志》载："在雅州府打箭炉厅东南大渡河上，其地旧无桥梁，河水迅激，不可施舟楫，行人从三渡口援索悬渡，至为危险。康熙四十年……抚臣熊泰奏言，距化林营八十里地名安乐，水平可建桥，以通行旅，遂造铁索桥。"又《古今图书集成》中《天全六番志》载："泸定桥，在泸水上，康熙四十五年所制铁索桥也。西炉附置戍守，税茶市而桥因以建。桥工费甚巨，以水势汹涌，其水达西炉，旧有皮船三渡……今皆废而集于桥。沈冷本天全部属，桥既成，檄天全工力修葺。"可见这座桥是在康熙四十年以后用"税茶市"的办法筹款建成的，桥成后，还强迫附近人民负修理之责。在建桥以前，大渡河有三个渡口，先是用皮船过渡，后来就"援索悬渡"。所谓"援索悬渡"，就是用藤索或竹索过江，绳索外套一大筒，人过渡时就捆在筒上，沿索"溜"过江去。这种溜索桥是四川省的一种原始悬桥。

泸定桥的修建，在当时确非易事。《小方壶斋舆地丛钞》载："康熙中修建此桥，曾于东岸先系铁索，以小舟载铁链过重，未及对岸辄覆，久之不成。后一番僧教以巨绳先系两岸，每绳上用十数短竹筒贯之，再以铁索入筒，缚绳数十丈，于对岸牵拽其筒，筒达铁索亦至。"其实，这就是溜索桥的遗意。

溜索桥不知起于何时，但悬桥在我国确有古老历史。《云南通志》载："景东厅津桥，跨澜沧江，两岸峭壁插汉，江流飞急，以铁索和南北岸为桥，相传汉明帝时建。"还有人说，在秦始皇时代，就有索桥了。《史记》云："燕太子丹质于秦，秦王遇之无礼，乃求归，秦王为机发之桥，故以陷丹，丹过之无虞。"据考证，这里所谓"机发之桥"，就是悬桥，桥的悬索可以放松，使人落水。由于山高水深，悬桥始于西北，当无疑义。《洛阳伽蓝记》载，比丘惠生于北魏孝明帝神龟二年（公元519年），奉使西域："从钵卢勤国向乌场国，铁索为桥，悬虚为渡，下不见底，旁无挽捉，倏忽之间，投躯万仞，是以行者望风谢路耳。"这座桥比泸定桥早约一千二百年，但已经用铁索了，可见我国古代少数民族的智慧。

泸定桥建成后,有咏桥诗云:"蜀疆多尚竹索桥,松维茂保跨江饶,几年频涉竟忘险,微躯一任轻风飘。斯桥熔铁作坚链,一十三条牵二岸……洪涛奔浪走其下,迢迢波际飞长虹。"在当时也还是盛事。

离这座泸定桥不远的地方,于1951年造了一座新的钢索悬桥,为康藏公路①之用。桥上有朱德委员长题的对联:"万里长征,犹忆泸关险;三军运戍,严防帝国侵。"从此,旧的泸定就过渡到新的泸定了!

卢沟桥

茅以升手抄卢沟桥碑文

卢沟桥在北京城西南十余公里永定河上。1937年7月7日,日本侵略军队在这里向中国驻军进攻。在全国人民抗日

---

① 今川藏公路。

热潮的影响下,中国驻军进行了抵抗,中国人民英勇的八年抗战就是从此开始的。

卢沟桥是一座永定河上的"连续桥"。桥长共二百六十五米,除两端桥堍五十三米外,桥身二百一十二米,分十一孔,每孔净空约十六米,桥宽约八米,桥高约十余米。桥面用石板铺砌,两旁有扶手石栏,各用石柱一百四十个,高一点四米,内嵌石板为栏。桥下为石墩,靠上游一面,筑成尖嘴形,以便分水破冰。

永定河,发源于山西,所经山谷,向少森林,且土性疏松,易被冲刷,因为河水挟沙特多,一至下游,地势平缓,又放沙淤积,垫高河身,遇有大水,即会横决泛滥,善淤善决,有如黄河,故古代有"无定河"之称。为了保障北京,清代康熙、乾隆两朝,曾在沿河两岸,筑有石堤束水,以防洪水威胁;在筑堤前,有人恐因此水势更猛,河底刷深,影响卢沟桥的基础,但事后并未发生问题,足见桥工坚固,经受考验。

卢沟桥的每个桥孔,均呈圆拱形,拱上为桥面,桥上载重,通过圆拱而传至拱脚的桥墩。每个桥墩,左右各有一拱,这一拱的脚就是下一拱的起点,因而全桥十一拱,连成一线,每一拱上的载重,就通过桥墩,而由全桥各拱所共同负担。这就把这十一个拱,连成一个整体,充分发挥了彼此互相的作用。就因这个缘故,上述的筑堤束水,才未发生事故。这就是"连续桥"形式的一个优点。由于多拱组成，卢沟桥的形式就叫做"联拱桥"。

永定河上的桥，为何叫作卢沟桥呢，原来这条河在唐代叫卢沟，据《畿辅通志》载："卢师为隋末神僧，能驯二龙，台其遗迹，唐书作卢思台，系在宛平县境内。卢沟系由河经卢思台而名，非仅取义于水黑曰卢也。其曰沟者，因由此入山，不通舟楫，如沟渠之类。"但永定河名卢沟的时间很短，为何桥名卢沟，沿习至今呢？很可能最早的卢沟桥，就是唐代修建的。

《唐书·苇挺传》："贞观十九年将有事于辽东，择人运粮。挺至幽州，漕渠壅塞，遂下米台（卢思台）侧。"《新唐书》云："自桑干水抵卢思台，行八百里，渠塞不可通，挺欲通漕至卢沟桥，达于宣大，以出蓟辽，而卢沟以上，山溪逼仄，不可通舟，故止下米于卢思台侧也。"可见唐代卢沟上已有桥，这桥才叫"卢沟"。不过这桥是何情况，是木桥或浮桥，是否即在今址，都不可知。许亢宗《奉使行程录》云："卢沟河水极湍激，每候水浅深，置小桥以渡，岁以为常，近年于此河两岸，造浮梁，建龙祠，仿佛如黎阳三山制度。"可见河上有浮桥是由来已久了。至于石桥，根据记载，那是到金代才建成的。

《金史·世宗本纪》："大定二十八年（公元1188年）五月诏，芦沟河（元明时，卢沟常作芦沟）使旅往来之津要，令建石桥，未行而世宗崩。大定二十九年六月，章宗以涉者病河流湍急，诏命造舟，既而更命建石桥，明昌三年三月（公元1192年）成，敕命名曰广利，有司谓，车驾之所经行，使客商旅之要路，请官建东西廊，令人居之……亦便于观望也。"这时所建石桥是何形状，未见记载。唯据意大利人马可·波罗所作游记，他曾于公元1292年，即在桥建成后一百年，亲眼见到此桥，大为赞赏。他说："在这条河上，有一座很好看的石桥，在世界上也许是无可比拟的。它有三百'步'长，八'步'宽，因而十个人不难并骑过桥。它有二十四个拱，支持在水中的二十五座桥墩上，墩是'蛇纹石'做的，砌工极好。"这就和现在所见的桥，大不相同了。因为现在的桥只有十一拱，而非二十四拱。不过，究竟马氏所见到的是否二十四拱，也有疑义，尽管他又特别指出，拱下有二十五座桥墩。北京中国历史博物馆，藏有一幅《运筏图》，绘出卢沟桥两岸情景。北岸没有宛平县城，还只是空旷的广场。两岸来往行人可分为两类，一类是骑马乘轿的官人和指挥行人的小吏，一类是背包挑担、推车运货的劳动者。官吏是阔面多须，头戴阔边盔状的帽子，身穿圆翻领或敞尖领的衣服。还有坐在人群中，监视居民的小头目。所有这些，根据专家鉴定，都说明这幅画是元人作品，与马氏游记同时，但这画中的卢沟桥就只有十一个拱，而非二十四个拱。

卢沟桥建成后，经过元、明、清三代修理就逐渐形成今天的式样。其中小修居多,大修只有一次。《读史方舆纪要》载:"元至正十四年(公元1338年)命造过街塔于桥上。"《明史》载:"卢沟桥正统九年(公元1444年)重建石栏,刻为狮形。"《畿辅通志》载:"卢沟桥……金明昌年间建,元、明两代屡有修筑,清康熙元年(公元1662年)发帑修筑,七年水溢,桥圯,东北十二丈重修,御制碑文,建亭于桥北。"《顺天府志》载:"卢沟桥……金明昌初建。明正统间重修,皇清至康熙八年重修,长里许,插柏为基,雕石为栏。"由于上年水溢桥圯,康熙八年的重修应当是较为彻底的,可能修成现在的形式。《畿辅通志》载:"雍正十年(公元1732年)重修桥面;乾隆十七年(公元1752年)重修桥面、狮柱、石栏、桥厢,五十年(公元1785年)重修桥面、东西陲,加长石道。桥东西长六十六丈,南北宽二丈四尺,两旁金边栏杆,宽二尺七寸。东桥坡长十八丈,西桥坡长三十二丈。东桥翅南长六丈,北长六丈五尺;西桥翅南北均长六丈,出土一尺四。"又云:"桥长六十六丈,虹十有一孔。寻常水宽约四五孔,夏秋水涨达七八孔,唯遇雨潦极盛之时,则十一孔均有水,若冬春水小,才只两孔过溜而已。"上文中的"东西坡"即桥堍,"南北翅"即桥头路面放宽处,"出土"即桥头路面距原来地面的高度。马氏游记中,对这"坡""翅"的细节,也有描写,好像观察得很细致,但对桥拱数目这样的重要资料,反而不符,未知何故。然而从他的游记中,也可看出,清代修理,只不过是恢复原样而已。

卢沟桥上的附属建筑,如金代所建东西廊和元代所修过街塔,今俱不见,清代所造碑亭则依然存在。最负盛名的石栏狮柱却自桥成至今,一直脍炙人口,名闻中外。马可·波罗游记中说"桥顶有一高大柱子,立在石龟上,柱底附近有一大石狮,狮上面另有一狮。由此向桥墩,每隔一'步'半,即有一根好看的柱子,上面都有狮子,在全桥的各柱之间,都有精工雕刻的栏板,共同构成美丽的奇观"。可见建桥时就已注意美化桥上的装饰,表现出当时的艺术风尚。其后明代修理,"重建石栏,刻为狮形",清代修理的"雕石为栏",重修"狮柱石栏",等等,大概都是加以补充,雕工精益求精。表现得最突出的是栏柱头的狮子,不但雕出的数量大,而且雕琢的艺术极高,画狮子形态,淋漓尽致。各个柱头狮子,无一相同,有多有少,有动有静。有的大小嬉戏,神态活现;有的交头接耳,恍同对话;更有昂首耸耳,好像在倾听水声人语。造型之妙,叹观止矣!而且,"卢沟桥的狮子数不清",成为北京谚语,由来已久,明代《帝京景物略》说:"石栏列柱头,狮母乳、顾抱负赘,态色相得,数之辄不尽。"清代《顺天府志》说:"栏上百狮,子母抱负,不可数计。"这个谜,到了1962年才被打破,经过北京市文物工作队的调查,卢沟桥的狮子,大小总共四百八十五个。

卢沟桥地处入都孔道，行旅当然稠密。上述《运筏图》中在桥两头，绘有许多酒亭客舍，檐前帘幌高悬，都足见那时的繁荣景象。元代蒲道源诗："卢沟石桥天下雄，正当京师往来冲。"（见《辛斋诗话》）陈高《卢沟桥晓月图》诗："卢沟桥西车马多，山头白日照清波。"（见《陈子上存稿》）杨奂诗："自有五陵年少在，平明骑马过卢沟。"（见《还山遗稿》）到了明代，更是兴旺。《戴司成集》载："桥上两旁皆石栏，雕刻石狮，形状奇巧，金明昌间所造，两崖多旅舍，以其密迩京师，驿通四海，行人使客，往来络绎，疏星晓月，曙景苍然，亦一奇也。"吴国伦《卢沟桥》诗："喧喧行路人，日昃去未已。"（见《詹甄诗稿》）张野过卢沟的《满江红》词："桥下水，东流急，桥上客，纷如织。"（见《古山乐府》）在清代，卢沟桥更成为"九逵咽喉"（《顺天府志》），并在此设关抽税，致民间有"卢沟桥，卢沟桥，雁过也拔毛"的谚语。桥南五里的长辛镇，为了便利过桥的客商，开了许多客店、马店、大车店等等的店，跟着繁盛起来，因而就改名为长辛店。

茅以升拓印卢沟桥碑上诗

同时，经过文人渲染，卢沟桥又成为风景胜地。据《明昌遗事》，金代起即有"卢沟晓月"之说。元代袁桷有《题卢沟烟雨图》诗云："驱马上河梁，园晕新雨纹。"（见《清客居士集》）赵宽有《题卢沟晓月图》诗云："长桥卧波鳌背耸，上有车马萧萧行。"（见《半江集》）《黄图杂志》云："元时卢沟桥畔有符氏雅集亭。"桥旁有亭，亭称"雅集"，可见那时文酒宴会之胜。到了明清两代，卢沟晓月，更是脍炙人口，题咏日多，如明王英诗"曙色微涵波影动，残光犹带浪花流"，顾起元诗"最是征人望乡处，卢沟桥上月如霜"（见《嫩真草堂集》），鲜于必仁《折桂令》曲"出都门，山甚空蒙，林影玲珑，桥俯危波，车通远塞，栏倚长空。起宿霭千寻卧龙，掣流云万丈垂虹，路杳疏钟，似蚁行人，如步蟾宫"。

卢沟桥又名古战场。《读史方舆纪要》载："金兵南迁，留太子守中都，辽军杀其主将以叛，福兴闻变，遣军阻于卢沟桥，使勿得渡。"又《元史》："天历初，上都兵入紫荆关，游兵逼都城南，大都兵与战于卢沟桥，败之。"又《明史》："建文中，李景隆谋攻北平，燕将请守卢沟桥以御之。"又《革除录》："李景隆征兵诸路，合五十万，引兵渡卢沟桥，遂围北平。"又《破梦闲话》："于是当桥之北规里许为斗城，局制虽小，而崇墉百雉，俨若雄关……并于崇祯丁丑，特设参将控制之。"此"斗城"即宛平县城，1937年"七七"抗战时，我守军由此上桥，抗击了日本帝国主义的侵略。

安平桥

安平桥在福建泉州晋江县安海镇的西南，跨越海湾，通往南安县的水头镇，为南宋绍兴二十二年（公元1152年）建成的石梁桥，又名五里桥。并非离开什么城市有五里路，而是它本身就有五里长，这在世界古桥中，恐怕是唯一的。泉州民间多年来传说"天下无桥长此桥"，却也当之无愧。在我国现代桥梁中，除去郑州黄河桥外，也还没有比它更长的。

安平桥所以要这样长，是因为要跨过一个海湾，从东面安海镇的海岸跨到西面水头镇的海岸。海湾通向台湾海峡，里面的船只虽不能远涉重洋，但在安海与水头之间，却是古代的唯一交通工具。泉州自南北朝起，就有了海外交通，到了唐代，更成为全国对外贸易的重要港口。安海镇古时名安海渡，原是个水陆码头。由于泉州繁盛，它就跟着兴旺起来。南宋赵令衿《石井镇安平桥记》（据《清源旧志》"安平桥在修仁里、石井镇、安海渡"）云："濒海之境，海道以十数，其最大者曰石井，次曰万安，皆距闽数十里，而远近南北官道所以出也……唯石井地居其中，而溪尤大，方舟而济者日千万计。"可见安海渡需要安平桥，同万安渡需要洛阳桥，同一迫切。

由于都是跨海，这两桥的修建，也同样艰难。安海到水头的海面，已经够宽了，同时，还纳入从西面来的注入海湾的河水。秋季又有台风。当山洪暴发而又加海潮袭击时，海湾里波涛汹涌，过渡都很危险，何况造桥。上述赵文云："飓风潮波，无时不至，船交水中，进退不可，失势下颠，漂垫相保，从古已然，大为民患。"清代陈万策《重修安平桥记》云："安平地压巨海，广衍数十里，南北往来市舶之区，泉之一大都会也。其西襟九溪之流，波涛萦折，以浚于海。"因此，安平桥建成后，万民争涌。赵文云："老壮会观，眩骇呼舞，车者徒者，载者负者，往者来者，祈祈舒舒，无所濡壅。"

《清源旧志》云："安平桥……界晋江南安溪，相望六七里，往来先以舟渡，绍兴八年（公元1138年）僧祖派始为石桥，镇人黄护与僧智渊各施钱万缗为之倡。派与护殁，越十四载未竟。二十一年（公元1151年）太守赵公令衿卒成之。其长一千三百四步，广三步有奇，疏为水道三百六十有二，自为记，榜曰安平桥。"赵令衿《石井镇安平桥记》云："……爰有僧祖派始作新桥，今派死不克竟……黄逸为倡，率僧惠胜……径始之日，人咸劝趋，即石于山，依材于麓，费缗钱二万有奇，而公私无忧，自绍兴之辛未（公元1151年）十一月，越明年壬申十一月而毕，榜曰安平桥。其长八百十有一丈，其广一丈有六尺，疏为水道者三百六十有二，以栏楯为周防，绳直砥平,左右若一,阮然玉路,崿然金堤,雄丽坚密,工侔鬼神。"据福建省文物管理委员会 1957 年 12 月调查:"桥现存长度两千零七十二米,而宽三至三点八米。设桥墩三百一十四座,全用花岗岩筑成。桥面直铺着四至七条石板,板的长短不一,长八至十一米,宽零点五至零点八米,厚零点三四至零点七八米。桥板两端的接头处又有横铺的石条,桥墩有三种不同形式。"桥墩皆用条石横直交错叠砌而成,一为长方形墩,为数最多,二为船形墩,两端成尖状,便于排水,设在水流较急而较宽的主要港道,三为半船形墩,一端为尖状,另一端为方形,也设于较深的港道部分。(以上见《文物参考资料》1958 年第 12 期)

从结构形式来说,安平桥几乎是完全模仿万安渡的洛阳桥的,两桥都在泉州濒海地区,都是所谓"简支式"的石梁桥,不过洛桥只长五百四十米,却早在一百年前就完工了。很可能,安平桥的建筑方法,也借鉴于洛阳桥,没有什么特殊创造。然而安平桥的长度为洛阳桥的四倍,工作量大得多,但在赵令衿时,却只用了一年时间就完成了。虽然在祖派、黄护、智渊等的提倡下,也许做过一些工程,而这一年的桥工成就,确实是惊人的,难怪赵令衿自夸"实古今之殊胜,东南所未有也"。《清源旧志》说,安平桥成后,"明年复有镇人请于公(赵令衿)曰,镇东南隅,复名东洋,其港深阔,愿复得桥。

公许之,不半载而成,长八百六十五步,分二百四十二间。较安海桥(即安平桥)为三之二"。规模为安平桥三分之二的东洋桥,居然在半年内就建成了,可见赵令衿的造桥队伍中,确实有卓越的工程师。自从东洋桥建成后,安平桥就又名西桥,但东洋桥不久就毁坏无存了。有个《重修安海桥募缘疏》说:"自东桥荡析,恻孤影以存羊,叹反复之无常,觉成亏之有数。"

因为桥太长,建桥时就在桥上造了五座亭子,以便行人休息。上述赵文云:"又因其余材,为东西中五亭以休。"《安平志》云:"桥之东西中半,凡为亭五,后废其二,唯东西中三亭存耳。东为超然亭,以祀观音,后火焚,今已筑城废,中为泗州亭,祀佛于其中,西在南安三十九都鸡幕山下,水陆坊。"根据近年调查:"桥亭现存五座,分设在桥的东、西、中等处,全系清代改建。东端桥亭叫水心亭(亦名桥头亭),中部桥亭也叫水心亭,因位于桥的中部,又称为中亭。中亭规模较大,面宽十米,周围保存有重修碑记十三座……亭的两侧还立着两尊石刻武士像,高一点四米,手执长剑……西端桥亭叫海潮庵……在中亭以东四百五十米和以西二百八十米处,分设路亭一座,建筑简单。"(见《文物参考资料》1958年第12期)《安海志·水心亭》节云:"水心亭即西桥中亭。附在桥中,丙申迁界毁,康熙丙子(公元1696年)复界后重建,居西桥之中，为晋（江）南（安）交界，故号曰中亭。"此外，在桥东端安海境，有八角七层砖塔一座，因塔身涂白，俗称白塔，传说是宋代建筑，今尚完整，在中亭以西三百米处，还有两个方形实心小塔，对立于桥两侧的海滩上，其风格与宋代石塔相近。

桥上原有扶栏望柱，故赵令衿文中有"以栏楯为周防"语，并附以诗云："玉梁千尺天投虹，直槛横栏翔虚空。"但现时一无存者，足见历代遭受破坏的严重。除去山洪与潮水的冲击和台风的侵袭外，还受地震影响。在晋江和惠安两县交界的洛阳桥，就因地震成灾而被损过三次。安平桥也很难幸免。《泉州府志》云："嘉靖三十七年（公元 1558 年）倭又四千余至惠安……倭分两支……一由清源山前寇南安，陷之。"其后两年，又一再犯南安，安平桥为南北孔道，也可能遭到倭寇的破坏。《重修安海桥募缘疏》中有"寇贼怯而不敢过，险同天堑"之语，可见桥是做过战场的，经过这样多的天灾人祸而能保存至今，安平桥的修理，当然是很频繁的了。

在中亭周围保存的十三座重修碑记中，最早的明天顺三年（公元 1459 年），最晚的清光绪十二年（公元 1886 年）。但根据记载，明初永乐二年（公元 1404 年）即有修缮（见《安平志》）。在这许多修理中，比较重要的有：（1）明天顺初年的，据陈弘《重修安平桥记》云："……逮倾圮而当南涯溪潮之处，毁断尤甚，乡人以木板代跨以渡……过者病焉，人咸乐输，遂先新水心亭,次及桥道,自北涯起,倾者砌,断者续,因复建亭于上,是岁十月兴工,越三年己卯(公元1459年)八月而讫。"(2)明万历年间,据颜嘉梧《水心亭碑记》云:"……陈弘鸠众重修……绵绵至今一百二十六年有奇,而日毁月损者……时有乡先生……筑城以捍其西,修桥以壮其趾,俱以未遑而先即世……躬执募缘之役,先新水心亭,次及桥道,自是岁(公元1600年)十一月兴工,越明年……庶几就绪。"(3)明崇祯十年(公元1637年)秋,郑芝龙再倡修中亭,翌年正月竣事,芝龙撰立碑记,竖在中亭对面桥边。其后,清顺治十八年(公元1661年),统治者为了消灭郑成功抗清的群众力量,迫令安海居民迁徙内地,把所有桥上建筑物,全部焚毁,中亭也在其内。(4)清康熙二十三年(公元1684年),安海人逐渐在废墟上重建家园,施琅复兴中亭。(5)清康熙四十六年(公元1707年),蓝理筑西埭,以海土填水心亭两旁,阔十丈,长一里许,筑屋百余间为市,重拆桥头十余坎,以断西塔路,别于桥旁筑土岸,接三陛门,以通新街市。(见《安平志》)(6)清雍正四年(公元1726年)时,因过去部分桥坎被水冲坏,暂以木桥代替,改木为石,翌年完成。张无咎《重修安平西桥记》云:"安平镇之里民以西桥倾圮具告,万民病涉,舆徒阗咽,望洋之众断断如也……民鼓舞竞劝,趋事赴功,不待桴鼓之督,而圮者整,断者续,不日而已落成,计所捐与所乐输凡及千缗。"

（7）清嘉庆十二年（公元1807年）重修桥道，历时一年多，翌年竣工，有徐汝澜撰《重修安平桥记》。（8）清同治五年（公元1866年）修建中亭，门口石柱上，刻有对联一副"世间有佛宗斯佛，天下无桥长此桥"。（9）民国十七年（公元1928年）有旅菲律宾闽籍华侨，捐资修葺桥道。

从以上修桥记录，得知几件值得注意的事。（1）不论石刻或文献，都未提到明代永乐以前修桥的事，而自桥成到永乐，已经二百五十多年了。也许有过小修理，但桥身坚固，应无疑义。（2）几乎每次修桥经费，都以募化乐输而来，说明民间对这桥的重视，同时也可见历来政府都不把修桥当作本身责任。并且，安平桥之所以能创建，就是由于群众力量。上述赵令衿的桥记中就曾说"斯桥之作，因众志之和，资乐输之费，一举工集，贻利千载，是岂偶然也哉"。（3）靠募捐来修桥，当然也不简单，其中必有很多热心人士，出了力量；但也少不了一些"急公好义"的士绅，希图在修桥碑文里留下名字。值得提出的，是这里面的和尚的贡献。根据赵令衿文及《清源旧志》，在建桥时就有僧祖派、僧惠胜、僧智渊等，而且僧祖派还是个工程师，因为赵文里说他"始作新桥"。在历代修桥工作中，和尚关系一定很多，不过上述修桥碑文中未曾提名而已。但是，桥上各亭内，多有佛像，可见佛教对这桥的影响是很大的。（4）桥上建筑，如五亭的损坏，好像比桥身还多,清顺治时全被焚毁不必说了,就是一般破坏也比较严重。在很多修桥文中,都说到"先新水心亭,次及桥道"。这也证明,桥身比亭子坚固。(5)从郑芝龙修桥及安海居民被迫迁徙,看出这座桥在郑成功抗清战役中的作用。(6)清康熙间桥上修了百余间房子做市场,可见那时桥上交通的繁盛。

根据1961年调查,安平桥的现状如下:由于自然变迁,桥下海滩,渐被泥沙淤塞,除中亭港、西姑港及水头港三港外,其余几乎全部淤成陆地。东首白塔以西的约长二百米的桥身,自清初填塞造屋后,到现在已成为小街道,中亭以东的海滩,也变为农田。桥的原来长度及桥墩数量,因之大为减少。据实测,现桥全长为两千零七十米,桥墩存三百三十一座,内方形墩二百五十九座,船形墩二十七座,半船形墩四十五座。

王世懋《闽部疏》云:"闽中桥梁甲天下。"所以能"甲",就因先有洛阳桥,后有安平桥。

古籍中提到的安平桥

安济桥及永通桥

安济桥和永通桥都在河北省赵县，都属于同一类型的石拱桥。安济桥即是中外闻名的赵州桥，在赵县南门外五里的洨河上，建成于隋代开皇末年至大业初年间。永通桥在赵县西门外清水河上，建成于安济桥后，但不晚于宋代。两桥的结构形式几乎完全一样，不过一大一小，因此安济桥又叫大石桥，永通桥又叫小石桥，小桥简直是大桥的雏形。

赵县洨河发源于山西，流经宁晋县，入滏阳河，平时水小，甚至常年无水，河身可做耕地，但发水时则成巨川。张孝时《洨河考》云："洨河发源于封龙山北之南寨村，两壁峰峦峻削，瀑布悬崖，水皆从石罅中流出。"《水经注》云："洨水不出山，而假力于近山之泉。今考洨河，实受西山诸水，每大雨时行，伏水迅发，建瓴而下，势不可遏。"《赵州志》云："当时颇称巨川，今仅有涓涓细流。唯夏秋霖潦，挟众山泉来注，其势不可遏，然不久变为细流矣。"《读史方舆纪要》："宋咸平五年，河北漕臣景望开镇州常山南河入洨水至赵州，以利漕，即此。"可见洨河是山泉及河水汇合而成，冬春干涸，夏秋泛滥。最近1954年大水，河身全满，洪峰距桥洞顶点仅一点七七米，然水过后，又成细流。永通桥下的清水河，夏季也很汹涌，明代王之翰《重修永通桥记》中有云："大雨霖霆，百潦会流，层浪浴天，惊涛拍岸，澎濑灏湫，横恣涯渚之间。"

赵州地处要冲，自古为南北官道所经，交通茂密。历代歌颂赵州桥的诗文中，都提到这点。唐代张彧《安济桥铭》："万里书传，三边檄奏，邮亭控引，事物殷富，夕发蓟壖，朝趋禁雷。"宋代杜德源《安济桥》诗："坦平简直千人过，驿使驰驱万国通。"元代杨鱼《安济桥》诗："五丁凿石极坚顽，陌上行人得往还。"明代鲍捷《安济桥》诗："洨河之水清且泳。来往征人急于蚁……任渠车马纷于织，往过来续无病涉。"明代张居敬《重修大石桥记》："赵为畿辅要区……四通之域也，往来肩摩毂击，而洨水汇四泉之流……假令津梁不通，则邮传檄奏之驰驱，货殖车舆之辐辏，将不胜争渡之喧。"永通桥上的繁盛，也大体近似。宋代杜德源《永通桥》诗："金道马尘奔驿传，玉栏狮影炽清晖。"明代王之翰《重修永通桥记》："地居九省之冲途，扼西京之要历……桥修而行人之往来，不厉不揭，而咸攸济也。"可见，由于洨河和清水河的阻隔，大小石桥的修建，在当时确有需要。而且，两桥的设计必须适应桥下水流的特性和桥上交通的条件。

安济桥只用一孔石拱，跨过洨河，跨度三十七点零二米，连南北桥埝，共长五十点八二米。我国石拱，一般都是半圆形，如卢沟桥，但这个桥的拱形却不是半个圆，而只是圆弧的一段,因而拱顶在拱脚水平线上的高度,不是圆的半径,而是大大小于半径,这个高度叫做"弧矢"。如按这桥拱形的圆弧说,圆的半径应当是二十七点七米,但因只用了一个弧,拱的弧矢就只有七点二三米,和拱的跨度相比,约为五分之一。这样的拱,叫做"平拱",以别于像卢沟桥的"穹隆拱"。拱圈的厚度为一点零三米,全拱一律。每道拱圈,用约四十三个石块组成,每块约厚三十多厘米。重约一吨。因为桥厚约九米,故拱圈有二十八道,每道自成一独立拱。板面亦石块铺成,分为三股,中间走车,两旁行人。在桥面和拱圈之间,拱顶两旁都有三角地带,这地带叫"拱肩"。一般拱桥的拱肩都是实心的,即是用石块填得满满的。但安济桥不同,在拱肩里还开了小的拱洞,拱顶每边有两个,全桥有四个小拱,伏在下面大拱的肩上。这就把拱肩敞开了,所以这种拱叫做"敞肩拱",安济桥是世界上最早的敞肩拱桥。两肩上的小拱是对称的,大的跨度约三点八一米,小的二点八五米;拱圈都是圆弧形,也用石块砌成,在全桥宽度内,也是二十八道。

永通桥的结构形式,完全模仿安济桥,也是用一孔石拱,跨过清水河,跨度约二十六米,弧矢约五点二米。桥宽约六点三米,无人行便道,由二十一道石拱组成。大拱上也伏有四个小拱,每个小拱也有二十一道拱圈,但是小拱对大拱的比例,不完全同于安济桥,小石桥的小拱比大拱,比起大石桥的小拱比大拱,要显得大一点,这正是设计的巧妙。

为何大小石桥都采取这种"单跨敞肩拱"的形式呢? 首先是洨河发水时,"建瓴而下,势不可遏",因而桥建成后要尽量减少对水流的阻遏,不但河中不应有桥墩挡水,而且拱桥两肩的三角地带,也不该形成挡水墙。单跨过河,就没有水中桥墩;两肩空敞,就减少阻水面积,比起"实肩拱",可增加过水面积百分之十六点五。其次,单跨过河,则全桥重量和桥上车马荷载,都集中到两个拱脚,引起对桥台基础的很大压力,但如拱肩里有四个拱洞,则可减少大拱圈上的重量五百多吨,相当于桥身自重的五分之一。洨河河床是冲积性的沙土,并非岩石,桥台建筑在这样的天然地基上,减轻拱脚压力,是有很大意义的。再其次,安济桥全部石料都是青白色石灰岩,从外地获鹿、赞皇、元氏各县石场采运,工料费当然可观,桥身开挖四个拱洞,就可节约石料二百多立方米,这从经济上说,是个附带的收获。

为何大小石桥都采用平坦式的拱桥,而不用穹隆式的拱桥呢? 首先是桥上的车马一定很多,而车马上桥,路面就不宜太陡。如用穹隆拱,路面就要做成阶台式,那么车子就无法上桥了,而北方运输却主要靠大车。现因用了平拱,桥上路面坡度就比较平缓,只合百分之七,即平行一百米,上升七米。近代公路的坡度,最大为百分之四,安济桥的坡度就算适当了。其次,穹隆式拱桥的优点是便于桥下走船,这在我国南方是相宜的,那里的水道多,主要运输靠船舶,而且船桅较高,陆上也是肩挑多而车辆少。大小石桥下的河流,虽也走船,但船不高,没有穹隆拱的需要。再其次,如果为了水流畅通,只用单跨拱而不用多孔桥,那么,平拱就比穹隆拱省工省料,更为经济。最后,只有用了平拱,才能把拱肩敞开,在大拱上加上四个小拱,而这在穹隆拱是不可能的。

为何大小石桥的拱圈都用二十几道独立拱,并排组成桥的宽度,而不用一道很宽的拱呢?首先是每道拱圈都能独立站稳,不受他道影响,一道损坏,不影响全桥安全。其次是施工时可以一道一道地砌,使桥逐渐加宽,假如砌成几道后,忽然发水,就可暂停工作,候水退再做。但如做一道全桥的宽拱,半途停工是很危险的。再其次,搭拱桥要用拱形木架做支座,拱圈狭了,拱架就省了,一道拱圈砌完,同一拱架可用来砌下一道,这样重复用很多次,比用一次桥宽的拱架,节约何可胜计。再其次是修桥方便,哪一道拱圈坏了,就修哪一道,不像全桥宽的拱圈,"牵一发而动全身"。但是,这样很多道独立拱圈拼成全桥一个拱,也有它的缺点,那就是各个拱圈如果不协调动作,就会失去整体作用。这个缺点正是独立拱的优点所引起的。必须克服缺点,才能发挥优点,因此,大小石桥就采用了许多办法,把这二十几道独立拱,联系成为一个整体的全桥拱。结果虽非十分满意，却已经过千年考验。

所有这些，都是大小石桥在桥梁结构上的特点。有了这些特点，桥梁作为交通工具的作用，就能充分发挥了。而且它还不妨碍桥下河流的畅通，并能抵御桥上各种自然界力量的侵蚀，以便长期服务。像这样的技术上的成就，而且远在隋代就已如此圆满地完成，这在我国桥梁史上是空前的。不但在我国，就是在世界上也是空前的，因为在欧洲，真正敞肩式的拱桥是到19世纪才通行的。至于石砌拱桥，在欧洲西班牙有一座建成于罗马帝国时期，是多孔穹隆式的联拱桥，据说距今已有一千八百年之久。然而中经战事，有六百年因毁损未用，其服务年限仍远逊安济桥。技术上的任何成就都不可能是偶然的，其中必有科学道理，为长期的实践所验证。安济桥的敞肩拱、平拱路面、并非独立拱等等的创造发明，不但为千年以上的实践所证明为有效，而且根据1955年的实测，拱圈两脚的沉陷是基本平衡的，并且为数无多，这是对任何桥梁最严格的考验。更当特别指出的是这桥的一孔跨度长达三十七米以上，这在当时世界上也是少见的，跨度大而拱圈薄，这桥如何能胜任繁重的陆上运输呢？若以近代的科学方法来验算，即可发现，这桥的构造是充分利用了小拱对大拱的被动压力的，这在当时，诚然是只知其然而不知其所以然，然而所以能知其然，那就无疑是根据实践的经验总结了。可见安济桥不但在桥梁技术上有贡献，就是在桥梁学术上，它的成就，也是大大超越时代的。

安济桥不但在技术和学术上有超时代的成就，而且在艺术上也是个光辉形象。它的弧形平拱和敞肩小拱的本身就是个艺术品，线条柔和，构造空灵，既稳重又轻盈，寓雄伟于秀逸。并且拱上桥面亦作弧形，不过弧径稍大，与拱弧天然配合。桥面两旁有扶栏望柱，栏板有蟠龙石雕，望柱上有狮首石像。更有斗子卷叶、多瓣花饰等等雕刻，多系隋代精品。一切美术布置都无堆砌现象。可以说，整座安济桥是个技术、学术和艺术的结晶品，也是个真、善、美的统一体。永通桥的情况也近似，连栏板浮雕都很美，是善于模拟的。

安济桥建成后，树立了榜样，风声所播，各地仿效者甚多，俨然形成一个"敞肩拱学派"。在赵州，除永通桥外，有济美桥，在宋村东北的洨河水上，建成于明代（公元1594年），其时正在修理大石桥。在山西，有晋城的景德桥，亦名沁阳桥，建成于金代（公元1161～1189年），有崞县的普济桥，建成于金代（公元1203年）。在河南，有济源县的望春桥，亦名通济桥，建成于金代（公元1177年）。在浙江，有余杭县的苕溪桥，建成于明代（公元1368年）。这些桥都是石拱桥，大拱上都有小拱，不过有的小拱在大拱脚的桥墩上，还非正式敞肩拱。

全世界敞肩拱学派的始祖是安济桥的总工程师李春。在我国历史和文献里,关于桥的记载是非常丰富的,有兴修文记、工程杂录、人物故事、抒情文艺等等,真乃洋洋大观。然而唯独对于桥梁本身最有密切关系,在设计施工上负有重责的技术工作者,总是缺略不书,认为无足轻重,实是无可补偿的一件憾事,想不到这件憾事倒不曾碰在安济桥上,因为唐代张嘉贞所作的《安济桥铭》的序文里,开宗明义的第一句话就是:"赵州洨河石桥,隋匠李春之迹也。"这就为我国桥梁史留下了极其珍贵的史料。李春应当列入世界名人的行列,为造福人类留念,为我国文化争光。纪念李春也就是纪念他所代表的我国古代的所有桥工巨匠,因为他的成就无疑是所有他的先驱者的经验总结,和在这基础上的发扬光大。北京大学在《金石汇目分编》中发现一项记录,在安济桥下曾有一块唐山石工李通题名石,上有"开皇□□年"(□为缺文,下同)字样,则是和李春同时修桥的,还有李通,但不知他在安济桥的"迹"是怎样的。

安济桥的建成,反映了当时社会经济的物质条件。从安济桥下发掘出的一些断碣残碑中可以看到有六块"修桥主"题名石,这些石块有的和桥上小拱的墩石相同,当是修桥时的同时作品。从此证明,修桥经费中必有一部分是"修桥主"

所捐赠的。古时把修桥补路当作做功德，同时也为自己祈福。甚至"布施勾栏"的三块名石上，也有"阖家同增福寿"等字样，捐钱的人多，说明社会尚有财力。根据这些"修桥主"题名石上的文字，俞同奎同志在《安济桥的补充文献》（《文物参考资料》1957年第3期）一文中，推论安济桥的建桥年份，认为是"开皇末年兴修，经过仁寿四年，到大业初年完成"，这就把建桥时期提前到公元605年左右了。

安济桥建成后，历代赞美歌颂的诗文，绵延不绝。见于《赵州志》著录的就有唐代张嘉贞的《安济桥铭》的序文："制造奇特，人不知其所以为……緘穿隆崇、豁然无楹……两涯嵌四穴，盖以杀怒水之荡突……其栏槛华柱，锤斫龙兽之状，蟠绕拏踞……"唐代张彧的《安济桥铭》云："虹舒电施，虎步云构……敞作洞门，呀为石窦……力将岸争，势与空斗……"宋代杜得源诗："驾石飞梁尽一虹，苍龙惊蛰背磨空……云吐月轮高拱壮，雨添春水去朝东。"元代刘百熙诗："半夜移来山鬼泣，一虹横绝海神惊，水从碧玉环中过，人在苍龙背上行……"明代祝万祉诗："百尺长虹横水面，一湾新月出云霄，恒山北接千峰秀，驿路南来万国遥……"又陆健诗："车马人千里，乾坤此一桥……"清代安汝功诗："天桥苍虬卷，横波百步长，匪心坚不转，万古作津梁。"对永通桥的称赞也不少。宋代杜德源《永通桥》诗云："并驾南桥具体微，石材工迹世传稀,洞开夜月轮初转,蛰启春龙势欲飞……"明代王之翰《重修永通桥记》云:"桥不楹而耸,如驾之虹,洞然大虚,如弦之目,旁挟小窦者四,上列倚栏者三十二,缔造之工,形势之巧,直足颉颃大石,称二难于天下……"上述诗文中,有的是在桥上刻石留迹的,如张嘉贞和张彧的两个石桥铭柱础残石,都在桥下发掘出来了,同时还挖出很多题有职衔的残石,大概都是想借"到此一游"而"留名千古"的,其中有的是唐代所刻,可见在那时,就已以过桥为荣了。还有一宋人石刻,于仰慕李春之余,表示了不甘示弱之意,他说:"不如其人之善得盛名于世,□桥也以石为桥,历海内不少矣,晋江之洛阳,庐山之三峡,备物□□。"抬出福建泉州的洛阳桥做对比,却也够得上南北辉映,然而洛阳桥却晚了四百五十年。

也正因安济桥地处要冲,它成为历代战争中争夺赵州时的受难者。争夺赵州的战役是很多的。就在上述唐代张彧的石桥铭序文中,他就说:"李公晟奉诏总禁戎三万,北定河朔,冬十月师次赵郡。"那是指公元782年,节度使王武俊攻赵州,李晟来解围事。在这以前,公元698年,武则天时"突厥将直攻定州,陷之,杀吏民数千人,又陷赵州";公元756年,开元十五年春,"史思明攻陷赵郡,夏四月,郭子仪、李光弼败之,进拔赵郡,九月,史思明复陷赵郡"。后来,公元910年,周德威攻朱温"将兵出井陉,屯赵州,自是镇定复称唐天祐年号";公元986年,"契丹大举入寇,攻赵州,大掠而去";公元999年,"契丹隆绪,入寇赵州,帝(真宗)自将御之";公元1105年,宋徽宗升赵州为庆源军,此后"金人入寇,赵地遂为战场";公元1213年,"蒙古分兵攻金河北诸郡,悉拔之,赵地人民杀戮几尽";公元1369年,"十一月,大将军徐达、副将军常遇春,率师下赵州"。(以上俱见《赵州志》)在这些战史中,虽未提到桥名,但大小石桥为进出南门、西门所必经,桥上争夺战是必不可免的。

根据记载,赵州又处在地震区域。根据《真定府志》,公元776年唐代宗十一年,"定州地震,冬无雪";公元777年唐代宗十二年,"恒州地震,宁晋地裂数丈,震三日乃止";公元1307年宋景祐四年,"定州地震";公元1808年元至大元年,"冀宁路地震";公元1508年明正德三年,"武强地震,白毛生";公元1538年明嘉靖十七年,"深州地震,自西北震起,一月间凡数十次倾圮庐";公元1626年明天启六年,"地大震,鸡犬皆鸣,振物有声"。较小地震,不及备载。虽未提到赵州,但同在真定府治,大震时必然波及。可见大小石桥,千年来所经地震,也非同小可。

战争和地震,对任何建筑物说来,都是不利的,大小石桥经过了千百年的战争和地震,所受损害,就可想而知了。此外,每年山洪暴发,水流冲击,势必震撼全桥,风雨摧残,冰霜冻结，更是经常剥蚀。每天过桥车马，践踏不停，不仅伤及路面，而且重车上桥，往往走斜线，下桥时，由北往南者，冲向东南方，由南往北者，冲向西北方，破坏了各道拱圈的联结性。因此，大小石桥千百年来经受了各种自然界和人为的侵凌，无可避免，甚至还有意想不到的残伤。从桥下挖出的石刻中，有一块上写"……从南第一虹下□一僧居焉，曰□心"，下有诗"安济桥通宫道□，□心洞里有闲人，长流不断东西水，□□□驰南北尘"。敞肩小拱，居然还有僧舍作用，当为李春所不及料，然而，穴居人能对桥无伤乎？

损伤既多，大小石桥的历代修缮，就史不绝书。先谈安济桥。最早一次，约在桥成后二百年。在桥下挖出的石刻中，有唐代刘超然的《新修石桥记》，作于贞元九年（公元793年），文云"郡人建石梁几二百祀，壬申岁（公元792年）七月大水，方割陷于梁北之左趾，下坟岸以崩落，上排箦又嵌欹，则修之为可"，可见"二百祀"中全桥完整如故，只是在上年大水中，北岸桥台西首，发现沉落现象，牵动上面拱圈开裂，这才急需修理。桥工坚固，于此可知。这在从桥下挖出的石刻，唐代崔恂的《石桥咏》中，也得到证明，诗云"代久堤维固，年深砌不隳……□俊起花叶，模写跃蛟螭"，都是形容这桥的完整壮丽的。这诗据俞同奎同志考证作于唐开元八年（公元720年），亦即在贞元大水之前。后来到了宋代，《宋史·方技传》云"僧怀丙,真定人,巧思出天性……赵州洨河,凿石为桥,镕铁贯其中,自唐以来,相传数百年大水不能坏,岁久乡民多盗凿铁,桥遂欹倒,计千夫不能正,怀丙不役众工,以术正之,使复故"。大概是靠外拱圈有侧倾现象,而联系拱圈的铁件,又有短缺,因而要把拱圈"正"起来,千夫不能,而怀丙一人正之,可见他并不是"役众工"来恢复凿铁,而是针对拱圈侧倾的自然原因,设法消除这个原因,使桥复故。他的"术"当然是技术,可惜不传了。再过五百年,从明代嘉靖三十八年(公元1559年)开始,对拱石和路面,做了多次修理,皆有碑文为证。张居敬《重修大石桥记》云"……世庙初(约公元1530年左右),有鬻薪者以舫运置桥下,火逸延焚,致桥石微隙,而腰铁因之剥削,且上为辎重穿敝",说明桥拱和路面,都需要修理,"以癸亥岁(公元1563年)……肩其役,垂若而年,石敝如前",并未修好,后来再继,"经始于丁酉(公元1597年)秋,而冬告竣,胜地飞梁,依然如故",为期不过数月,想所修不会太多,孙人学《重修大石桥记》云"桥之上辙迹深处,积三十年为一易石……重为修饰,以永其胜……嘉靖四十二年(公元1563年)"。说明上述张文中所谓"肩其役"的成绩,不过是"修饰"而已,所以"石敝如前"。翟汝孝《重修大石桥记》云"嘉靖己未(公元1559年)……注意郡人修葺……缉其桥之中路,得八尺许",从此先修路面,"于嘉靖壬戌(公元 1562 年）冬十二月兴工，至次年癸亥四月十五日告成。所修南北码头及栏槛柱脚，锤斫龙兽，复加旧制，且增崇故事形象，备极工巧。焕然维新，境内改观矣"。经过四个多月时间，又修南北码头（想是桥堍）和栏槛，大概都是表面工作，并未加强桥拱。就这样，再过八十年到明末，据传说，拱桥的西面靠外五道拱圈坍塌，又过一百多年到清代乾隆间，拱桥的东面靠外三道拱圈坍塌。西面五道，后来不知何时修复。因此，直到 1949 年解放时，大石桥的二十八道拱圈，尚有二十三道存在，其中二十道，可能依然还是隋代李春的遗迹，并未经过修理。

关于永通桥的修理，见于文献的不多。从明代王之翰的《重修永通桥记》得知当大石桥于嘉靖壬戌兴工，癸亥告成时，小石桥也在同时修理，因为"铺石磨砻，栏杆斜倚，行者病其坎壈，而居者叹其倾颓"，也就是路既不平，扶栏也要倒，于是"扶簿募缘，以与四方共嘉惠，不一月得钱如千缗，取石于山，因材于地，穿者起之，如砥平也，倚者易之，如绳正也，雕栏之列，兽伏星罗，照者彩也，文石之砌，鳞次绣错，巩其固也，壬戌之秋，癸亥之夏，为日三百，而大功告成"。三百天内一千缗，就把大功完成了，所修工程的内容，可想而知。大概拱圈二十一道，也和大石桥同样坚固，无须修缮。

从大石桥的"修桥主"和小石桥的"募缘簿"看来，开始建桥时固然有部分工款来自捐赠,到了后来修桥时,恐怕全部经费,都是"随缘乐助"。由于路面和扶栏,都是有目共睹的,因而化缘比较容易,同时对栏板雕柱,也不妨"备极工巧,焕然维新"。出钱多的人,或是修桥记中留名,或是立柱刻石纪念,都是求之不得的。出力而不出钱的,当然也大有人在,在这里面值得注意的是和尚。关于重修大石桥的张居敬文内有"令僧人明进缘募得若干缗"语,孙人学文内有"僧悟成募缘"语,这两位和尚,除了化缘而外,不知曾否参加了修桥工作,如同宋僧怀丙那样。和尚修庙,往往是化缘兼施工,他们中尽有卓越的工程师。不过大石桥里还有道士遗迹,明代翟汝孝修桥文内有"曩桥之穹隆处悬一石约丈余",一道人修好,"神实相之",好像是个"仙迹"。

由于大石桥的巧夺天工,自宋代起就盛传桥上有仙迹。杜德源诗云:"休夸世俗遗仙迹,自古神工役此工。"世俗传说:桥是鲁班修,修完后,张果老骑驴,柴荣推车,同时上桥,使用仙法,运来五岳名山,把桥压得摇摇欲坠,但鲁班用手在桥东侧使劲顶住,居然把桥保住了,因而桥上留下几处"仙迹",即驴蹄印、车道沟、手印,还有张果老斗笠印及柴荣膝印。这些"仙迹"都在桥的东侧,路面三分之一的范围内,因而罗英同志在《中国石桥》里说,为了让开"仙迹",就要在桥面当中行车,这于保护桥梁是有益的。大石桥也因"仙迹"而名气愈大。京剧《小放牛》唱词有："赵州桥,鲁班爷修,玉石栏杆圣人留,张果老骑驴桥上走,柴王爷推车轧了一道沟。"把柴荣称作王爷,恐怕这种唱词在宋代就有了,后来小石桥也挤进大石桥的神话里去了,据说当鲁班修大石桥时,他的妹妹修小石桥,两人竞赛,要在一夜之间完成,不料妹妹到底敌不过哥哥,快到天亮,小石桥还差得远,幸亏有天神过境,打抱不平,帮助妹妹施法,这才使她获得胜利。然而妹妹曾否留下"仙迹",倒未听过。《小放牛》唱词中"玉石栏杆圣人留"的传说,恐怕起于唐代,《朝野佥载》里有一故事："赵州石桥,其工磨砻密致如削,望之如初月出云,长虹饮涧。上有勾栏,皆石也,勾栏并为石狮子,龙朔(公元 661～663 年)中……盗二石狮子去,后复募匠修之,莫能相类者。"由于原来雕刻精美,后人很难模仿,就把它的创作权,推给"圣人"去了。

安济桥建成后,为人民服务了一千三百五十多年,虽经历代修缮,从未彻底。只有在中国解放后,它才得到新生。人民政府重视文物保护,于 1953 年起即着手大修。那时东边外侧的三道拱圈,早已塌毁,邻近两圈也有很大裂缝,有随时坍塌可能。西边外侧五道拱圈曾经坍塌重砌,石质犹新。为何西边的拱圈比东边的倒得多,倒得早呢? 据说是因为千百年都是进城的重车多,出城的重车少,因而桥上向西北方的冲击力,大过向东南方的冲击力。经过了彻底调查,做出了复旧设计,在 1955 年到 1956 年间,将全桥修竣。由于在河底挖出了一千五百多块积石,其中有原来桥上的栏板雕柱,因而桥面和扶栏,都可完全恢复原状。修竣后的大石桥,从远处看去,仿佛就像一座近代化的拱桥,但它却是一件一千三百五十多年的古代文物,为国家重点保护了!

主要参考资料

1. 罗英《中国石桥》,1959 年,人民交通出版社。

2. 梁思成《赵县大石桥即安济桥》,《中国营造学社汇刊》第 5 卷第 1 期。

3. 俞同奎《安济桥的补充文献》,《文物参考资料》1957 年第 3 期。

4. 李全庆《赵州桥修复工程竣工》,《文物》1959 年第 2 期。

5. 福建省文物管理委员会《福建安海宋代安平桥调查记》,《文物参考资料》1958 年第 12 期。

6. 福建省文物管理委员会供给的调查资料。

7.《古今图书集成·职方典》。

原载 1963 年《文物》第 9 期

介绍五座古桥

——珠浦桥、广济桥、洛阳桥、宝带桥及灞桥

在 1963 年第 9 期的《文物》上，登载了我写的关于我国古代桥梁的一篇短文，原名《五桥颂》，发表时改为《重点文物保护单位中的桥》，因为所写的五座桥，在那时国务院已经公布为第一批全国重点文物保护单位了。现在又介绍五座在我国历史上都曾发挥过巨大作用，在科学技术上都有过重要贡献的古桥。这五座桥是：四川的珠浦桥，广东的广济桥，福建的洛阳桥，江苏的宝带桥及陕西的灞桥。

珠浦桥

珠浦桥在四川省灌县西北，岷江分为"外江"与"内江"的"浦"上，其下即"都江堰"的"鱼嘴"。原为竹索桥，长三百二十米，1964 年改为钢索桥，长二百四十米。

这座桥不在公路干线的孔道上,桥上不能行驶汽车,但于来往灌县的行旅及都江堰在内江与外江岁修时的过江运输,极为需要。都江堰范围内,有二王庙及伏龙观,[①]解放前,川西各县人民来此观光的很多,桥上交通,最称繁盛。

珠浦桥和都江堰是分不开的。都江堰是川西成都大平原上一套完整的农田灌溉系统的总称,这个灌溉系统已有两千二百多年的历史。都江堰本身,原为岷江中的一个沙洲,将江分为内外两股,其内江分出后,在灌县城西为一山所阻,折入岷江,与外江合并。山名"灌口山",[②]秦代李冰任四川太守时(公元前251年)将这山凿开一个缺口,让内江流出,与成都平原各河川汇合,组成一个灌溉系统。李冰开凿灌口山之处名"离碓",故《史记·河渠书》中云:"蜀守冰,凿离碓。"因凿山而开出的缺口,名"宝瓶口"。灌口山也名为"金灌口"或"玉垒山"。为了稳定内外两江的分流,岷江沙洲两岸,各筑起石堤一道,名内外"金刚堤"。两堤在沙洲分水尖端相遇,形成沙洲的鱼嘴。内江经过宝瓶口的流量,有一定限度,过此限度,则超出的流量及所挟泥沙,在未到宝瓶口前,即从飞沙堰漫溢,而在沙洲尾流入岷江。通过各种控制设施,内

---

① 二王庙为纪念李冰及其子二郎的庙宇,伏龙观为纪念李冰的专祠。
② 灌口山之名始于汉代,《汉书·文翁传》:"景帝末(约公元前141年)为蜀郡守。"《元和志》:"汉文翁穿湔江灌溉,故以名山。"

外两江得到适度调节,防洪引水,保证了岷江的航运及灌溉的功能。自李冰创始,两千多年来,都江堰水利工程,逐渐改进,直到解放后,才得到系统的更大的发展。

珠浦桥何时创建,尚未查得记载,唯知在宋代(公元960～1279年)时,名"平事桥"(根据灌县文物保管所1965年简介)。桥名珠浦,因"浦"为"大水有小口别通者"(见《风土记》),而"珠浦"则是因与宝瓶口、金灌口、玉垒山等配合而被赐以嘉名。这桥在明末(公元1628～1643年)与都江堰同时被毁,到清代嘉庆八年(公元1803年)方重建,改名为"安澜桥"。

《四川通志》云:"其制两岸榘石为穴,键石为笼,夹植巨木,屹砥湍流,编竹绳跨江,横阔一丈,离水面五丈,长一百二十丈。"《灌县志》云:"桥在县西二里……崇德庙(即二王庙)前,旧有索桥,即珠浦桥也,久废,设义渡以济往来,每当夏秋水溢,常有覆溺之患。嘉庆八年……仿旧置建立,长九十丈,高二丈二尺,阔一丈,名安澜桥。"

"编竹绳跨江"的竹索桥,其结构如下:以竹丝编成竹缆,粗如碗口,陆续接长,横跨全江,其两端绕系于横卧大木碾,转动木碾时拉紧竹缆,以免下垂过度。大木碾安置于木笼内,木笼位于两岸石岩中所凿的石室。竹缆十根平列,上铺木板为桥面,可以行人,两旁各有较细竹缆六根,作为栏杆。

由于桥底竹缆太长，下面用木排架八座及石墩一座承托，将桥分成九孔，全长三百二十米，一孔最大跨度达六十一米。每座木排架用大木桩五根，打入江底，中用横木连接，下有石块堆砌，其两边木桩较长，形成斜柱。石墩一座，位于都江堰的鱼嘴上，内有石室，亦有大木碾，可以拉紧竹缆，其作用与两岸的大木碾相同。

这座桥，以竹为缆，以木为桩，都是就地取材。与都江堰的水利工程相似，用竹笼装石，筑成堤堰，用竹木绑成三脚架的"杩槎"，放在水边，堆上黏土，成为临时挡水坝，费省效宏，简单易行，足见历代劳动人民的巧思高艺。

竹材的强度甚高，几与钢铁相近，但易受气候影响，雨淋水浸，容易伸长；气候干燥，又易收缩；因而使用时间，受了限制。竹索桥必须随时察看，经常检修，并规定三年大修一次。珠浦桥附近有竹林，用新竹换旧竹，旧竹除可利用者外，以之出售，料价可抵修桥工费，是个自力更生的维修方法。

竹索桥的缺点，与一般索桥相似，缺乏刚性，人行其上，最怕桥身摇晃。《黔书》中描述盘江铁索桥云："然絙长力弱，人行桥上，足左右上下，絙輒因之而升降，身亦为之摇撼，眩晕不克自持，乘车马者至此必下，且不容二人接武而行，必待前者陟岸，后者始登，若强而相蹑，震动愈甚。"珠浦桥虽用竹索，行人尚未如此惊恐。1964 年，随着都江堰的发展，将竹索改为钢索,承托缆索的排架木桩,改为钢筋混凝土桩,桥身更形稳定。由于在都江堰鱼嘴上兴修发电站,桥身缩短为二百四十米。珠浦桥化险为夷,万民欢颂。唐代诗人杜甫有咏竹桥诗,赞美建桥者云:"伐竹为桥结构同,褰裳不涉往来通。天寒白鹤归华表,日落青龙见水中。顾我老非题柱客,知君才是济川功。合欢却笑千年事,驱石何时到海东。"他未能料到社会主义的伟大,我们已经"驱石"(此引秦始皇"鞭石成桥"的神话,实际上是将大石块放在下有小轮的木板上,人推石块前行,当作"鞭石")建桥到"海东"了。

竹索桥修建时,有一段民间故事:在都江堰旁的内外两岸上,旧桥早毁,唯凭舟渡,夏季江水暴涨,渡客常多覆溺,即幸而安渡,也受渡船勒索。有塾师何先德,关心人民疾苦,倡议修建竹索桥,并参加工作。不意在索桥将要完工时,竹索断毁于风雨之夜,官僚们怕何先德揭发弊端,借口他的过失,将他杀害了。群情激愤,拥护其妻何娘子,继夫遗志,出面负责施工,终于将桥建成。人们为了纪念他们夫妇,就将建成之桥,改名为"夫妻桥"。这故事流传很广,后来就被编成川剧,搬上舞台。

广济桥

广济桥在广东潮州市东面韩江上,旧名济川桥,通常称为湘子桥。解放后,1958年大修,改建成通行大型汽车的桥梁。

这座桥创建于宋代,迄今已近八百年。全长五百一十八米,中有一段,用船只连为浮桥,可以解开,让出航道,成为可分可合的活动桥,是我国桥梁史上的一个特例。后来这座活动桥改为固定桥。

广济桥为福建、江西、广东三省交通要道,具有经济上、军事上的重要性。

韩江由两大支流汇为巨川,东支汀江,源出福建长汀;西支海江,源出广东安流、平远等地,两支集合于三河坝,曲折南流,经过潮州,在澄海、汕头入海。韩江流量,变迁很大,枯水与洪水,相差几达一百倍。全江流域的地平,高低悬殊,所经山谷,多系经过风化的砂岩、页岩,因而水流湍急,所挟泥沙,淤积于中下游,江水盛涨时溃决堤防,久旱枯涸又阻塞航运,既破坏农田,又妨碍交通。这江原名“意溪”,又名“恶溪”“鳄溪”。唐代韩愈(公元767~824年),因谏阻迎佛骨,于元和十四年(公元819年)被贬为潮州刺史。因江水汹涌,又有鳄鱼,故他在《潮州谢表》中有云:"通海口,下恶水,涛泷壮猛,难计程期。"他在此又写了有名的《祭鳄鱼文》。他这年三月到潮,十月调任。后来,这"恶水"就改名为韩江。潮州境内的笔架山,因他尝登览,就改名为韩山,他所手植的橡木就改名为韩木。

韩江是很重要的水运航道,上游水急滩多,不利舟行,但自大埔至汕头,则水深面阔,可通大船,甚至海洋巨舰,也可由汕头直驶大埔,既长且宽的大木排,也可由上游顺流而下。由于韩江流经闽、粤富饶地区,在潮州的渡江需要,自古已然,自南宋以后,更形迫切。但是要在这壮猛急流上造桥,而往来船舶又如此之多,这座桥究竟如何造呢?

据《潮州府志》云:"广济桥,在城东,跨韩江上,广二丈,长一百八十丈,旧名济川。西洲创于宋……增筑为十。东洲……增筑为十有三,久之洲坏桥断。宣德中……垒石为墩二十有三……造舟二十有四为浮桥,更今名。……嘉靖间……桥南北皆甃石栏而圬以灰,岁金桥夫四十四名、渡夫十名司守。"文中所谓东西洲,指东西段的桥墩。从这里看出,这桥的修建,是分段进行的,由宋至明的三百年间,工程未断。

根据记载,桥工分三段进行。西段靠城关,先行动工,自宋代乾道六年(公元1170年)开始,历时五十七年,建成桥墩九座,后在明代正德元年(公元1506年)扩建一墩,故西段共有桥墩十座。东段从对岸开始,自宋代绍熙元年(公元1190年)开始,历时十六年,也建成桥墩九座,后于明代宣德十年(公元1435年)扩建五墩,故东段共有桥墩十四座。两段共建桥墩二十四座,但后来因沙土淤积及修整江岸的原因,两岸各去了两墩,现在共存桥墩二十座。

在东西两段石桥之间,尚有将近一百米宽度的河道,未曾筑墩。据明代姚友直《广济桥记》中有云:"中流惊湍尤深,不可为墩,设舟二十四为浮梁,栏楯铁链三,每链重四千斤,连亘以渡往来。"

所筑桥墩,全部都是石砌,大小不一,形态各异,南北两端,均作尖形,石块与石块之间不用灰浆,但凿有卯榫,使相契合,然都庞大异常,闻所未闻,其宽度自六米至十三米,长度自十一米至二十二米,各墩宽度相加,总和达二百零七米,占全桥总长百分之四十。水中桥墩所占河道流水面积,更在百分之四十以上。这当然抬高了水位,增加了流速,并刷深了河床,大大增加了中流筑墩的困难。所谓西段九墩,费时五十七年,东段九墩,费时十六年,并非一次修筑所费的时间,而是修成复坏,断断续续所费的全部时间。桥墩宽度增加,又压缩了船舶过桥的通道,迫使较大船只及下放木排的航行,都集中于东西两段尽头桥墩之间的河道,并使"中流惊湍尤深,不可为墩"。桥工如此困难,而不放弃,虽旷日持久,终抵于成,足见这桥对人民生活的重要性。明代姚友直在所作《广济桥记》中反映了这情况:"其途通闽浙,达二京,实为南北要冲。其流急如马骋而汹涌,触之者木石俱往。水落沙涌,一苇可渡,水涨沙逸,数里旷隔,虽设济舟,日不能三四渡,咫尺之居若千里,士女不得渡,日夜野宿,以俟其便,军民病涉,莫此为甚。自宋至是,因循不能修复者殆百余岁,凡登途而望者,莫不痛恨,以为斯桥不复,终古苦涉矣。"

"设舟二十四为浮梁",将东西两段石桥,连为一体,是一很高明的设计,然而因此妨碍了中段河道上的航行,发生了一个极大的矛盾,因为那时从汕头来的大船,仍要上驶大埔,东西段的桥墩中间,均不能通过。于是得一解决办法:浮梁中的船舶,既是由铁链连系,则解开铁链,将船移动,即可让出河道,暂通水运,然后再拉紧铁链,使船归位,依然成一浮梁。这样就使整个广济桥成为一座可分可合的活动桥。这是一种"想当然耳"的说法,未见确证,初见于罗英所著的《中国石桥》(1959年出版)。甚至用船舶搭成一座浮梁,究竟创造于何时,亦未见记载,但明代姚友直的《广济桥记》中有云"中流不能为梁者,仍设浮舟,系以铁缆",可能在那时以前,即以浮舟代梁了。

全桥桥面,除浮梁上铺木板外,其东西两段各墩之间,均用巨石做梁,拼成桥面,石梁尺寸,异常庞大,最小者高一米,宽零点八米,长十二米;最大者高一点二米,宽一米,长十八米。

　　广济桥的创建,费时之久,固所罕见,而历代修理的频繁,也很惊人。除小修不计外,明代大修即有五次:(1)宣德中(公元1426～1435年),重修西段十墩,东段十三墩,又造舟二十四为浮梁,并将浮梁上木板改为石梁,以期稳定,因这时大船不往大埔,止于潮州,无解舟通航的必要;(2)弘治中(公元1488～1505年),桥"复圮于水",重修一次;(3)正德中(公元1506～1521年),"大水坏梁,易以石,相继重修";(4)嘉靖间(公元1522～1566年),除修桥外,并于桥上"立东西亭,南北增造石栏杆";(5)万历间(公元1573～1620年)重修,"桥上木屋,群集成市"。到了清代,重修次数更多:(1)顺治间(公元1644～1652年)因遭兵祸,石梁仅存十分之一,全部修复;(2)顺治十年(公元1653年)焚毁大修;(3)顺治十一年(公元1654年)"以大木架梁修理";(4)康熙十年(公元1671年)重修;(5)康熙十六年(公元1677年)"八月十四日夜,西岸桥下吼声如牛,石墩忽倒其一",隔了三年修复;(6)康熙二十四年(公元1685年),"易木以石";(7)康熙五十九年(公元1720年)"水决东岸石墩,没者二",其后隔了四年修一墩,再隔四年又修一墩;(8)道光二十二年(公元1842年)大水冲毁东岸九墩,陆续修复,"中部仍用浮船十八只",可见在这以前,浮船已非二十四只了。后来,在抗日战争期间,于公元 1939 年,在中段浮梁处,取消船只,改建悬索吊桥,据云只通车一次即废,现吊桥铁塔犹存。

广济桥的修理,为何如此频繁呢? 当然由于建筑关系,然而也是灾难过多所致:(1)洪水泛滥,冲毁桥墩多次,至于大水漫桥的次数更多,公元 1911 年,全桥即为大水淹没。(2)飓风为害,海水上溢,致桥梁受损。自明弘治至清乾隆年间,发生飓风三十三次,平均每七年一次。其后亦数见不鲜。(3)地震。自明永乐至清乾隆年间,有二十三次,平均每十五年一次。至公元 1918 年更遇七八级地震,桥墩发生裂缝及倾斜现象。除天灾外,尚有兵祸,清顺治二年、十年均遇兵火焚毁。1939 年,抗日战争中,遭受敌机轰炸。在各种灾难侵袭下,桥墩损坏日多:(1)裂缝。所有桥墩都不免,轻重不一,有的块石走样,有的坠落。(2)孔洞。除裂缝外,尚有大小不一、深浅不同的孔洞。由于块石为波浪冲击吸出,或为地震离心力抛出,或为木排撞击脱落,最大孔洞,可容一只小船进出。(3)倾侧。地震影响较大,如公元 1918 年地震后,东段各墩,都有向上游偏欹的现象。

经过八百年的"服役",广济桥在解放时,已经残破不堪。1958 年大修,全桥整旧更新,呈现今日雄姿。在大修时,也重视文物的保护,所有旧桥墩全部保留利用,并予以彻底修整,增加了强度。

现在的广济桥，全长五百一十八米，分为三段，靠市区的西岸部分，长一百三十七点三米，计七孔八墩，每孔跨度，自八米至十七点五米不等；东岸部分，长二百八十三点四米，计十二孔、十二墩、一桥台，每孔跨度，自九点四米至十二点九米不等。两部分桥墩，都是石砌，为长轴六边形，上下游作成尖形，以利水流。桥面下为原来石梁，上为钢筋混凝土路面。东西两段之中间部分，长九十七点三米，原系船搭"浮梁"，现改为三孔钢梁，每孔跨度三十四点七米，上铺钢筋混凝土桥面，下由"高桩承台"的桥墩支承。全桥桥面原宽五米，现为七米。桥上车道两旁有人行道，外为预制混凝土的回纹花格栏杆及灯柱。

历年来桥上常有各种建筑物，作为点缀，如亭屋、茶亭、碑亭、石塔等。明代姚友直《广济桥记》中云："桥之上乃立亭屋百二十六间……作高楼十有二……"由桥西亭而东为楼五，凡楼屋五十间，由浮梁而东为楼七，凡楼屋七十六间，可见规模的宏大，后来都毁了。清代雍正三年（公元1725年），在浮梁两边桥墩上，铸置两只铁牛，后来一只坠河中，故潮州当时流传一首民歌："潮州湘桥好风流，十八梭船廿四洲，廿四楼台廿四样，两只铁牛一只留。"（梭船为浮梁之船，原为二十四只，后减为十八；洲为桥墩，原为二十四座，后为二十座；

廿四楼台,因"高楼十有二",每楼分东西两个部分。)

广济桥又名湘子桥,成为通称。据传说:湘子是唐代韩愈的侄子。韩愈在韩江上造桥,累年不成,韩湘子下凡来助,施展"仙法",一夜间就成功了,故以命名来纪念他。这完全是无稽之谈。首先,广济桥创建,在韩愈死后三百年;其次,韩湘子是韩愈的侄孙而非侄子,从未到过潮州;而且他并非好道学仙之人。从宋代朱熹(公元1150~1200年)的《考异》起,经明清两代学者考证,都已辨明得很清楚了。根据《唐书·韩愈传》和他的年谱、碑铭及诗文集,韩愈有三个哥哥,均早死,长兄韩会和次兄韩介的双桃子名老成,即愈所写《祭十二郎文》中的"十二郎"。十二郎也早死,遗两子,长名韩湘,次名韩滂。韩愈自潮州迁到袁州(今江西省宜春县)后,才将两个侄孙接去,不久滂死,愈又往长安,湘随去,于长庆三年(公元823年)成进士,时年三十,累官至大理寺丞,未见有学道的记载。不知何故,唐末《青琐高议》及《酉阳杂俎》中,都记了韩湘的"聚土覆盆""盆中开花"的神话,后来愈说愈广,愈广愈奇,宋末《续仙传》中将韩湘子列为"八仙"之一,到了元代,就有剧本《八仙庆寿》,韩湘子更成为一个著名的神仙了。至于他的名字联系到广济桥,更不可解。也许是由于韩愈曾任潮州刺史之故。然而愈在潮州,不过六个多月,而且当时并未为潮人所重,所谓韩江、韩山、韩木等等尊韩名称,都是唐代以后的事。是否因愈为大文豪,引以为荣,又值韩湘子名气日盛,再加一桥名来尊韩,均尚有待于考证。

洛阳桥

我国古桥中,民间传说最广的,几乎通国皆知,北方有赵州桥,南方有洛阳桥。福建省在1961年4月,公布洛阳桥为省级重点文物保护单位。

洛阳桥位于福建泉州洛阳江的入海尾间上,为福州至厦门之间的重要孔道。桥长八百三十四米,桥面宽七米,桥栏高一点零五米,桥面现为钢筋混凝土筑成,可通行重型汽车。

洛阳江,发源于戴云山,流经安溪、南安、泉州,东入海口泉州湾。《泉州府志》:"洛阳江在府城东北二十里,实晋江、惠安二县夹界之江也,群山逶迤数百里,至江而尽。昔唐宣宗(李忱)(公元847~860年)微行,览山川胜概,有类吾洛阳之语,因以名。"《闽书》:"宋《淳祐郡志》引沈存中(梦溪)《笔谈》云,'水以漳洛名甚众,洛,落也,水落于下谓之洛,旧号洛洋。'《九域志》作乐洋江,以落为乐,误矣。后又以洛阳名之,而好事者又云,唐宣宗出家时,至此,谓风景大类西洛。则不根为甚。按宣宗尝至同安夕阳寺,西洛之说,似亦未必无据者,古谶'洛阳沙平,泉南公卿'。则洛阳之名已古矣。"《金石萃编》谓:"然欧四门已有洛阳亭送别诗,岂肃代之世,预知有宣宗语耶。"可见洛阳江得名故事,并无定论。泉州有《浯江郡志》云:"晋南渡时,衣冠士族,避地于此,故又名晋江。"如果东晋时避地来泉的士族,已经多得能使江以"晋"名,那么,南宋以后,从开封、洛阳来的士族更多,岂不会将"洛洋江"改名为"洛阳江"的故事,更加渲染吗?洛阳这个名字,因历史悠久,好像有魔力,不但国内重视,而且传闻海外,据说日本京都,仿洛阳规划,至今有"东洛""西洛"之称。

洛阳江入海口处,水阔五里,正当交通要道,不仅北宋南渡以后,闽南人口骤增,经济发展,而且泉州自南朝即为出海港口,到唐代更成为对外贸易的四大港口之一,因而过渡的人,日益增多,而风险也愈甚。《泉州府志》中有云:"万安桥未建,旧设海渡渡人,每岁遇飓风大作,沉舟被溺而死者无数。"因名渡口为"万安渡"以祷之,而建桥需要愈形迫切。然而江流湍急,海潮汹涌,水面广阔,泥沙之深莫测,建桥又谈何容易。

《福建通志》载:"宋庆历初(公元1041年),郡人李宠始甃石作浮桥,皇祐五年(公元1053年),僧宗己及郡人王实、卢锡倡为石桥,未就。会蔡襄守郡,踵而成之。"创始作浮桥的人,文中说是李宠,但在《读史方舆纪要》中,则作陈宠,未知孰是。但不论李或陈,在史书中均未查到他的传略。蔡襄本人在桥成后，写了一篇《万安桥记》，因在万安渡上故名万安桥，也可见洛阳桥的名称，到了南宋以后，因南渡的人日见其多，才逐渐盛行起来。蔡襄的桥记，后人为之刻在碑石上，现存桥南的蔡襄祠内。因蔡襄是有名的书法家，历来传说碑文是他的亲笔（也有人说是他的曾孙蔡梀写的），于是称颂云"当与桥争胜"。碑石上的刻文如下："泉州万安渡石桥，始造于皇祐五年（公元1053年）四月庚寅，以嘉祐四年（公元1059年）十二月辛未讫工。垒趾于渊，酾水为四十七道，梁空以行，其长三千六百尺，广丈有五尺，翼以扶栏，如其长之数而两之。靡金钱一千四百万，求诸施者。渡实支海，去舟而徒，易危而安，民莫不利。职其事者卢锡、王实、许忠，浮图义波、宗善等十有五人。既成，太守莆阳蔡襄为之合乐宴饮而落之。明年秋，蒙召还京，道由是出，因纪所作，勒于岸左。"文虽不长，但记下了几项重要事迹。一是造桥时间共六年八个月；二是桥长三千六百尺，有四十七孔，桥宽十五尺；三是造桥经费一千四百万，由募捐而来；四是负责造桥的人共十五位，内五位列名，两人是和尚，三人大概是募捐和管事务的，其余十人名未列，想是设计和施工的桥匠。和尚二人想系为募化和监工的，其中义波和尚，特别有功，至今桥上有他的"真身庵"。那时及稍后的福建桥工，几乎无一桥无和尚参加的。文中"垒趾于渊"即建筑桥墩，为造桥的关键工作，惜文中未详。但蔡襄在桥成后的隔年,"道由是出",可见桥成无恙,筑墩的方法是可靠的。是什么方法呢?《福建通志》记桥的长和宽后说:"以蛎房散置石基,益胶固焉。"又《名胜志》中叙万安石桥文中,亦有"桥下种蛎固其基"之语,《通志》又云"元丰初(公元 1078 年),王祖道知州事,奏立法,禁取蛎房",则桥成后二十年,蛎房之效益著,故立禁取之法。

究竟蛎房对桥基的作用何在呢?

罗英在所著《中国石桥》中,由于参阅古籍,广集传闻,并征询蔡襄后裔,认为洛阳桥的建筑方法如下:在江底随着桥梁中线,满铺大块石,并向中线两侧,展开至相当宽度,成一条整体的横过江的矮石堤。但洛阳江水深流急,所抛石块,小的易被水流漂失,大的平时尚能稳定,但潮汛遇风,仍可漂流入海。如何将石块胶成整体,成一大问题,因那时石灰浆在水中不能凝结。后来多方探寻黏合材料,看到在这江里"盛产牡蛎,皆附石而生,初如拳石,四面伸展,渐长至一二丈,崭岩如山,俗呼壕山"。因而就想出,在所抛石块上,繁殖蛎房。试之果然生效,就用作桥基,在上面建筑桥墩。用大长条石齿牙交错,互相叠压,逐层垒砌,筑成桥墩。墩的上下游两头,俱作尖形,以分水势。桥墩间距离不一,视所采用的石梁长度而定,两墩间净孔,约在一丈五六尺上下。沿岸开采的石梁,预先放在木浮排上,等到两邻近桥墩完成后,即趁涨潮之时，驶入两桥墩间，俟潮退，木排下降，石梁即可落在石墩上。石梁上铺桥面，旁立桥栏，全工告竣。这样筑成的桥墩底盘，即现代桥工中的所谓"筏形基础"。现代桥工的"筏形基础"方法，运用还不到一百年，而我国桥工在近九百年前就已运用，可算我国桥工的一个贡献。

关于洛阳桥的兴建，神话很多，录其一则。《闽书》云："是桥也，坦夷如官道，可三里余，扬州'廿四'，空悬月夜，吴江'飞压'，仅表虹垂，兼以忠惠（蔡襄）记文，书笔并精千古，于是好事者竞传桥异，谓桥未兴时，深不可趾，忠惠为檄，使隶投之海而告之，隶叹息曰，茫茫远海，何所投檄，买酒剧饮，醉卧小艇上，醒而起视，则檄已换，第书一'醋'字。忠惠公曰，神示我矣，当二十一日酉时潮退可趾也。"又传此隶名夏得海（下得海），桥上曾有"夏将军庙"。关于"醋"字神话，传说最广，许多洛阳桥记载中，几乎无一不提此事，并且一般都解释为"二十一日酉时"，"廿""一""日""酉"合起来，恰是"醋"字。这是一种附会的说法，酉时已经天黑，如何能施工？按宋代历法"建寅"，以正月为寅，则酉为八月，上引《闽书》云"二十一日酉时"，其"酉时"或为"酉月"即八月之刊误，而八月二十一日正是大潮趋弱之时，适合施工需要。是何种施工呢？并非江底抛石之工。据明代姜志礼《重修万安桥记》中有云："昔忠惠公以二十一日安桥，今须事也，及是日，滔天之水果自东来(此文有误,应是弱潮),石梁遂上。"说明利用潮水,为的是上"石梁",而非下石基,即罗文所说,石梁在木排上,随潮水涨落,安装在桥墩上的工作。至于一般施工,每日皆可为之,不像用浮排上石梁,是要选择适当的潮水涨落的。

关于"蔡状元起造洛阳桥"的故事,流传已久,但经考证,发现一些问题,值得讨论,今试为解答。

(1)蔡襄与洛阳桥。根据宋代欧阳修所作《端明殿学士蔡公襄墓志铭》、明代徐燉所作《蔡忠惠公年谱》及蔡襄本人诗文集,他两度为泉州太守,第一次赴任在至和三年(公元1056年)二月初七日,闰三月诏移知福州,六月离泉,八月初四日赴福州;嘉祐三年(公元1058年)再知泉州,七月初一赴任,五年秋,召拜翰林学士权三司使,六年二月就道。洛阳桥工程,从公元1053年四月到公元1059年十二月的六年八个月中,蔡襄在开工后三年才来过不足六个月,隔了两年再来,离完工也只有十六个月,这样看来,好像他对桥工并无多大关系。可是,为何说这桥是他造的呢?《宋史·蔡襄传》载他"徙知泉州,距州二十里万安渡,绝海而济,往来畏其险,襄立石为梁,其长三百六十丈,种蛎于础以为固,至今赖焉"。这记载应当可靠,不过不够详细;但在《福建通志》中找到了答案,这里说:"皇祐五年,僧宗己及郡人王实、卢锡倡为石桥,未就,会蔡襄守郡,踵而成之。"文中"倡""未就""会""踵"

"成"等字,说明了在皇祐五年开工,但没有成就,等到蔡襄来,才把工程接过来,提出施工方案,将它完成。什么施工方案呢? 一就是"种蛎于础以为固",二就是利用潮水,浮运石梁。

"蛎房"是牡蛎的壳聚凝而成的,在汕头海边,久有利用蛎房筑堤的方法。1970 年根治海河时,在海边就遇到坚实的蛎房成岩,长二百二十一米,宽一百五十米,厚两米,开挖困难,严重影响到施工进度,就是一证。在洛阳桥,不是用它代替石块,而是用以胶凝石块。

还有关于和尚问题。《通志》中说是"僧宗己",而蔡襄写的桥记中说是"浮图义波、宗善",可见开工之始,宗己并无大用,蔡襄来泉后,义波、宗善才能发挥作用。

(2)蛎房和潮水的利用。这和蔡襄本人的经历有关。《宋史·蔡襄传》云:"蔡襄字君谟,兴化仙游人。"但蔡本人在他写的桥记中云,"太守莆阳(即莆田)蔡襄为之合乐宴饮而落之。"究竟蔡是仙游人还是莆田人呢? 这两地相距不远,但莆田靠海,两地的生活习惯是有些不同的。蔡襄的父亲是莆田人,母亲是仙游人,蔡幼年常住仙游,不过莆田是他出生之地,他对沿海劳动人民的丰富经验应有所闻,知道海潮涨落及蛎房作用,后来便应用到桥工上了。

(3)关于醋字神话。《泉州府志》中有一段记载,说它是同蔡锡有关,而不是同蔡襄有关:"宣德间(公元 1426~1435年),蔡锡知泉州,先是洛阳桥圮坏,故石有刻文云'石摧颓,蔡再来',至是锡捐俸修之。海深不可趾,锡檄文海神,遣卒投之,卒醉卧海上,寤视檄面,题一醋字,锡曰酉月廿一日也,至期潮果不至,桥成,民祠于蔡忠惠祠畔。按明《列卿传》亦以移檄海神为蔡锡事。"这段资料很可贵,因为很多讲蔡襄造桥的文中,都要提到这个"醋"字神话,而其可靠性,实有问题。首先,蔡襄当时在泉州是很有名声的,他提出的施工方案,群众不会反对,无须借助于神话。其次,洛阳桥上的修桥碑文中,凡提到这个神话的,都写于蔡锡以后,而非以前。再其次,各文中,对这醋字的解释不一致,有的说是"酉时",有的说是"酉月",而酉月正是蔡锡的解释。从情理推测,蔡锡是修桥而非造桥,施工方法尽可依照前例办理,可能是关于利用潮水一事,发生疑问,故蔡锡编制出这神话,甚至假托说是以前蔡襄的故事,以期动听,因而这故事就传开了。由于蔡襄的名气大,后来传说蔡锡的故事,提到这个姓蔡的泉州太守如何修洛阳桥,很容易把它当作蔡襄的神话。故备此一说,以待续究。

(4)蔡襄的状元。蔡襄是状元,在民间传说极广,然而在"官书"里,他并不是。他生于北宋大中祥符四年(公元 1011年),死于治平四年(公元 1067 年),在天圣七年(公元 1029

年)中进士,年仅十八岁。在唐代,进士中的第一名,称为状元,但在北宋,各省省试的第一名就称状元,即是后来的所谓会元,而蔡襄都不是。到了南宋,恢复廷试第一为状元的制度,但加了一条特例,即中进士时,名列中优等,后来做官立功多次者,也可称状元。蔡襄是北宋进士,曾以龙图阁学士身份做过开封府知府(与包拯同),又升端明殿学士,因而做官是有"功"了,人们就把他当作南宋进士,称为状元。

以上是附带谈到同蔡襄有关的几个问题。

当然,洛阳桥这样艰巨工程的成功,最根本的还是由于不少桥工巨匠的贡献,不能把它只归功于蔡襄。据《泉州府志》云:"石匠四至,各呈其艺,有献石狮子者,其发玲珑有条理;又一人所献,其口开处小容指,有珠圆转在口中,殆不可测。"当然"四至"者中,不乏桥工巨匠。参加桥工的劳动人民,胼手胝足,倾沥血汗。虽然桥记不载其名,但其功德巍巍,与桥并存,这是任何人都不能抹杀的。

洛阳桥告成后,举国传闻,《八闽通志》载诗二章,其一曰《洛阳桥》,诗云:"一望五里排琨瑶,行人不忧怆海潮,憧憧往来乘仙飙,蔡公作成去还朝,玉虹依旧横青霄,考之溱洧功何辽,千古万古无倾摇。"历代诗人赞美者尤多,如明代徐㷸诗:"路尽平畴水色空,飞梁遥跨海西东,潮来直涌千寻雪,日落斜横百丈虹。"明代凌登名诗:"洛阳之桥天下奇,飞虹千丈横江垂……约束涛浪鞭蛟螭，雄镇东南数千里。"录此以见一斑。

更动人的是将造桥故事搬上舞台。我在幼年时曾看过一出京剧，就叫《洛阳桥》，是个灯彩戏，演出三百六十行中各行艺人，手持各种彩灯，象形工具，手舞足蹈，兴高采烈地过桥景象；并在唱词中道出他本行业因桥受益的感激心情。我当时也为这桥的故事所感动。后来事隔多年，在这桥上读到清代方鼎所作的一篇碑文《重修宋蔡忠惠公祠记》，文中有云"余少于氍毹中见里开演蔡端明洛阳桥事，岁必数四"，碑记写于乾隆二十八年（公元1763年），才知道舞台上演出这类戏，由来已久。以桥为名的剧曲，原本很多，但像这样专演造桥故事的，却极罕见。

洛阳桥的最大影响，仍然在桥工方面。各地闻风兴起，看到这样困难的巨大桥工，居然胜利完成，不但增长了知识，而且也壮大了胆量，于是在泉州首先掀起了"造桥热"。仅在南宋绍兴年间（公元1131～1162年），便有十座石桥的兴建，其中有的比洛阳桥还要雄伟，如晋江县安海镇的跨越海湾的安平桥，长达五里，致有"天下无桥长此桥"之誉。《闽部疏》中有云"闽中桥梁甲天下"，推崇洛阳、安平两桥，为全省桥梁增光。

一座桥建成后，不但要胜任过桥运输，还要能抵抗天灾人祸,洛阳桥在这方面是经受了考验的,据统计,关于地震,自公元1290至1722年的四百三十二年中,成灾五十二次,平均八年一次,桥被损一次;关于飓风,自公元1343至1752年的四百零九年中,成灾二十一次,平均十九年一次,桥被损一次;关于水患,自公元1066至1760年的六百九十四年中,成灾五十七次,平均十二年一次,桥被损坏三次。受天灾的损坏,平均约一百年一次,而在公元1596至1722年的一百二十六年中,损坏达六七次之多。这些数字,当然还不完全,但已可看出灾害的严重了。此外还有战祸,如《读史方舆纪要》载"明嘉靖三十七年(公元1558年)官兵败倭于此""昔延平王郑成功,曾据此桥,大战清兵"。《闽部疏》云"洛阳桥……近两岸一山,皆大石,倭乱时,城其上而楼之,扃钥甚固,倭不能过洛阳之南"。

根据《泉州府志》及历代修桥碑文记载,洛阳桥修建至今,历时九百多年,先后修理和重建达十七次之多,其修理时间相隔最长的约一百七十余年,平均约五十年修一次。其中较大的有:宋代绍兴八年(公元1138年),因飓风侵袭,遭受毁坏,做了首次修理,但距建桥时间已八十年;明代永乐六年(公元1408年),因地震、飓风和水患,大修;宣德初年(公元1426年),桥址下陷,潮至漫桥,将桥墩增高三尺;景泰四年(公元1453年),"桥梁断其三间",为之修复;隆庆元年(公元

1567 年),石基损坏数处,重修,仍严厉禁止取蛎;万历三十五年(公元 1607 年),地大震,桥梁圮,址复低陷,进行大修;清代康熙二十年(公元 1681 年),重修;雍正八年(公元 1730年),因历经多次飓风、地震和水灾而桥崩,至此修复;乾隆二十六年(公元 1761 年),重修;道光二十三年(公元 1843 年),桥倾欹大修;咸丰十年(公元 1860 年),因洪流冲击,桥墩有倾圮之虞,修理一次。此后未见大修,桥身日见损害,到了公元 1932 年,已经破坏不堪,几乎不能行车,由当时驻在闽南的第十九路军,将桥大加修理,恢复原来桥墩,在上面添筑矮墩,提高路面,在矮墩上建筑钢筋混凝土横梁和桥面,原来墩上石梁,仍予保留,于是全桥改观,可通大型汽车。抗日战争时,又遭日本侵略军破坏,1946 年用军用钢梁,架设两孔,暂维交通。解放战争时,国民党反动派大加破坏,桥身被炸毁数段,桥上文物损坏十之八九,幸桥墩无恙。洛阳桥经受了几百年的灾难,依然屹立,足见创始工程的伟绩。

解放后,洛阳桥历经修整,面目一新,现在实况如下:从北岸惠安县境起,有石垒桥堤一段,桥由堤接出,经过一个小岛,名中洲,继续南展,达于南岸晋江县境。桥长八百四十三米,砌出水面的船形桥墩四十六座,墩上桥梁、桥面和桥栏,全部用钢筋混凝土筑成,桥面宽七米,桥栏高一点零五米。桥上附属文物及建筑如下:(1)亭两座,一为中亭,另一为西川甘雨亭,均在中洲上,中亭内有修桥碑石十二座及摩岩两方;西川甘雨亭原为祈雨之地,内有"天下第一桥"横幅。(2)佛塔五座,全为石构,计有四种形式,均筑于桥旁扶栏外。(3)石刻武士像四尊,分别立于桥的南北两端,均戴盔披甲,手持长剑。(4)纪念建筑物三所,一为蔡忠惠公祠,在桥南街尾,曾奉祀蔡襄,重修多次,内有蔡书《万安桥记》碑一座,另明、清碑记八座,又有明代蔡锡石像;二为昭惠庙,在桥北街旁,内有"永镇万安"匾额,所奉祀者可能为"镇海之神";三为真身庵,在桥北洛阳街靠海处,为了纪念造桥和尚义波而修建的,此地原系他所住的茅舍。(5)有关洛阳桥修建的碑记二十六座,分布在桥中亭周围及桥南蔡襄祠和桥北昭惠庙等地。一座桥的兴建及修理的石刻碑文,有二十六座之多的,恐怕在国内桥梁中为仅见。这不但说明这桥的兴建艰难,修理繁复,更重要的是这桥与人民生活,该有多么密切的关系。

宝带桥

长江以南地区,河流纵横,自古就发展了的水上运输,为经济繁荣的一大因素。为了便利陆路交通,桥梁当然也随之增多。东南水乡桥梁之盛,远远超过其他地区,特别在人口密集的都市,更是如此。如江苏苏州,载于图籍的桥梁,计有三百五十九座之多，其中最宏伟的即是宝带桥，全长三百一十七米，建于公元 816 至 819 年。

宝带桥在苏州市南，距葑门六里，与运河平行，位于运河与澹台湖之间的玳玳河上，玳玳河通连运河，经吴淞江入海。这桥为苏州至杭、嘉、湖陆路所必经，因跨诸湖之口，又是航路通往运河及吴淞江的一个关口，行旅繁忙，交通上极为重要。

据《苏州府志》载："去郡东南十五里，有宝带桥，唐刺史王仲舒捐宝带助费创建（约公元 816 年），故名……宋绍定五年（公元 1232 年）……重建……明正统间（公元 1436～1449 年）……重修……清康熙九年（公元 1670 年）大水冲圮，十二年（公元 1673 年）重修，咸丰十年（公元 1860 年）又毁，同治十一年（公元 1872 年）工程局重建，北堍建有碑亭。"

又据明代陈循《修宝带桥记》云："苏州府城之南半舍，古运河之西，有桥曰宝带。运河自汉武帝时开，以通闽越贡赋，首尾亘震泽东墙百余里，风涛冲激，不利舟楫，唐刺史王仲舒始作塘堤，障于河之西岸，今东南之要道是也。然河之支流，断堤而入吴淞江，以入于海，堤不可遏，此桥所为建，仲舒鬻所束宝带以助工费，因名。元末修葺之功不续，桥遂坍毁，有司架木以济，至今百有余岁……正统十年（公元 1445 年）秋，为桥长一千二百二十五尺，洞孔下可通舟楫者五十三，而高其中之三，以通巨舰，冬十一月落成。"

为了保障运河不受太湖风涛的冲击，在西岸筑了塘堤，留一缺口，以通支流，又在支流上造桥，以利交通，所造成的桥，维持了四百年，到了南宋末期才重建，也可见这桥的创建质量是很高的。重建后一百年，由于维修不继，桥又倒坍，用木架便桥，又维持了一百多年，方建成五十三孔的石拱桥，其中有三孔特高，"以通巨舰"。这在当时可说是十分巨大的工程。桥成后再过二百年，又为大水冲圮，三年内修复。再过二百年又毁，十二年后重建。这时（公元1872年）建成的桥，大概一直遗留到今天，又是一百年了。

在清康熙十二年（公元1673年）重修后，至咸丰十年（公元1860年）又毁三间，桥的形式仍是够雄伟的。英国人马戛尔尼在所著《乾隆英使觐见记》一书的1793年11月7日从镇江往杭州的运河道上日记中云："七日礼拜四晨间抵常州府，过一建筑极坚固之三孔桥，其中一孔甚高，吾船直过其下，无需下桅……已而又过三小湖，乃互相毗连者，其旁有一长桥，环洞之多，几及一百，奇观也。"其同伴摆劳氏《中国旅行记》中有云："此种世间不可多见之长桥，惜于夜间过之……后有一瑞士仆人，偶至舱面，见此不可思议之建筑物，即凝神数其环洞之数，后以数之再三，不能数清……"他们见了宝带桥的惊叹之情，有如13世纪意大利人马可·波罗见了北京南边的卢沟桥一样。

今天所见的宝带桥,系石拱形式,有桥洞五十三孔,其中三孔特高,很像明代(公元1445年)修建的那座桥。那座桥的前身是木架便桥,便桥前身便是经历了五百年的最初创建的桥。那座创建的桥,是否也是五十三孔并有三孔特高呢,这当然有待于考证。木便桥以前的五百年中,四百年是唐代的桥,一百年是宋代的桥;宋代的拱桥现存者尚无中间突起的例,然而那是石拱,则无疑义。元代一位僧人名善住,在过此桥时,曾有一诗:"借得他山石,还将石作梁。真从堤上去,横跨水中央。白鹭下秋色,苍龙浮夕阳。涛声当夜起,并入榜歌长。"所谓苍龙,指的就是拱。至于唐代,已先有赵州桥(隋代建),将宝带桥修成石拱,很有可能,甚至为了便于行船,对桥洞做出特殊设计,将拱圈抬高,亦非意外。这样说来,宝带桥的现在的形式,可能在唐代就早有规模了。我国古代桥梁具有高度技术造诣,有待考察研究者甚多,此为一例。

现存的宝带桥,全长约三百一十七米,为联拱石桥。所谓联拱,即是各拱在拱脚相联,全桥成一整体。石拱共五十三孔,每孔跨径(拱脚至拱脚的水平距离),除第十四至十六中间三孔(由北端数起)处,平均为四点六米。第十四孔与第十六孔的跨径,各为六点五米,第十五孔跨径最大,为七点四五米。第十四至第十六孔的顶点高出其他各拱,从第十三孔起，拱身逐渐隆起，至第十五孔拱顶为最高峰。第十三孔至十七孔，拱上桥面成弓形曲线，在第十二至十三孔及第十七至十八孔间，桥面各有一段反曲线。桥宽四点一米，桥堍为喇叭形，桥端宽六点一米。上桥下桥两端各有石狮一对，桥端北头约两米处，有石塔一座，高约三米。桥之北端向北约六米处，有碑亭一座，内有张松声的碑记。全桥体态雄伟，外形壮丽，不愧古人所谓"长虹卧波，鳌背连云"。

石拱桥以石拱为骨干，所有桥上载重，均通过石拱而传达至桥墩。石拱隆起为弧形，当然便利桥下行船，但其主要作用在充分发挥石料的强度。弯拱这种形式的建筑构件，是我国的一大发明，古文献中所谓"囷""窌""窦""甕"等等，就都是拱。拱有整体浇成，如钢筋混凝土拱，和分块拼成，如砖拱、石拱及钢拱。石块砌成的拱，是我国较大石桥的最普遍结构，因而砌拱的方式也是多种多样的。大体可分两类，一是将石块砌成一片片的单独拱圈，并将这些片的单拱拼合成为整体拱圈，如赵州桥；一是将石块砌成与桥同宽的一条条的长石，并将这些长条石按弧形砌成整体拱圈，如卢沟桥。宝带桥则两法兼用，将全桥拱圈，用与桥同宽的长条石，将整个拱圈分成若干隔间，在每个隔间内，用块石砌成一片片的弧形短拱，各片合拢，拼成短拱，与长条石一起，拼合成为整体拱圈。宝带桥的拱石，尚有一特点，即每两石块之间，均用榫头及卯眼拼接，因而在受到压力时，可以微微移动，将不平衡之力，自行调整。由于榫卯具有铰接作用，用这种块石砌成的拱，名为"多铰拱"。同时，砌合这些石块时，不用灰浆，成为"干砌"。

拱圈的两端拱脚，砌在两个桥墩上，每个桥墩，支持两个拱圈的拱脚，两拱之间，形成一个三角地带。各拱圈的上面为平坦的路面。在路面与拱圈之间的空腹，须用沙土填满，因而在空腹的两侧，都要砌墙来挡土，这种桥墩上三角形的石墙，名为"肩墙"。因而空腹内的沙土，上有路面，下有拱圈，两旁有肩墙，四面包围，如将沙土填塞很紧，则四面都有压力，可将路面上载重的一部分，直接传递至桥墩，来减轻拱圈上的担负。这种沙土的受挤压力，名为"被动压力"。

上述的多铰拱及被动压力的利用，都是我国劳动人民根据实践经验的创造，成为修建拱桥的优秀传统，为世界各国所罕见者，我国石拱桥的广泛应用，宝带桥是其一例。

有的古桥可用来行驶汽车，但宝带桥因有突起部分，则形不便。1933 年修筑苏州至嘉兴的公路时，于旧桥西面造了一条木便桥，名新宝带桥。1935 年修筑苏州至嘉兴的铁路时，又于旧桥东面建成一座铁路桥。抗日战争中，铁路桥全毁，宝带桥南头，也有六孔被炸，公路桥大部破坏。解放后，新宝带桥先行修复，旧宝带桥于 1956 年冬恢复旧观。

宝带桥当然是为了交通而设,然因地处胜境,同时也点缀了湖山,江南好风景,得此益增佳趣。明代王宠诗云:"春水桃花色,星桥宝带名,鲸吞山岛动,虹卧五湖平。"又袁裘诗云:"分野表三吴,星桥控五湖,济川恒壮业,驱石望雄图。"由此联想到,苏州有很多名桥,风光旖旎,远近传闻,最著者如唐代张继的《枫桥夜泊》诗,很概括地衬托出苏州桥的诗情画意,宝带桥亦在其内。

灞　桥

灞桥是我国历史上最古老、久负盛名而又相当宏伟的一座桥,虽然它现在已非原样,它位于陕西省西安市东十公里的灞河上,河宽四百米,是西安东去的一个必经孔道。而西安又是汉唐故都,因而很早就有了桥,但受河水冲击,屡毁屡建,桥式变化很大,1955年重建,才能稳定下来。算来,这个灞桥之名已有两千年历史了。

灞河又称灞水,为"八水绕长安"的最大一条水,是渭河的一个支流,发源于蓝田县南的秦岭。北魏郦道元著《水经注》"渭水"条中云:"霸者,水上地名也,古曰滋水矣,秦穆公霸世,更名滋水为霸水,以显霸功。水出蓝田县蓝田谷,所谓多玉者也。西北有铜谷水,次东有辋谷水,二水合而西注……霸水又左合浐水，历白鹿原东……霸水又北会两川，又北经王莽九庙南……霸水又北入于渭水。"群流所经达一百余公里，皆汇于灞水以入渭。可见这条河的规模是不小的。据陕西省的《蓝田县志》称："灞水，在县境，古滋水也，其水北流，会浐水，北入于渭，其源出县东南秦岭倒回峪，西北流九十里，入万年县界。旧河岸阔六十丈，今河水涨溢，岸颓为谷，其阔盖不啻数十倍矣。"可见这河逐年涨溢，泛溢为灾，河流阔度，自旧河六十丈，扩展到现时的两倍，至于文中所说"数十倍"，想是指洪水泛滥时的情况而言。

由于靠近帝都，往来行旅，日益增多，灞河上的桥，大概来历很古，然而何时初建，尚无文证。陕西省《西安府志》述中渭桥云"秦始皇跨渭作宫（阿房宫），渭水中贯，以象天汉，横桥南渡，以法牵牛"。这座桥，唐杜牧在他的《阿房宫赋》中云"长桥卧波，未云何龙"，称它是"长桥"，想必不小。在那时已能在渭河上建长桥，灞河相隔不远，又是交通要道，能否在秦始皇时，也建了一座"长桥"呢？陕西省《咸阳县志》中述西渭桥云："汉名便桥，唐名咸阳桥。盖其地夏秋时以舟渡，秋深则作桥，桥成则舟废；冬春二时水涨，则舟行而桥废，故斯地以渡名，复以桥名。"这肯定了在汉代时渭河上夏秋以舟渡，秋深则作桥，故名此桥为"便桥"。灞河上可能是同一情况，汉代就有便桥，然而从汉代的何时开始呢？《水经注》中对灞水又云"水上有桥,谓之灞桥"。其释文引《汉书·王莽传》云"地皇三年(公元22年)二月,霸桥火灾,数千人以水沃救不灭。晨焚夕尽,王莽恶之,下书曰,甲午火桥,乙未立春之日也,予以神明圣祖,黄虞遗统,受命至于地皇四年为十五年正,以三年冬终,绝灭霸驳之桥,欲以兴成新室,统一长存之道。其名霸桥为长存桥"。可见灞河在公元23年冬至节,确实有了木桥,至于夏秋时,是否也和西渭桥一样以舟渡,那就待考了。但《初学记》中有云"汉作灞桥,以石为梁",又《读史方舆纪要》云:"汉灞陵在古长安城东二十里,南北两桥,通新丰、赵店。"灞河在桥址流向是从西北往东南,中有沙洲,故桥分南北。大概在王莽以后的东汉年代,灞桥就从木梁改为石梁了。

《西安府志》云:"灞桥在府城东二十五里,隋开皇二年(公元582年)造。"所谓造,当然不是修理。汉代石梁,经过三国和南北朝时代的大动荡,一定是毁而复修,修而复毁,不知几多次了。《府志》又说这桥"唐景龙二年(公元708年)仍旧",就是说,隋代所造的灞桥,到了唐代,经历了一百多年而依然如故,证明隋代所造的桥应当是一座石桥,这石桥在唐代盛世,想必维修得很好。唐代定都长安,将都城大加修整,《府志》说那时西安是"代为帝都,世称天府,城郭宫室之巨丽,市井风俗之阜繁,实为秦中之奥区神皋也",可想对这灞河津梁，也必是加意修饰的。故《府志》说，在桥旁两岸"筑堤五里，栽柳万株，游人肩摩毂击，为长安之壮观"。因此灞桥两堠，至今北有灞桥镇，南有柳巷村，保存着唐代遗风。

《府志》"灞桥"条文内又云："汉时送行者，多至此折柳赠别，取江淹《别赋》句，又呼为'销魂桥'。"文内"折柳赠别"，取自《三辅黄图》一书，下段呼为销魂桥，来源于江淹《别赋》句："黯然销魂者唯别而已矣。"但江淹是南朝梁代人（公元444～505年），离汉代数百年，显然将两事混在一起了。《开元遗事》载"灞陵有桥，来迎去送，至此黯然，故人呼为销魂桥"，《西安府志》载有西安"十二景"，其中之一为"灞陵风雪：灞陵桥边多古柳，春风披拂，飞絮如雪，赠别攀条，黯然神往"。因此"灞桥折柳"，汉唐以来，成为离别伤怀的同义语，在历代诗文史，时时见到，灞桥之名，借以传播。唐代大诗人李白的《忆秦娥》词中，就有"年年柳色，灞陵伤别"之句。其他唐人诗句，如岑参的"初行莫星发，且宿灞桥头"，李商隐的"朝来灞水桥边问，未抵青袍送玉珂"，郑谷的"秦楚年年有离别，扬鞭挥袖灞陵桥"，罗邺的"何处离人不堪听，灞陵斜日袅垂杨"。这类诗多不胜书，再录宋代陆游两句："风雪灞桥吟虽苦，杜曲桑麻兴本浓。"再录一段故事：唐代雍陶，送客至情尽桥（四川简阳沱江上），问其故，左右曰，送迎之地止此，陶命笔题其柱曰"折柳桥"，并为诗云"从来只有情难尽，何事名为情尽桥,自此改名为折柳,任他离恨一条条"。

唐代以后,这桥时有兴废,宋赵顼(神宗)(公元 1068 ~ 1085 年)时重修,至元代至元三年(公元 1337 年)大修,明代成化六年(公元 1470 年)增修,旋以沙土壅塞,遂废。清代康熙六年(公元 1667 年)以桥工难就,复设舟渡,水落则济以木桥。三十九年(公元 1700 年)又造桥,甫三年即圮。乾隆二十九年(公元 1764 年),有三郡士民输金,复建石墩木桥,为水洞二十四、旱洞十二,越五年桥复坏,仅存桥墩五座,于是援前例,定冬春搭浮桥,夏秋设舟渡之法,因夏秋山洪暴发,沙逐水流,致河床淤积,河面渐宽,增加了搭桥的困难。

清代道光十三年(公元 1833 年),陕西省集资重修,鉴于康乾间随修随圮的经验,采用了所谓"石轴柱"桥基的方法,建起石梁木梁结合的石面桥,经历了一百二十多年,屹立水中,仍然坚固,也是一个桥工杰作。此桥结构,具载《灞桥图说》一书中,其大要如下:桥长一百三十四丈,分为六十七孔,各孔间跨径由四米到七米不等,共有石轴柱四百零八根,分六柱为一门,每门底有石碾盘六具,下为木桩,石轴柱上,平砌石梁,横加托木,两排石轴柱间,安放木梁,横铺板枋,其上两边砖墙,中填灰土,然后是石板桥面及石栏杆。

这座灞桥,于 1955 年改建为新式钢筋混凝土板桥,全长八百三十六点二米,计六十四孔,每孔跨径六米,桥宽七米,两旁人行道各宽一点五米,桥上可行驶重型汽车。所有桥墩,仍利用原有石轴柱重建,用钢筋混凝土柱加高。

在桥下游一百七十米处为铁路桥,彼此平行,长四百四十一米,由十六孔二十五米跨度的钢梁构成,完工于 1934 年,恰在旧灞桥大修的百年以后。

新灞桥是建筑在旧灞桥的基础上的,尽管面目全非,却是渊源一脉,经过两千年的几度沧桑,旧桥并未无声消逝,而是获得新生,更形宏传,再次证明社会主义制度的无比优越性。

附注:本文写就,承陈其英老友,详加校阅,指出漏误,并出示珍藏史料,因而得以修正补充,书此志谢。

原载 1972 年《文物》第 1 期

灞桥与升仙桥

我国古代桥梁构造技术是
宝贵的民族遗产

——《桥梁史话》前言

桥梁是一国文化的表征,我国文化悠久,自古以来建成的桥梁,不计其数,其中多有划时代的杰出结构,对世界人类做出了巨大贡献。这类桥梁都是我国宝贵的民族遗产,有的现在尚继续为人民服务,有的经过加固,仍能发挥作用,有的已随岁月消逝,无遗迹可寻。所有它们的修建经过、科学成就、艺术创造等等,都有记录成书的价值,用来研究其优良传统,古为今用,并在国际上交流,发扬祖国文化。这就是要为我国劳动人民在历史上的功勋,编写一部《中国桥梁史》。

桥梁史并非流水账,而是生产发展的写照。生产对桥梁提出要求,同时也给它以物质条件。桥梁构造的演变总是和生产发展相适应的。一座桥的兴废更要接受政治、军事、经济、文化等等的影响。因而写桥梁不但写它本身,还要写它的背景，这就牵涉到很多复杂问题了。

就是写桥梁本身也不简单，最棘手的是技术资料不足。我国历史文献中，虽然不乏有关桥梁的记载，比如各省各县的地方志中都有当地桥梁的一些资料，在《永乐大典》或《四库全书》里，更可看到不少有关桥梁的诗词歌赋和各种传记的文章，但它们都有一共同特点，就是缺乏关于桥梁的技术资料，更不重视桥工巨匠的辛勤业绩。历史文献中所缺，正是桥梁史中所应补。如何从实地调查及博览群书来补充，是完成桥梁史的一个最大难题。

况且我国古桥的数量，大得惊人，必须就散布在我大好河山中的千千万万桥，选出代表作，不厌求详，加以介绍，来显示我国古桥的全貌。再加古桥的类型特别多，所有近代桥梁中的最主要形式，差不多都有，而且有的还比国外的早得多，应当把这许多丰富多彩的桥梁形式，搜罗得比较齐全，并加以评述。只有这样，才能将古代劳动人民通过建桥工作所进行的各种斗争的成果，垂之久远。

作为一部完整的《中国桥梁史》的前驱。本书用史话的形式，先就一些久负盛名的古桥，从政治、经济、文化、科学、技术、艺术等各方面，予以介绍，并推论其发展演变的过程，然后陆续提出其他具有特点的桥，化整为零，既使读者先睹为快，又不妨碍全书的系统性，是对编辑史书工作的一个创举。

从《史话》中看来，我国古桥的构造，在科学技术上，有很多成就在世界上为先进，有过辉煌的历史。不但桥梁如此，我国的其他科学技术，在古代"往往远远超过同时代的欧洲，特别是在 15 世纪以前，更是如此"。（英国人李约瑟在他所著的《中国科学技术史》序言中语）鉴古知今，我国人民勤劳勇敢，对今后桥梁事业的发展，是具有充分信心的。

正当全国人民响应党中央的号召，向四个现代化大进军的时候，这本书的出版，当不失为鼓舞人心之一助！

1977 年 11 月

《中国古代科技成就》前言

　　正当全国人民努力奋斗,争取在本世纪末实现四个现代化,赶超世界科学技术先进水平的时候,出版这本介绍我国古代科学技术成就的书,是否会有不急之务的感想或疑问呢? 水有源,树有根,科学技术也有继承发展的问题。毛主席曾说:"中国的长期封建社会中,创造了灿烂的古代文化。清理古代文化的发展过程,剔除其封建性的糟粕,吸收其民主性的精华,是发展民族新文化提高民族自信心的必要条件。"列宁也说:"马克思主义……并没有抛弃资产阶级时代最宝贵的成就,相反地却吸收和改造了两千多年来人类思想和文化发展中一切有价值的东西。"因此,这本书的出版,正是为了鉴古知今,来加强我们当前为了那宏伟目标而奋斗的信心。从这本书的内容看来,我国古代的科学技术,历来就是处于世界上的前列,有过惊人的辉煌历史,只是在近二三百年前，才开始走下坡路。正如英国人李约瑟在他所著的《中国科学技术史》的序言中所说："中国的这些发明和发现往往远远超过同时代的欧洲，特别是在 15 世纪之前，更是如此（关于这一点可以毫不费力地加以证明）。"

科学技术的成就，并非纸上谈兵，而应该是确实无疑地表现在活生生的各种事实上。如果一项科学创见或技术发明，不能最终反映到人民生活上来推动历史前进，那就不能算是成就。这本书所介绍的成就，都可以在我国历史上得到验证，都可以算是当之无愧的成就。

首先，几千年来，我国除短暂时期外，政治上始终统一，尽管民族众多而未分裂成欧洲那样；更不像罗马帝国或蒙古帝国，只是盛极一时，以后就衰亡下去。我们中国和他们不同。我们中华民族上下五千年，屹立于大地，而且日益繁荣昌盛。主要原因之一，正如本书的内容所体现的，就是由于有我们自己的科学文化的辉煌成就。

在今天的世界上，我国国土并非最大，但是人口最多。这不能说只是由于地理条件如气候、土地、资源等比较优越的缘故，因为有同样优越条件的国家，人口都比我国少得多。这应当主要归功于我国古代的农业和医药科学的成就。当然，其他经济方面和文化也有重大影响。

在国内人口增长的同时，我国海外华侨人数也很多，到今天已有三四千万人，散布在世界各地，主要在东南亚一带，在当地做出多方面的贡献。他们依靠祖国的文化，形成团结的力量，这文化里就有科学技术，是华侨立足海外的一种凭借。

说到华侨，不由得想到两千年来，我国科学技术在国际文化交流中的地位和影响。汉代张骞出使西域，开辟了"丝绸之路"，通往中亚、西亚各国，唐代鉴真和尚去日本，明代郑和下西洋，不少的探险先驱，都带去了祖国的科学技术。当然，这也同时开辟了我国吸收外国文化的途径。

从 17 世纪耶稣会传教士来到北京以后，西方的科学技术开始输入我国。到了清代末期，封建统治者崇洋媚外；五四运动后又有人提倡"全盘欧化论"，结果西方的科学技术就逐渐占领了我国的文化阵地。直到解放以后，由于毛主席的领导和周总理的关怀，我国古代科学技术才逐步恢复到它应有的地位。因而全国各地都特别重视出土文物的发掘、整理、研究、展览等工作，并对古代遗留下来的建筑、桥梁、古迹等，贯彻执行了国家重点保护的方针。从大量的古代文物中，可以验证我国古代的科学技术对我国悠久文化所起的重大作用。

新中国成立不久，1950 年，我中央人民政府就颁布法令，规定古迹、珍贵文物图书和稀有生物保护办法，并且颁发古文化遗址的调查发掘暂行办法。二十八年来，出土文物的数量之多，价值之高，都非常惊人，使我们对我国各民族的文化遗产，有了广泛和深切的认识，特别是对古代科学技术，能亲眼看到它成就的伟大。比如河北满城西汉刘胜墓中的"金缕玉衣"；湖北江陵凤凰山西汉文帝年间的古墓里有非常完整的男尸一具，外形和内脏的保存都胜过长沙马王堆汉墓里的女尸；陕西岐山、扶风交界处发掘出西周大型建筑遗迹；陕西咸阳发掘出秦始皇时代宫殿遗址；广州市发掘出秦汉造船工场遗址等等。数不胜数的两千年前的遗物中，哪项没有科学技术的贡献呢！

当然，从这许多文物和遗址中，首先接触到的是当时手工艺的水平，在某些方面，两千年前劳动人民的技巧，竟可以同今天的技巧相比。手工艺表现在物质上的成品，必定牵涉到各种材料的生产和利用，如铜铁、玉石、木料等等，在材料的制成和使用中，结构和装配等等里面就都有技术，有属于冶金工程的，有属于采矿工程、土木工程、机械工程的，甚至还有化学工程。我国两千年前就有了这样的技术，这是很可以引为自豪的。至于科学水平，这是表现在技术上的，技术之所以成功，必定暗合科学道理，这就证明了，当时劳动人民的生产实践已经掌握了自然界里物质运动的一些规律。

在秦汉以后的文化高潮推动下，我国的科学技术逐步发展,如本书中所介绍。更可贵的是,在自然科学方面,如天文、数学、物理、化学、地学、生物学、医学、药物学等等,有的成就超过西方一千年,如祖冲之的圆周率,以及气象学、地震学方面,也处于世界的最前列。至于技术,对人类的贡献就更多了。如指南针、火药、印刷术等等都是我们祖先发明的。在各种工程上的成就更是数不胜数。所有古代科学技术的成就,都是我国人民几千年来勤劳勇敢、机智奋斗的结果。我国人民有无穷的智慧和力量,富于创造性,不但表现在政治、经济、军事、文艺方面,也表现在科学技术方面。

由于长期的封建统治,我国人民的聪明才智,几千年来,并未得到充分发展,特别表现在科学技术上,否则,成就一定会更加伟大。也因为这个缘故,古代科学技术上有过特殊贡献的学者、技师和劳动人民,大都是默默无闻的。然而名虽不传,他们的功绩是不朽的。他们的辛勤成果,犹能重见于今日。

现在看来,我国古代的科学技术,是否都已经陈旧,不值一顾了呢?如果把今天的新科学、新技术好好分析一下,往往可以看到旧科学、旧技术的痕迹,因为新的总是从旧的发展而来的。从整体看来,当然已经面目全非,但是从它组成的部分或零件来研究,穷源探本,往往能看出它的脉络所在。即使是从西方传来的东西,也会发现有的部分原来是从我们传过去的旧东西里继承来的。从实践来的旧技术，有的形成传统，到今天还有它一定的价值，所以还能古为今用。最突出的例子，河北省的赵州桥可以算一个。它已经有一千三百多年的历史，但是它的"敞肩拱"技术，今天桥梁工程上还在广泛应用，并且在它的基础上，发展出新型的"双曲拱"。由此可见，在科学技术上，优良传统是很可宝贵的。

我国古代的科学技术，在当时的世界上是领先的。在科学技术的竞赛场上，我国是得过锦标的。我国有过这样的历史，在今天的同一竞赛场上，对世界的先进水平，我们是能够赶上并且超过的。我们不但有信心，而且有能力。这本书就是有力的见证。

1978 年 3 月

中国的古桥和新桥

引　言

中国是一个多山多水的国家,山河壮丽闻名于世界。可是,巍巍群山,滔滔江河,对交通来说,却是一种障碍。中国有一条有名的长江,横贯中国中南部。它江宽水深,风起浪作,南北交通不易,古时被人们喻为难以逾越的"天堑"。在中国西南地区,峰峦重叠,万水千山,交通更是闭塞,唐代诗人李白曾以"蜀道难,难于上青天"的诗句来形容那里交通的困难。

但人类是有征服自然的本领的。中国人民勤劳、勇敢,在同自然界长期斗争的实践中,有过无数伟大的创造。在古代,劳动人民和匠师们,为了便利行旅交通和发展运输事业,曾经建造了许多各种构造和形式的跨江越谷的桥梁,其中有像赵州桥那样的杰作。在中华人民共和国建立以来的三十多年中,中国的桥梁建设呈现了蓬勃发展的新局面。过去从来没有被征服过的号称"天堑"的长江江面,已先后架起了四座巨型现代化桥梁。其中南京长江大桥,规模宏伟,技术复杂,在桥梁科学技术的许多方面达到了世界先进水平。南京长江大桥的建成,西南铁路线上许多桥梁的建成,说明任何高山险水再也不能阻挡我国强大的建桥队伍的胜利进军了,自然界的险阻,在人民的智慧和力量面前低头了。

中国的桥梁建筑有很悠久的历史。原始的桥梁仅在水流平缓的河道叠置石块,或以树干搁置河道,以便通行,随后逐步发展到利用绳索、竹、木、砖、石等材料架桥。远在周代(公元前11世纪~前771年)已有浮桥和简单石桥的出现,公元1世纪,在灞河(在今陕西省)上出现了以石为梁的长桥。公元282年时,在河南洛阳已建成有石拱桥。此后,石拱桥和石梁桥逐渐推广到全国各地,隋、唐以后,桥梁建造得越来越多,越来越好了。公元605年,在河北省赵县(古代曾叫赵州),杰出的石拱赵州桥问世了。公元1050年,在福建泉州建成了著名的石梁洛阳桥,这两座名桥至今仍巍然屹立。若就古代建成的石拱、石梁桥统计一下,为数在百万以上,大小木桥及悬桥尚不在内。13世纪时的意大利旅行家马可·波罗远道来中国,他在游记中说,中国是一个"多桥的国家",

仅杭州一地就有桥一万二千座（这是传抄之误，实际不足一千座）。他盛赞北京附近的卢沟桥，"在世界上也许是无可比拟的"。可见在 13 世纪时，中国的桥就既多且好了。中国解放后的新桥，吸取了现代先进技术并继承、发扬了古桥的优良传统，从而开拓了自己的发展道路。

中国古桥一瞥

石 桥

赵州桥，又名安济桥，位于河北省赵县城南的洨河上，约建成于公元 594 至 605 年，是一座造成后一直使用到现在的最古的石桥。

洨河平时虽是涓涓细流，但在夏秋大雨季节，河水暴涨，急流汹涌，因而造桥很是不易。据记载，这座桥是我国隋代的一群造桥工人建成的，其中为首的一位名叫李春，他是我国桥梁史上的一位杰出的工程师。李春他们造的赵州桥有很多创造性的特点：(1)它是一座独孔的石拱桥，跨度达三十七点零二公尺，在当时是最长的。它的拱圈不是普通的半圆形，而仅仅是圆弧的一段，形成一个扁弧，因而拱圈上的路面不是隆起而是平直的，便于车马通行。（2）敞肩拱是它的另一个非常突出的优点。就是在桥的两个拱肩（石桥两端、桥面与拱圈之间的三角地带称拱肩）和桥面之间各建两个小拱，使拱肩敞开，这种"拱"就称为敞肩拱。这种结构，不但可以减轻桥的重量，节省建筑材料，还可以分流洪水，减少水的冲击，延长桥的寿命，并且使桥型更加壮观。（3）拱圈是石块拼成的，但它的拼法很特别。它是由并排二十八道石拱圈构成的一个整体，每道石拱圈都能独立发挥作用，一道坏了，不影响全桥安全。（4）全桥结构匀称，壮丽美观，和四周景色配合得十分和谐，桥上的石栏石板也雕刻得古朴美观。从这些特点可以看出赵州桥的高度技术水平和稀有的艺术价值。赵州桥在一千三百多年的漫长岁月里，经过千百次洪水的冲击、地质变迁、车辆重压，依然屹立如故，这在世界桥梁史上也是罕见的。

赵州桥是在我国古代无数木桥和砖石桥梁的基础上发展和创造出来的，它是我国桥梁建筑史上杰出的范例，表现了我国劳动人民的高度智慧和技巧。赵州桥的传统，不但为我国一千多年来的石拱桥所继承，也为我国现代的钢筋混凝土拱桥所应用，而且有了许多新的创造。

洛阳桥，又名万安桥。这是我国东部福建省泉州的一座名桥。泉州是宋代对外贸易的一个重要港口，经济繁荣，交通日盛，11世纪至13世纪时这里造桥很多。洛阳桥位于泉州的洛阳江入海处。一条江的入海处必然江面辽阔，海潮凶猛,并有飓风袭击,平时过渡也很危险,何况修桥。这座桥建成于公元 1059 年,原来长一千二百公尺,宽五公尺,有桥孔四十七个,每孔上架平直的石梁七根,每根大约高五十公分①,宽六十公分,长十一公尺,石梁上即是路面。经过历代修理,现存桥长八百三十四公尺,桥宽七公尺,桥墩二十七座。墩上桥梁、桥面与桥栏,已部分改用钢筋混凝土筑成。当时修桥还无机械,只有简单工具,在这海口风浪俱作之处,这座桥是如何建成的呢?当时的一些桥工巨匠发挥了智慧和力量,他们先在海底抛石,铺满桥址,形成一道水下的石基,然后利用浅海里蛎房(一种介属动物,附石而生,得海潮而活,排泄一种胶质黏液)的繁殖,把石基胶固,使成整体,在这石基上堆砌大石块建筑桥墩。洛阳桥这种基础,就是近代筏形基础的先声。他们安装石梁的方法也很巧妙,用大船装载整根石梁,利用潮水涨落,把石梁搁置到桥墩上。经过六年的艰苦努力,终于把这座桥建成了。桥成后经过多次的飓风、地震及洪水的袭击,并无多大的损坏,足见它构造的坚固。

安平桥,这是泉州的又一名桥。建成于公元 1152 年。在泉州安海镇,跨过海口到水头镇,计长五里,因而又名五里桥。平常所谓三里桥、五里桥,多系指离城远近而言,但这座

---

① 公分:厘米。

桥却是实长五里,故民间传说"天下无桥长此桥"。现在实测,桥长两千零七十公尺,桥墩三百三十一座,堪称稀有的宏伟结构。这桥仿照洛阳桥,采用石墩石梁式,其施工方法也可能相同。但洛阳桥造了六年,而这桥长两倍半却只用了一年时间,由于桥太长,桥上修了五座亭子,以便行人休息。

宝带桥,地处江南水乡的江苏苏州,那里的桥自古以来就很多。直到现在民间还有谚语"一出门来两座桥"。在众多的桥梁中,最负盛名的要算宝带桥了。它横跨在南北大运河旁边的玳玳河上,正当澹台、历山诸湖山,湖光山色十分优美,远远望去,它的五十三个桥孔宛如颗颗明珠接连起来,真像一条宝带。宝带桥全长三百一十七公尺,为石拱桥,共五十三孔,其中三孔特别高耸,可通大型船舶。在隆起的三孔上,桥面形成弓形弧线,弧线两头,各有小段反曲线。这桥最初创建于公元816年,后于公元1232年重建,从那时起定为今日形式。

卢沟桥。中国古桥在世界上出名最早的要算北京附近的卢沟桥了。前面提到13世纪时马可·波罗在他的游记中就曾称道过此桥。卢沟桥建成于公元1192年,也是石拱桥。桥长二百六十五公尺,宽约八公尺,由十一孔石拱组成。卢沟桥的狮子柱最为脍炙人口。栏柱头上雕刻的狮子有四百八十五个,雕刻艺术极高,狮子造型生动,体态各异。

卢沟桥的闻名于世,还不仅仅是它的雄伟壮丽和雕刻精湛的石狮子,而在于它是中国人民抗日战争的光荣纪念地。1937年7月7日晚上,日本帝国主义突然向卢沟桥的宛平中国驻军开炮,开始了对我国的大规模武装侵略。中国人民从此燃起了抗日的烽火,在中国共产党领导下,经过八年艰苦奋斗,终于打败了日本侵略者,取得了抗日战争的最后胜利。卢沟桥也因此名垂史册了。

悬　桥

泸定桥。有些古桥,以它著名的历史,反映了中国人民不屈不挠的战斗精神。除上述卢沟桥外,坐落在我国西南四川省大渡河上的泸定桥,同样令人永志不忘。这是一座长一百零三公尺、宽三公尺的铁索悬桥。1935年5月,毛主席率领中国工农红军长征到达这里,敌人把桥上木板烧掉了。英勇的工农红军,冒着浓密的硝烟弹雨,攀着铁索爬了过去,消灭了对岸桥头的敌人,把红旗插在敌人的碉堡上。这就是有名的红军强渡泸定桥的英雄事迹。当看到那用铁索和木板组成的泸定桥时,就不禁会想起毛主席的《长征》诗:"金沙水拍云崖暖,大渡桥横铁索寒。"泸定桥建于公元1706年,桥身用铁链九根系于两岸,悬挂空中,上铺木板,形成桥面。桥两旁各有铁链两根,用作扶栏。两岸各有碉堡一座,内有树立的铁桩。后面有横卧铁锚,牢系过河的九根铁链。这种铁链桥摇曳空中,很不稳定,过桥时感到危险。解放后,于附近造了一座新的钢索悬桥和一座双曲拱桥,通行重载汽车,旧泸定桥就成为珍贵的革命文物了。

珠浦桥。中国西南一带悬索桥很多,而且渊源甚古,晋代法显和尚于公元399年,唐代玄奘和尚于公元621年,曾先后往印度取经,在他俩的记述中都提到途中所见悬挂于山石间的悬索桥。高原地区河道,流经峭壁深渊,造桥筑墩异常困难,悬索桥不需水中支柱,当然是最好的结构形式。但铁索铁链在往昔内地都不易得,于是就地取材,因材致用,而有藤索桥或竹索桥的产生,体现了群众的智慧。在四川省灌县有一座竹索珠浦桥,位于岷江上。共长三百三十公尺,分为九孔,用大木建立木排架,作为桥墩。最大一孔跨度达六十一公尺。木排架上悬挂竹缆十根,竹缆两端锚碇于两岸石崖凿出的石室中。竹缆上铺木板,并有压板索两根,压着木板使它不会移动。左右两旁各有竹索五根,上下并列,作为扶栏。桥的修理也很巧妙,因附近有竹林,每年抽换一部分竹缆,既不影响交通,又能为桥长期维修。这座桥的修建早在千年以前,后来桥损,改为渡船过江,公元1803年重建。1974年因桥址兴建水闸,于下游一百公尺处,依原样改建钢索悬桥。

木 桥

虹桥。除了竹桥,中国古代的木桥当然最多,但因木质易损,无法保留到今天,只可从历史文献中略知一二。值得提出的是中国名画《清明上河图》中的虹桥。这幅画是宋代一位名画家张择端的手笔,它绘出了当时北宋都城汴京(河南开封)的繁华景象。图中画出了一座桥,桥下许多来往的船只争着过桥,桥上行人熙来攘往,但最引人注意的是,这幅画将这座木桥的结构很细致地描绘出来了。原来这是一座木拱桥,由五片拱架组成,上铺木板为路面。这五片拱架虽然构成弧形,但每片拱架却是由许多短的、直的木料拼接而成的。其拼接之法是将每根木料中心点放在一根横木上面,一端放在另一根横木下面,因而将这木料夹起,让木料的另一端伸出去,形成所谓"伸臂梁"。许多伸臂梁,由桥的两岸节节伸到桥的中心,这样就构成一片拱形木架了,然后将五片拱架联系在一起,成为整体,以期稳定。这种桥在当时叫做"飞桥",即近代所谓"伸臂拱桥"。

以上介绍的一些桥都是为了交通运输而建造的。此外,还有在名胜地区便于游人散步、浏览风景的各种小桥,它们的造型优美,体态轻盈,平添自然佳趣,成为风景的点缀。这类桥格局虽小,结构却也不简单。北京颐和园内有两座别致的园林桥。一是玉带桥,全部用汉白玉石砌成,桥面呈双向反曲线,上桥和下桥时使人眼前展开了不同角度的景致。另一座是十七孔桥,桥上路面呈一弧形,各拱高度与此配合,构成一幅美丽画图。其他如浙江杭州西湖的九曲桥,山西太原晋祠的鱼沼飞梁等也很有名。

古桥的卓越成就

首先,最明显的是桥梁形式的多样化,几乎具备了现代桥梁所有的主要形式。这就是:(1)梁桥,以平直的纵梁为桥身的骨干,如洛阳桥、安平桥;(2)拱桥,以弯曲的拱圈为桥身的骨干,如赵州桥、卢沟桥;(3)索桥,以悬挂的缆索为桥身的骨干,如泸定桥、珠浦桥;(4)伸臂桥,以伸出的构件为桥身的骨干,如虹桥。

第二,中国古桥结构精致,施工巧妙,符合科学原理。这反映在石拱桥建造上更为鲜明。中国古桥中最多的是石拱桥,"拱"在中国发明最早,也用得最普遍,可以说,石拱桥就是中国桥的象征。石拱桥越修越多,质量也就随着提高,日益趋于科学化。

古代拱桥的砌筑方法很是讲究,符合科学要求。譬如,任何材料在承受载重或遇气候变化时,没有不变形的,因此在一座桥内要避免不同材料的相对变形。古桥所以全用一种材料而不夹以他种,用石料亦系来自同一石场,并且砌拱不用灰浆,就是这个原因。有的石拱还用榫头嵌入卯槽的方法形成"铰"形结合,更是考虑到全拱的整体变形。

第三,中国古桥在艺术上的成就也是很高的。中国古桥造型优美,线条匀称,并力求与四周环境相配合。

新中国的桥梁建设

1949年中华人民共和国成立后,中国的桥梁建设事业从此开始了一个历史新纪元。

解放后,人民政府在开始全国规模的建桥规划的同时,对中国古桥进行了认真的整饰、修缮和保护。1961年国务院还公布了泸定桥、卢沟桥、赵州桥、安平桥等为重点文物保护单位,并指示说:"我国丰富的革命文物和历史文物,是世界人类进步文化的宝贵财产。切实保护这些文物,对于促进我国的科学研究和社会主义文化建设,起着重要的作用。"许多有名的古桥经过维修恢复了青春。它们一方面在交通上继续发挥原来的作用,同时,在它们身上也体现了我国造桥的优良传统,为今后的桥梁建设提供借鉴。

三十年来,我国的交通事业日益发达,桥梁建设出现了迅速发展的局面。目前,除西藏外,全国各省、市、自治区都有了铁路,铁路通车里程到 1974 年止比解放前增长了三倍多,公路通车里程增加了八倍多。路多,当然桥也多。解放后,广大建桥工人和工程技术人员,吸取了古代劳动人民丰富的实践经验,在新建的铁路、公路线上建造了大量的新式桥梁。

我国西南的四川、云南、贵州三省,解放前没有一条标准铁路。现在,这个地区已建成成渝、宝成、川黔、黔桂、贵昆、成昆、湘黔等铁路。形成一个环形铁路网,并通过其他铁路干线,同我国西北、华北、中南、和东南沿海等广大地区相通。根本改变了过去交通落后的状况。这些线路穿山越谷,跨江过河,方桥、高桥、长桥连续出现。成昆路为例,它穿过大小凉山、横断山脉和金沙江、大渡河等自古以来的天险地区,一个隧道接着一个隧道,一座桥梁接着一座桥梁,桥隧相连,蔚为壮观。全线架起了六百三十五座桥梁。平均每一点七公里就有一座大型或中型桥梁,沿线有好几种我国最大跨度的铁路桥梁,其中金沙江大桥的钢梁跨径达一百九十二公尺。

西藏,地处称为"世界屋脊"的青藏高原上,解放前全境没有一公里公路,如今建成了以拉萨为中心的公路交通网,共有一万五千八百多公里的公路通了车,奔流在喜马拉雅山脉和冈底斯山脉之间的雅鲁藏布江是世界上最高的一条河流,现在在这条江上已矗立两座现代化的公路桥梁。为了沟通西藏和内地的联系，英勇的中国人民解放军和各族人民，以大无畏的革命精神，和高原上恶劣的气候环境做斗争，劈开奇峰峭壁，穿过高大的山脉，跨过湍急的江河，修筑了川藏、青藏和新藏等著名的公路。以全长两千余公里的川藏公路为例，它翻过了十四座大山，这些大山一般都在海拔四千公尺以上，跨过大渡河、金沙江、澜沧江、怒江等十条大河，它的桥梁工程十分艰巨，仅在金沙江到澜沧江的悬崖深谷，惊涛急流上就架起了一百多座大小桥梁。

解放前万里长江无一桥，全国第二大河黄河仅有两座铁路桥、一座公路桥。现在，长江上已架起了四座大型铁路桥梁，在黄河上已建成的铁路桥有十座，公路桥二十余座。1969年1月，我国吸收国内外的建桥经验，建成了具有世界先进水平的南京长江大桥。它的建成进一步锻炼和考验了我国建桥队伍的战无不胜、攻无不克的攻坚能力，并极大地促进了钢铁、水泥、结构和桥机制造等一系列与建桥有关的材料和机械制造工业的发展。

长江"天堑"变通途

长江是我国第一大河，也是世界上最长的河流之一。它发源于号称"世界屋脊"的青藏高原，流经青海、西藏、云南、四川、湖北、湖南、江西、安徽、江苏和上海十个省、市，在上海入海。烟波浩瀚的长江全长五千八百公里，流域面积宽广，达一百八十万平方公里，是我国工农业生产比较发达的区域。过去由于长江的横隔，曾给中国南北交通造成了极大困难。

武汉长江大桥。1955年9月大桥工程正式开工，1957年10月就落成通车。大桥从汉阳的龟山，跨过长江，伸到武昌的蛇山。它和汉水铁路桥、汉水公路桥连接，把武汉三镇联成一体，沟通了中国的南北交通。武汉长江大桥是一座雄伟壮丽的铁路公路两用的双层桥。总长一千六百七十公尺，其中正桥一千一百五十六公尺，两岸引桥共五百一十四公尺。双线铁路在下层，六车道、十八公尺宽的公路在上层，上下层每侧各有二点二五公尺宽的人行道。桥下净空（指通航水面至桥身钢梁底的空间高度）很高，可通航大型轮船。在正桥、引桥相接处的桥台上矗立两座桥头堡，系六层高楼，各层有厅堂，可供游人休息及各种活动之用。

南京长江大桥。1961年在南京建设另一座长江大桥，于1968年建成了。这是我国自行设计和施工的最大的现代化桥梁。南京长江大桥也是一座双层铁路、公路两用桥。铁路桥和公路桥都由正桥和引桥组成。正桥两端有桥头堡，高七十多公尺，与武汉大桥相似。铁路桥全长六千七百七十二公尺，江面上正桥长一千五百七十四公尺。铺设双轨，南北两岸过江的列车可以同时对开。屹立在惊涛骇浪中的九个桥墩，像一座座擎天柱，凌空托起了用高强度合金钢制成的、跨度达一百六十公尺的巨大钢梁。桥下即使在涨水季节也可通航万吨轮船。公路桥面在上层，全长四千五百多公尺，其中引桥长两千六百多公尺。平坦宽阔的桥面可容四辆大卡车并行，人们走上大桥的公路引桥，首先要经过长长的一段双曲拱桥。它南岸连续十八孔，北岸连续四孔，具有浓厚民族特色的双曲拱，使大桥显得分外壮丽多姿。

南京长江大桥是比武汉长江大桥规模更宏伟、技术更复杂的一项桥梁工程。如此长的桥身，如此深的基础，堪称世界上的一个伟大建筑。这座桥的建成标志着我国桥梁科学技术进入了一个崭新的阶段，是我国桥梁科学技术的一个新的里程碑。

大河上下凯歌声

黄河是中国的第二大河，发源于青海省境内，经过青海、甘肃、宁夏、内蒙、山西、陕西、河南、山东等省区，最后流入渤海，全长四千四百八十多公里。在这么长的一条河流上，解放前仅有两座铁路桥，一在天津至浦口的铁路上，靠近济南。一在北京至武汉的铁路上，靠近郑州。另一座公路桥建在兰州。而且这三座桥还都是外国人修造的。解放后，人民政府就积极筹备修建跨过黄河的新铁路桥和公路桥。现在,黄河上已建有铁路桥十座,公铁两用桥一座,公路及农用桥不下二十余座。

新建郑州黄河大桥。位于旧铁路大桥下游半公里。全长两千八百九十九公尺,七十一孔,每孔跨度四十公尺,系双线铁路桥,钢板梁结构。全桥七十二个桥墩基础均采用两根三点六公尺直径的钢筋混凝土管柱。

潼关黄河大桥。这座桥连接南岸的陇海路和北岸的南同蒲,对西北地区的交通极为重要。桥全长一千一百九十四点二公尺,二十三个实体钢筋混凝土圆形桥墩托起了钢桁梁,显得轻巧美观。

北镇黄河大桥。位于山东省北镇地区,是座公路桥,全长一千四百公尺。桥墩基础采用钻孔灌注桩基,桩径达一点五公尺,最深达一百零七公尺,创造了这种桩基罕见的深度纪录。大桥的建成对鲁北地区工农业生产及交通运输的发展起了重要的促进作用。

甘肃靖远黄河大桥。地处青、甘、宁一带的黄河上游地区,近年来随着工农业的发展,也兴建了不少的桥梁。靖远黄河桥是在甘肃境内跨越黄河的公铁两用桥,全长三百五十点三公尺,公路和铁路路面并列布置。

条条江河落长虹

中国幅员广阔,河流众多。除了长江、黄河外,所有江河,随着铁路、公路的大规模兴建和农田水利事业的发展,无一不需要建桥。在解放后的二十多年间,在其他江河上新建了许多各式各样的桥梁。其中较为有名、较具特色的有:广州珠江大桥(公铁两用桥)、南昌赣江大桥(公铁两用桥)、贵州乌江大桥(铁路桥)、河北永定河上的永定河七号桥(铁路桥)、福建乌龙江公路大桥、吉林前扶松花江公路大桥、重庆朝阳公路大桥等。

此外,在澜沧江、怒江、大渡河、金沙江等流域的西南边远山区,昔日山险水急,无人问津的地方,现在铁路穿山越谷,公路网络纵横,也修建了许多形式新颖、规模较大的桥梁。如横跨澜沧江的永平、永保公路桥,越过怒江的腊猛红旗公路桥、安宁河大桥(铁路桥)等。这些桥不但建造技术上都具有各自的独特之点,且桥依山势,蔚为佳景,更显得江山如画。

传统拱桥放新彩

松树坡石拱桥、"一线天"石拱桥。石拱桥是我国千年传统的桥梁结构形式,过去用于走车马的石路,后来用于走汽车的公路,解放后则大量用于走火车的铁路。以成渝铁路为例,建成的石拱桥有三百二十四座之多。宝成铁路上,从黄沙河至成都的一段就有一百七十五座石拱桥。这些石拱桥都在原有的传统上,进行过各种技术革新,以期适用于大跨度及重荷载。宝成铁路上有一座松树坡石拱桥,全长一百二十一公尺,由两孔三十八公尺跨度的大石拱组成。中间桥墩高达五十公尺。更值得注意的是,这座桥位于半径为三百公尺的曲线上,而且路面坡度高达千分之二十八。因此,桥上行车用电力机车索挽。成昆铁路上也修建了许多石拱桥。其中"一线天"石拱桥跨径达五十四公尺,全长六十三点一四公尺。此桥发扬了古代赵州桥传统,在大石拱的两肩上各有三个小拱。桥高二十六公尺,两岸悬崖陡立,有"一线天"之称。

长虹大桥、九溪沟桥。公路上的石拱桥,解放后也得到大发展。1961 年云南省建成长虹大桥,一孔石拱跨度为一百一十二点五公尺,此桥也发扬了赵州桥传统,在大石拱的两肩上,各有五个小拱,拱圈半圆形,跨度五公尺。大拱拱圈为悬链曲线,拱脚至拱顶高度为二十一点三公尺。拱上公路面宽七公尺,走两排汽车。1972 年在四川省建成九溪沟桥,跨径达到一百一十六公尺,这是目前跨径最大的一座石拱桥。

湔江人民大桥、浒湾石拱桥。为了就地取材并充分利用石料强度，减少拱圈石的数量和石料加工所费巨大劳动力，各地在工程实践中采用石肋拱代替板拱，采用天然卵石修建卵石拱都取得了可喜的成绩。1967年四川省修建的湔江人民大桥为十八孔，每孔跨度为二十公尺的卵石拱。1971年河南省修建的浒湾大桥，为石肋拱，为加强拱肋稳定性，肋间加设有石盖板。此桥主拱成"冂"形断面，独具风格。上述两桥都能通行六十吨拖拉机，建成后运营情况都极为良好。

新卢沟桥、前河双曲拱桥。1970年11月在北京附近旧卢沟桥下游半公里处建成一座新卢沟桥，这是座双曲拱的公路桥，全长五百一十公尺，分十七孔，每孔跨度三十公尺，拱桥拱圈由十根拱肋并列组成，每两肋上铺一系列拱波。拱肋为钢筋混凝土，拱波为无筋混凝土。这座桥仅仅用了四个月时间，就全部建成了，说明这种拱桥的优越性。1969年河南省建成前河双曲拱桥，单孔跨度达到一百五十公尺，为目前最大跨径的双曲拱桥。

长沙湘江大桥。更宏伟的双曲拱桥是湖南的长沙湘江大桥。湘江是湖南省的最大河流，经洞庭湖注入长江，流域面积占全省面积三分之一。长沙位于湘江东岸，东西交通为湘江割断，1972年9月在此建成公路桥一座。桥全长一千二百五十公尺，系赵州桥式的双曲拱桥，共十六孔，每孔大拱跨度七十六公尺，拱高九点五公尺。全桥为钢筋混凝土建筑。

公路面为四车道。宽十四公尺,两旁人行道各宽三公尺。长沙的湘江对岸为有名的风景胜地岳麓山,桥上往来行人甚多,故人行道特别宽。这座桥只用了一年时间就完成了。

宜宾岷江大桥。在总结双曲拱桥经验的基础上,群众又创造一种箱形拱。1971年四川省宜宾修建的岷江大桥,全长五百三十三点七五公尺。主拱圈由六片拱箱组成,每片拱箱施工分为开口箱、盖板和现浇混凝土三部分,前两部分为预制构件。这种拱桥施工程序大体与双曲拱相同,稳定性好,抗扭刚度大,整体性强,适合于大跨度拱桥的采用。

1978 年

中国的古桥和新桥(增删稿)

《中国古桥技术史》前言

　　我国地大物博，文化悠久，自古以来建成的桥梁不计其数，其中多有超时代的杰出结构，成为我国宝贵的民族遗产。我国史书文献浩如烟海，如《永乐大典》《四库全书》《古今图书集成》等，皆足以彪炳千古。可惜的是，在这浩瀚的书丛中，与桥有关的史、地、诗、文之选录，虽很丰富，而与桥有关的科学技术的记载，却寥若晨星，异常缺乏，但这正是我国文化的一种重要象征。

　　自本世纪以来，在国内外闻有介绍我国古桥的论文或专著。国外如 1937 年英国福尔梅约著的《中国桥梁》一书，国内如 1957 年唐寰澄著的《中国古代桥梁》一书，都记述了中国古代桥梁。但比较全面介绍我国桥梁历史的，当推 1959 年出版的罗英著的《中国石桥》和《中国桥梁史料》。我在《史料》这本书的《序言》中说："作者很谦虚地把他这本书称作

'史料',希望他这筚路蓝缕的工作能成为一部桥梁史的'先行官'。"

英国李约瑟在他的巨著《中国科学技术史》(1971 年出版)第四卷第三册《土木工程及水利》中,有一专章介绍我国古桥(中译稿约三万字),参考上述及其他上千种的中外文资料,广征博引,按桥梁结构分段叙述,引证翔实,极有见地,令人赞佩。但对中国古代桥梁史而言,尚非全璧。

1964 年第二届全国人民代表大会第四次会议时,我和梁思成同志联名提出"请政府组织委员会,编写《中国桥梁史》案"(第 124 号)。大会对此案的审查意见是:"由国务院交交通部会同文化部等有关部门研究。"同年 11 月 7 日全国人大常委会办公厅在《提案办理情形报告》中云:"据交通部报告,关于编写中国桥梁史问题,曾拜访了茅以升代表,并先后和有关部门进行了磋商,现正由交通部会同文化部、建筑工程部、铁道部等有关部门进行此项工作,准备先组成中国桥梁史委员会,然后再订编写方案。"当时我深为交通部如此负责的态度所感动。可惜的是,十年动乱,编写桥梁史的事也搁置下来。在此期间,国内各种报纸刊物,已开始登载有关我国古桥的史话,引起广大群众的兴趣与注意。我也曾在《文物》期刊的 1963 年 9 期和 1973 年 1 期中,介绍了由国家公布保护的十座古桥。各出版社以古桥为选题的读物,亦日渐增多,其中1977年成书的科普作品《桥梁史话》,受到读者的好评。同时,我也写了一本《中国桥梁——古代至今代》,译成日、英、法、德及西班牙五种文字,向国外介绍。然而,对于《中国古桥技术史》的编写,依然未能忘怀。

我感到很意外而又最兴奋的事,终于到来。1978年春,中国科学院自然科学史研究所决定组织人编写《中国古桥技术史》并派副所长段伯宇和黄炜同志来约我为主编,承告拟约请有关部门组成编写委员会,由各协作单位聘请一位专家,组成编写小组,按计划进行,预定四年完成。我虽自惭力薄,然对此得偿宿愿的使命,不得不欣然接受。

《中国古桥技术史》编写委员会于1978年9月28日成立,组成了编写组,于1978、1979、1980年开过三次编写会议,经过四年多的努力,全书告成,至堪庆贺。

本书各章的编写,以中青年专家执笔为主,并结合各协作院校有关科系的科研项目,配合进行。同时设立了资料小组,搜集文献资料、沟通情报,提供执笔人参考和为重点调查提供线索。并经交通部通知全国各省市交通、公路部门做了大量采访查报工作;各地文物部门也为本书提供了不少资料。

参加编写组的同志名单如下:戴竞、黄京群、徐承炉、金恒敦、薛镕、钟用达、黄梦平。以上各位对本书的完成,都大力支持,并提供了宝贵意见,至感欣幸。

编写组执笔人的名单如下:陈从周、唐寰澄、韩伯林、张驭寰、谢杰、沈康身、杨高中、潘洪萱、郑振飞、胡达和、瞿光义、夏树林、汪嘉铨、张尚杰、许宏儒。

全书统稿负责人:唐寰澄。

本书各章执笔人对所担任务,皆不辞辛劳,于业余时间,夜以继日,如期完成。主编人在此表示敬佩。

编写组经常得到自然科学史研究所负责同志的指导和支援,工作进行十分顺利;科学史所有关的研究组也对本书作了有益的审议与协助。在编委会及编写组筹建期间,黄梦平同志担负了筹备工作,对本书的开端做出了贡献。在本书编写过程中,唐寰澄同志协助主编人承担了全书文稿的统稿、编辑工作。许宏儒同志担任编写组秘书,对组织编写,收集资料,完成全书,做了大量工作。编委王玉生同志参加了定稿会议,进行了全书发稿前的校订工作。

对这次编成的《中国古桥技术史》,有以下几点需要说明:

(1)这是一部技术史而非通史,因而侧重于科学技术,而有关政治与经济方面的联系,不能求其完备。

(2)这是一部中国古代桥梁的技术史,下限至1881年铁路桥梁出现时为止。因为近代及现代的桥梁史包括铁路及公路的桥梁,而这两部分桥梁,铁道部与交通部都已编有画册,并已分别着手编写铁路桥梁与公路桥梁近代、现代史。

(3)本书系集体创作,从技术专业讲,执笔者皆研究有素;但从内容取舍、考证详略及文章体裁等方面,则不免各有所好。为了统一全书、避免各篇矛盾起见,由编写工作会议推定副主编唐寰澄同志负责统稿,将全书各章节文字和内容校阅、调整、补充,再由主编人定稿,如是往复数遍,以求完善。

(4)作为史书,本书有一特点,即进行了很多次实地考察。遇有古籍传说中错误遗漏之处,多经执笔者亲往实地调查,辛勤跋涉,补足其缺略,且有所发现,故本书之成,是有较为翔实的根据的。在此,还应铭志在考察过程中所得到的各省公路部门各级行政组织的大力协助与指导。特别应当提出并表达本书各专家同表感谢的,是交通部潘琪副部长为促成本书的完成,由交通部协拨专款,并对资料采访方面,给予多方便利。

(5)过去桥梁书籍,附图多于实物照片,为编写本书,作者们在各地考察中拍摄了大量照片,选入本书图版及插图五百余幅,图文对照,足以补充文字之不足;而各式桥型,姿态多样,留此古桥风貌,以遗后之学者。

写历史书,非仅笔墨之事,首先要查资料,有了资料,还要鉴别真伪是非，往往一字之差，意义为之左右。写技术史更多一探微索隐问题，而且本书所写各桥，大半仍留存人间，随时可以验证。因此，本书之能于短期告成，可知所有参加工作的同志们所费心血，非比一般，作为主编，我向全体同志表示敬佩。希望今后过若干时间，再作一次修改增补，以期更好地发挥其作用。

为了振兴中华，全国人民皆在努力于四化建设。荟集若干人力，耗费若干年月，从事于古代桥梁的技术史研究与撰述，是否有其必要，应当有所说明。表面上看来，为了提高我国的科学技术水平，势非引进国外的先进学术及设备不可。以桥梁为例，则近代的斜拉与刚构桥，皆为我国古代所无。至于新兴材料，如各种预应力结构所用者，更是闻所未闻。然而，任何新技术皆有其旧技术的根源，往往旧事物中孕育着新事物的萌芽。如各种斜拉创意，可于古代吊桥中求之；预应力的设想，可于古代施工方法中求之。近代文明渊源于古代文明，此世界之所以能进步。我国桥梁技术，过去曾在世界上领先，现在追述其历史，探索其所以领先之故，正是为了从中得到启发，来发扬光大其固有传统，为促进四化建设之一助。我相信，世界上没有不能造的桥，我国桥梁技术，将于此中再度显现其领先的可能性。

《中国古桥技术史》概论

　　我国古代桥梁,具有悠久历史和卓越成就。几千年来,历代人民通过辛勤劳动和反复实验,并且勤于改进,勇于创新;于是在长河急流,风涛鼓荡之间,架起了一座座坚固美观的长桥,飞跨两岸,畅通无阻。当近代铁路、公路交通还未出现以前,作为历史产物的古代桥梁,在各个不同的历史时期,基本上是随着社会经济的发展而发展,成功地完成了历史所赋予它的使命。

　　建筑依赖于生产力水平,建筑的发展是科学技术进步的具体表现。桥梁是一种既普遍而又特殊的建筑物。普遍,因为它是过河跨谷所必需,而河流峡谷则是遍布大地,随处可遇的。特殊,因为它是空中的道路。道路处在空中,它的结构就复杂了。由于复杂,就需要适当的材料和特殊的结构,并且还要有合乎科学和艺术的设计进行施工,才能建成,才能保持永久。这些条件，在科学昌明的今天，国家设有专科育才的制度，又有新品种的建筑材料、各种功能的大型机具和先进技术的情报交流，当然是得心应手，功效迥异。但在古代则条件大不相同。每一成果的取得，都是经历了漫长岁月和付出百倍于今的努力。

我国古代的桥工匠师，也和世界上其他民族一样，创建了形式多样的桥梁，虽然创建年代互有先后，但其基本形式则大致相同，足以说明桥梁建筑的规律性和科学性以及事物发展的必然结果。在中世纪，西方旅游家，曾赞誉我国为多桥的古国，他们以亲身的经历，翔实的文字，叙述了在我国所见到的长桥结构，往往是在其他国土上所罕见的。由此影响所及，引起了世界工程界的重视，在世界上享有盛名。这无疑是我国历代劳动人民对人类文明做出的杰出贡献，也是我国古代灿烂文化的一个组成部分。这一历程，即在现代化的今天，仍在继续前进，虽然时代不同，功效迥异，但事物进化原是永无止境的。

国外学者对我国古代桥梁的研究，已有一些专著，国内工程界在解放前后也开始了一些研究。用今天的科学水平，探讨过去的技术成就，鉴古知今，古为今用，是有极其深远的意义的。

第一节　古代桥梁的起源

早在远古时代，自然界便有不少天生的桥梁形式，如浙江天台山石梁，跨长六米，厚约三米，横跨飞瀑之上，梁上可通行人，是天台山风景之一。至于天然侵蚀成的石拱，也有很多，如广西桂林的象鼻峰和江西贵溪的仙人桥。《徐霞客游记》记江西宜黄狮子岩石巩（即拱）寺："寺北有蠹崖立溪上……是峰东西横跨，若飞梁天半，较贵溪石桥，轩大三倍。"树木横架便成木梁桥；藤萝跨悬，是为悬索桥。人类从自然界天生的桥梁得到启发，在生存的过程中，不断仿效自然，以解决行的问题。

我们的祖先，由原始游牧而进入定点聚居。随着生活、生产资料的日臻繁盛，逐渐完整地创建了宫室坛台，城廓道路，车舆舟楫，早期的建筑群，便成为部落聚居经营的特征。桥梁，这个闪耀着劳动人民创造智慧的产物，也初具规模，并且日益成为人民日常的政治、经济、生活中不可缺少的重要建筑物。

《文物》杂志 1976 年 8 月载，《河姆渡发现原始社会重要遗址》。考古家认为距今六千至七千年，在文化发达较晚的杭州湾以南的宁绍平原，已有了带有榫卯的木梁柱建筑构件,并发掘出一些小件饰物,其中有一个骨匕的木鞘,厚度均匀,上下平直,弧度一致,外壁两头缠有多道藤篾类的圈箍。由此可知,新石器晚期劳动人民智力发达,已经相当成熟,对木、竹、藤等材料性能已具备了多种组合利用的工艺。据此,则木梁木柱桥的物质条件早已具备了。

1954年,我国在陕西西安半坡村发现了新石器时代的氏族聚落,位于浐河东岸台地上。已发现密集的圆形件房四五十座,中间最大的建筑面积约一百平方米。在部落周围,挖有深宽各约五至六米的大围沟,这条沟当年可能有水,估计是为了防御封豕长蛇和异族侵略的设施。其出入之际,势必有桥,时约在公元前4000年左右。

史前和之后的原始桥梁,由于当时的材料、工艺等种种原因,不可能在风雨侵蚀、洪波荡突的漫长岁月中长久保存下来。但每一事物,总是新陈代谢、踵事增华,后者继承又发展前者,不断地由低级演进为高级,由简单而日臻完善。

一、桥梁字义释证

关于早期桥梁的情况,我们先从文字的释名进行考证。

桥、梁两字,在古代是异名同义的两个单词。汉许慎《说文解字》(下略为《说文》)作这样的解释:"梁,水桥也,从木水,刃声。"又"桥,水梁也,从木,乔声"互为通释,未说明其字义。清段玉裁《段氏说文解字注》(下略为《段氏注》)曾予以较详细的解释,意义便较清晰。其"梁"字注为:"梁之字,用木跨水,则今之桥也。《孟子》'十一月舆梁成';《国语》引《夏令》曰'九月除道,十月成梁';《诗·大雅》'造舟为梁'皆今之桥制,见于经传者,言梁不言桥也。若《尔雅》'隄,谓之梁';《毛诗》'石绝水曰梁'谓所以偃塞取鱼者,亦取亘(即横)于水之义,谓之梁。凡毛诗自'造舟为梁'外,多言鱼梁。"

《礼·王制》:"鱼梁,水堰也,堰水为关空,承之以笋以捕鱼,梁之曲者曰罶。"

对于桥的注释是:"水梁,水中之梁也。梁者,宫室所以关举南北者也。然其字本从水,则桥梁其本义而栋梁其假借也。凡独木者曰杠,骈木者曰桥。大而为陂陀者曰桥。古者挈皋(即桔槔,井上汲水的横木架)曰井桥。《曲礼》'奉席如桥衡',读若居颐反,取高举之义也。"

此外,还有同义的专门名词。如《尔雅·释宫》:"石杠谓之徛。"《说文》释"徛"称"举胫有渡也"。《段氏注》为:"聚石水中以为步渡彴也。"且又于"榷"字条说明"步渡彴"是:"然则石杠者谓两头聚石,以木横架之,可行,非石桥也。"又如《广志》:"独木之桥曰榷,亦曰彴。"《说文》释"榷"是:"水上横木,所以渡者。"然则,"徛""榷""杠""彴"指的都是木梁桥,或精确地说是独木桥。《广韵》《集韵》《韵会》注"矼"字:"古双切,音江,聚石为步渡水也,通作杠。"那就是今天所谓的踏步桥或汀步桥,是不加木梁排列较密的一个个跳墩子。

又,楚人称桥为圯,这些不同的名词,近代已不再沿用了。

根据段氏之说,可以知道:

(1)"桥"和"梁"是同义异名。"梁"这一专用名词,在文字应用上略先于"桥"。虽然,桥字出现也不晚。最近实物如1976年出土秦安陆令吏喜(公元前262~前217年)墓竹简上有"千(阡)佰(陌)津桥"字样。

(2)"梁"和"桥"都从木,最早的桥梁应是木制的。梁字的古文有写作"**㴪**"(漆)的。《段氏注》:"水阔者,必木与木相接,一其际也。"按照字形,二木之间有一横,似即指水阔而用有水中墩的多孔木梁桥。

(3)按《尔雅》和《毛传》的注,"梁"的字义不仅专指架木跨水的桥,还包括筑土或砌石绝阻断水所成横在水中的隄(通堤),即所谓堤梁。这和今称山梁或鼻梁含义相同。堤梁,从今日桥梁的含义来判别,不能称之为桥,那只是土或石堤上兼可以走人而已。

(4)"桥"和"梁"二字虽是异名同义,但桥字又特有其含义。徐锴《说文解字系传通释》注"桥":"乔,高而曲也,桥之为言趫也,矫然也。"联系上文"大而陂陀者曰桥",则"桥"字又形象地显示有坡度而中高的形状,与"梁"字仅训为跨水或绝水者有所不同。

二、桥梁最早记载

史前桥梁,无可查证。但我国有悠久的文字记载的历史,载于典籍的桥梁,比较早的如——

《史记·殷本纪》记帝纣:"厚赋税以实鹿台之钱,而盈巨桥之粟。"

《史记·周本纪》在武王伐纣既革殷命(公元前 1122 年)之后说:"命南宫括散鹿台之财,发巨桥之粟,以赈贫弱萌隶。"

关于巨桥,东汉许慎注以为是:"巨鹿,水之大桥也,有漕粟也。"虽然尚有别种注法,如邹诞生云:"巨,大;桥,器名也。"可是《水经注》"漳水"条:"衡漳水北经巨桥,邸阁西,旧有大梁(桥梁之梁非堤梁之梁)横水,故有巨桥之称。武王伐纣发巨桥之粟以赈殷之饥民。"可见郦道元也认为这是当年漳水上的大桥。因此在商、周之间公元前 12 世纪已不乏有名的桥梁。至于其桥式,因漳水是较阔的河流,推测为多孔的木梁桥,地点在今河北省曲周县东北。

考之《禹贡》,逾河、逾洛,津梁无定所。周有盟津,春秋有茅津、棘津、宋桑之津,津之有名,始见于记载。津,即渡口,由无定所的津渡,演进到固定的码头,是一个进步。黄河、洛水都是大河,其所谓津梁,乃是今日的渡口或浮桥。

《诗·大明》："亲迎于渭，造舟为梁。"朱注："造，作梁桥也。作船于水，比之而加板于其上以通行者也。即今之浮桥也……造舟为梁，文王所制。"时在公元前 12 世纪，是浮桥最早的记载。《左传》记鲁昭公元年（公元前 541 年）："秦公子铖出奔晋……造舟于河。"公子铖在秦国无法安身，随从资财，"其车千乘"，靠舟渡要用很长时间，于是搭起浮桥，使车辆连贯通过。这二座虽然都是属于临时一次搭用的浮桥，但由此可说明纪元前五百多年，人们已掌握了在长河大川中架设浮桥的技能。

《史记·秦本纪》记昭襄王五十年（公元前 257 年）"初作浮桥"，张守节注："此桥在同州临晋县东，渡河至蒲州，今蒲津桥也。"桥址即今山西省永济县西接陕西省朝邑县东境。"初"作浮桥，乃是指第一次在黄河上修建起固定式的浮桥。

拱式结构的出现，是人类的一大发明。从原始天然的侵蚀型石拱，到有意识地砌筑拱券，其间有很长的发展过程。根据历年发掘墓葬的调查，西汉末期（公元前 1 至 2 世纪）出现了多边形砖拱和圆形筒拱结构。如《考古》1973 年第二期载麻弥图庙一号墓墓室内为砖券顶，年代为汉武帝元朔二年（公元前 127 年），则拱的创始，必然更早。在我国古代建筑技术的术语中，砌拱称之为"发券"或"卷瓮"，这"卷瓮"二字是极有意思的。

《礼记·儒行篇》记有"蓬户瓮牖",即为用破坛子砌在土壁中作窗户。因此,"卷瓮"的方法只要用陶瓮出现之后,便有其可能,而陶瓮土屋则在公元前五六千年便有了,虽然这还不能称为真拱,也并不是桥。

有明确记载的石拱桥,见《水经注》"谷水"条记旅人桥:"凡是数桥,皆垒石为之……朱超然与兄书云,桥去洛阳宫六七里,悉用大石,下圆以通水,可以受大舫过也。题其上云,太康三年(公元282年)十一月初就功,日用七万五千人。"目前还没有查到比这更早的文字记录。

索桥是桥梁的基本形式之一,我国最早使用的都是竹索桥。

《四川名胜记》引西汉扬雄《蜀记》称四川成都的七星桥为:"桥上应七星,秦李冰所造(公元前256~前251年)。按七星桥者,一长星桥,今名万里;二贞星桥,今名安平;三玑星桥,今名建昌;四夷星桥,今名笮桥;五尾星桥,今名升仙……"

竹索,古写作"笮",亦通"筰"。唐李吉甫所写《元和郡县志》(以下略称《元和志》)称:昆明,"本汉定筰,凡言筰者,夷人于大江水上置藤(或竹索)桥,谓之筰"。又记四川成都内外江:"大江一名汶江,一名流江。经县(成都)南七里。蜀守李冰穿二江,成都中皆可行舟,溉田万顷。蜀中又谓流江为悬筰桥水。"

竹索桥发源于西南,在秦通蜀之前,这样的桥梁早就存在,公元前3世纪只是最早的记载。

以上所述是桥梁中几种主要结构形式:梁、舟、拱、索的最早记载,早期桥梁的出现,是以后长历两三千年桥梁发展的先驱。

至于《拾遗记》记:"尧命禹疏川奠岳,济巨海,鼋鼍以为桥梁。"《集仙录》说:"(周)穆王命八骏吉日甲子,鼋鼍以为梁,以济弱水而宾王母。"《述异记》有:"秦始皇作石桥于海上,欲过观海日出处……"《风俗记》传说:"织女七夕当渡银河,使鹊为桥。"这些都只是神话,但也反映出人民的意愿和丰富的想象力。

第二节 古代桥梁的几个发展阶段

我国是一个封建统治时间最长久的国家,据科学技术史研究工作者认为,从战国到秦汉,我国古代科学技术的发展,已经形成具有自己特色的体系,到宋元时期更达到一个高峰,在很多方面曾居于世界领先地位。在这期间,中国不断汲取其他国家的先进成果,也把自己的成果东向传入日本,西经阿拉伯辗转传入欧洲。我国桥梁技术的发展和流传,情况也大致如此。

我国幅员辽阔，山河壮丽，江河流域面积在一千平方公里以上的水系，达一千五百多条。秦岭、淮河以南，水流密如结网，加之人烟稠密，交通频繁，道路必须依赖桥梁为联系手段。桥梁直接关系道路的通塞。当近代铁路、公路还未出现，桥梁建筑还未成为近代科学的专门工艺之时，我国古代劳动人民也和世界其他民族一样，以坚韧不拔的毅力，勇敢智慧的创新精神，反复实践，摸索出一套符合事物发展规律的造桥技术，创造出形式多样的跨空长桥，标志着中华民族文化的特征。今天硕果仅存的若干古桥，经历了千百年天灾人祸的长期考验，仍然继续发挥其应有功用，它们绝不是孤立幸存的个体，而是千百万座古代桥梁，通过科学实践逐步发展成长的代表之作。因此，它们是一份极可珍贵的遗产；是我们今天古为今用的研究对象。古代桥梁发展的分期，大致可试分为四个阶段。

一、西周、春秋时期

第一阶段以西周、春秋时代为主，包括西周以前有历史记载的时代。这是古代桥梁的创始时期。

原始公社生产力非常落后，随着生产资料私有制的逐渐形成而进入有阶级的奴隶社会，生产力逐渐提高。西周公私田的制度，出现了封建土地私有制萌芽，促进了农业生产。又取山泽林薮之利，手工业和商业也活跃起来。这时期虽然已有了若干著名的桥梁，但一般情况下还做不到遇水架桥的程度，桥梁的建造尚不普遍。《诗经》上有很多涉水的诗句。如《诗·郑风》有："溱与洧方涣涣兮。"朱注是"冰解而水散之时也"，于是便有"子惠思我，褰裳涉溱""子惠思我，褰裳涉洧"等诗句。

对于稍深一些的河流，《诗·谷风》称："就其深矣，方（竹或木筏）之舟之；就其浅矣，泳之游之。"至于长江大河，如汉水、长江，则《诗·周南》："汉之广矣，不可泳思，江之永矣，不可方思。"

《孟子·离娄》："子产听郑国之政，以其乘舆济人于溱洧。孟子曰'惠而不知为政，岁十一月徒杠成，十二月舆梁成，民未病涉也……焉得人人而济之。故为政者每人而悦之，曰亦不足矣'。"所谓舆梁，宋孙奭《孟子音义并疏》说："舆梁者，盖桥上横架之板，若车舆者。"或有解作通车舆的桥梁。子产的未能在溱洧上造桥，是不知为政呢，还是力有所不及？

《国语》有为政不力、失职的记载："周定王使单襄公聘于宋，遂假道于陈，以聘于楚……道茀（音弗，路上长草）不可行也。侯不在疆，司空不视涂，泽不陂，川不梁……故先王之教曰'雨毕而除道，水涸而成梁'。"所谓"司空以时平易道路"，就是利用农闲季节，也恰好是枯水时期，召集民工，修筑道路桥梁，使人免去褰裳涉水之苦。

春秋时期的桥梁记载多语焉不详,且加上"梁"有木梁和堤梁的不同含义,使很多单独称梁的记载分不清是何种结构。可以分析的如《诗·谷风》和《诗·小弁》有"毋逝(去)我梁,毋发我笱",因为梁和笱同说,可见为捕鱼的石堤。而《诗·曹风》记"维鹈在梁,不濡其翼",捉鱼鸟在梁,便可能是堤梁。《诗·卫风》"在彼淇梁"注:"石绝水为梁。"那便是淇水上用石堆砌的堤梁。《春秋》鲁襄公十六年(公元前557年):"会于溴梁。"《尔雅》:"梁莫大于溴梁。"注:"溴水名梁堤也""然则以土石为堤,障绝水者名梁,虽所在皆有,而无大于溴水之旁者"。但是连筑堤也算在内的桥梁,建筑是比较简陋的。这一时期,正是逐步地摆脱渡河"深则厉,浅则揭"的原始状态,进而创作永久性的桥梁,是桥梁的筚路蓝缕的阶段。

二、秦、汉时期

第二阶段以秦、汉为主,包括战国及三国,为古代桥梁的创建发展时期。秦在统一中国后,虽然国祚短促,但为以汉族文化为中心的统一事业,做出了巨大贡献。汉承秦制,进一步巩固了中央集权的封建政体。东汉是我国建筑史上的一个灿烂发展时期,发明了人造建筑材料的砖,创造了砖结构体系及以石料为主体的石结构,而演进为新的拱券结构。在建筑艺术造型方面,又融合了佛教东渐的宗教色彩。秦、汉两代，大兴土木，阿房、未央，巍峨壮丽，大大开拓了建筑群的总体设计规模和个体建筑的高大形象。这时修建在都城的渭桥、灞桥，不仅长大宽广，而且饰以勾栏，植柳成荫，实用功能与艺术美化，交融而结成一体。

春秋末战国初期，我国在冶炼技术方面，同时发明了生铁和冶炼铁，出现了铁器。铁的出现，又大大推进了建筑方面对石料的多方面的利用，为大量建造石桥，提供了物质条件。首先在原来木柱梁桥发展的基础上，增添了石柱、石梁、石桥面等新构件。而石料用于桥梁建筑的重大意义，则在于由此而石拱桥应运而生，在实用、经济、美观各方面都起着划时代作用。石拱石梁的大量发展，不仅减少了维修费用，延长了桥梁使用寿命，还提高了结构理论、施工技术的科学水平。晋太康三年的七里涧旅人桥，并不是一个突然的出现，而是石拱建造技术已经相当成熟的作品之一。汉代画像砖"裸拱"——无拱上建筑的裸露拱券的出土，证明汉代已有石拱。秦始皇大修天下驰道，规模宏伟。筑路当然必须同时修桥，才能四通八达。《述征记》云："始皇东巡，弗行旧道，过菏水，率百官以下人提石以填之，俄而梁成。"这是说由于"弗行旧道"，所以临时提石成梁。传说又有始皇在海中立柱建桥的故事，海上架桥，在当时的条件下是不可能的，但足以说明在有可能的条件下，是有遇水架桥的愿望和相应措施的。

实际上,战国时期已大规模造桥,除前述李冰七桥外,《史记·滑稽列传》记战国魏人西门豹为邺城令(今河南临漳县):"发民凿十二渠……十二渠经绝驰道。"汉赵充国治军屯田,建桥七十二座,都说明当时修建桥梁已十分普遍。

《汉书·薛广德传》述广德谏阻文帝御楼船,称"乘船危,就桥安",则在当时人们的生活经验中,已确认从桥上通过,不仅便利,并且安全。由此,也可说明当时的造桥技术,已给人以安全可靠的信念。在这一时期里,汹涌宽阔的黄河上,架起了第一座蒲津渡浮桥;在四川产竹之乡出现了竹索笮桥。标志着公元纪年初期,梁、拱、吊,桥梁的三大基本体系,已在我国形成。

三、唐、宋时期

第三阶段以唐、宋为主,包括两晋、南北朝、五代,为古代桥梁发展的全盛时期。隋、唐国力,较之秦、汉更为强大。原因是自东汉以降,逐步对江南开发经营,在经济上有了极大的发展。隋代结束了南北分割、兵祸频仍的混乱局面,唐、宋两代取得了较长时间的安定统一,民营工商业发达,运河驿路畅通,航海技术进步。自东晋至南宋,大量汉族人南迁,经济发达远远超过了黄河流域。隋、唐在当时称为文明发达的强盛大国。指南针、活字印刷术和火药武器都发明于宋代,对世界发生了巨大的影响。建筑、造船、天文、水利等方面,也都取得了重大成就。在著名的古建筑中，隋匠李春首创的敞肩拱赵州石桥，北宋牢城废卒发明的叠梁结构虹桥，北宋蔡襄主持建成的筏形基础、植蛎胶固的泉州万安桥，潮州海阳县的石梁结合浮桥开关活动式的广济桥，在世界桥梁史上享有盛誉。在这段时期里，石桥墩砌筑工艺不断改进，日臻完善，为兴建长大桥梁铺平道路。从此石桥飞跃发展，数量上、质量上都达到历史的高峰。北宋木工喻皓，写成《木经》三卷。李诫编写的《营造法式》，内容包括土木工程技术、建筑设计规范和估工算价的规定，总结了古代劳动人民长期积累的丰富经验。这两部书虽然不属于桥梁建筑的专著，但由此可以察知工程技术上的一般法则，仍不失为可珍贵的最早技术文献。

四、元、明、清时期

第四阶段，元、明、清三代，对古桥的建造及修缮，由于驿路和漕运的发展，都各有过贡献。首先是对金代明昌三年（公元1192年）建造的永定河上的卢沟桥，根据《马可·波罗游记》记载，在元代更加修整一新。这个游记中还说到中国各地的桥梁既多且美，举出杭州、泉州等地为例，足见元代的桥梁是受了宋代桥梁很大的影响。明代桥梁不似宋、元繁盛，然亦有江西南城万年桥，贵州盘江桥等艰巨的工程。清代桥梁事业很多有别于前朝，一是对一些古桥进行了修缮和改造,因而延长了寿命;二是在川滇一带兴建了不少索桥,如泸定桥等,提高了索桥技术;三是大量发展了桥梁艺术,特别是公私园林中的小桥,其艺术性之高为前代所未有;四是为所造的一些大小桥梁留下了施工说明,有的比较完整,如灞桥、浐桥、文昌桥、万年桥等,从中也可看出宋代喻皓《木经》及李诫《营造法式》的影响。

在清代末季,我国桥梁历史上发生了一次技术革命。公元1881年我国第一条营业铁路从唐山到胥各庄开始通车,从此引进了新式的、前所未有的铁桥、钢桥、钢筋混凝土桥等各种以新式材料及结构所建成的各式桥梁,及为了造桥制成的各种施工机具。这在我国是"铁路桥梁"的开始。随着铁路的发展,公路也开始在湖南出现了,而在大城市中也需要新式桥,因而也引进了"公路桥梁"。

新式的铁路桥梁和公路桥梁都是从国外引进的。为了分清古代与近代桥梁的史实,本书即以铁路桥梁的出现为断代的标志。即在公元1881年以前建成的桥为古代桥,以后建成者为近代桥,包括铁路桥、公路桥及公路、铁路两用桥。

第三节　古代桥梁的卓越成就

桥梁的功能,是为了解决跨水或越谷的交通,建成架空道路来沟通经济动脉。凡是道路上的运输工具及行人都要能在桥上畅通无阻。一方面，它是一定历史背景下的产物，受到一定物质条件的制约；另一方面，它在时代前进、政治需要、经济发展的影响和促进下，不能停滞不前。我们对古代桥梁在历史上所取得的成就，可从下列几个方面来探讨。

一、早期桥梁概况

恩格斯在《自然辩证法》一书中说道："随着自然规律的知识的迅速增加，人对自然界施加反作用的手段也增加了。"他又在《反杜林论》中提出："自由不在于幻想中摆脱自然规律而独立，而在于认识这些规律，从而能够有计划地使自然规律为一定的目的服务。"我国古代桥梁的演进历程，情况正是如此。

远古时代，桥梁以最原始、最简陋的形状，出现于部落聚居的建筑群中。从此，随着社会进化，一系列的新要求就不断地提出，而每前进一步，都必须解决一系列的复杂问题，付出巨大的努力和代价。桥梁又与其他建筑物不同，它要在洪波巨浪里奠基，要在寒风暴雨中操作，要使用大量人力和器材，而一有疏虞，便可能前功尽弃。有的甚至随建随坏，长年累月不能竣工。也有新桥刚刚落成，便遭遇巨浸洪水，毁于一旦。而桥梁如毁坏，则不仅损及财产，而且丧失人命。因此，桥梁是无时无刻不在经受着严峻考验的。

早期的桥,多数只能建筑在地势平坦,河身不宽,水流平缓的地段,技术问题比较容易解决。早期桥的形式,不外聚石培土,或是木梁柱式的小桥;而在水面较宽、水流较急的河道上,则采用另一种方法——造舟为梁,如公元前11世纪,周文王"亲迎于渭,造舟为梁"。但那时候的浮梁,与后世的浮桥有所不同。那是一次使用,时间短暂,不须经受长期风涛鼓荡的考验。可是后世长期性的浮桥,恰正肇始于此。由于长期,于是船只锚定,船与岸的联系,随波上下而发生的长度伸缩以及河中行船通过等等问题,都必须一一妥善解决,显然问题就比较复杂了。浮桥,是最早出现于我国古文献中的一种桥型,也是在黄河上最早出现的第一座长桥。浮桥一直沿用至今,原因有二:一是它在军事上仍有实用价值,二则是在经济不够富裕的地区,可暂时借以维持目前交通。另外浮桥还是一种过渡手段,在古代有不少的桥梁,是经过浮桥、木桥、石桥三个循序而进的阶段,借以掌握水文资料,为建筑永久式桥梁做好准备。所以浮桥在历史上,有其一定的贡献。

　　梁桥出现的时代最早。早期的绝水为梁,有的垒石,有的培土,虽然也达到跨河越谷的目的,但它还不具备桥梁的本质,因为它不是架空飞跨。但另一方面,也可说这种早期的梁,是道路向桥梁转化的一种过渡形式,通过它,使人们产生了梁和路两种不同的概念。真正的桥,应以跨水行空为标志。而柱、墩，则又为桥梁构造的先决条件。在我国，梁柱式木桥，出现最早，在公元前6世纪，就有了木梁打桩立柱的文字记载，在技术上已有相当基础。由木柱而石柱而石墩，其演进迹象极为显著。由于木柱不能经久，石柱形式单薄，于是体积重、大的墩体就相继而出现。桥梁使用寿命的久暂，关键在于基础是否巩固。基础工程技术，是我国历代名工巨匠们的攻关重点，下基要深达岩层的观念久已形成。后汉《阳渠石桥铭》"攒立重石，累高固距"，隋《沣水石桥碑》"龟柱通泉，龙梁接汉"，都形象地描述了基础的坚固厚重。《旧唐书·李昭德传》："昭德始累石代柱，锐其前厮杀暴涛，水不能怒，自是无患。"又《宋书·河渠志》记留守向拱重修天津桥，"甃巨石为脚，高数丈，锐其前以疏水势，石纵缝以铁鼓络之，其制甚固"。"甃巨石为脚"，即是桥墩，"桥墩"这个名词，出现较晚，实际石墩的采用，至迟也在晋代以前。《水经注》有云"凡是数桥（指睪门、阳渠、旅人等桥）皆垒石为之，亦高壮矣"，可为明证。于此，可见石墩的创建和不断改进以及"分水尖"所起的缓冲护墩的作用，前后经历了七百多年。墩的形式，由上大下小改为垂直，又改为下大上小，这是由于最早想用挑出伸臂加大净跨，最后则吸取长期的经验教训，采用下大上小加强墩身；同时对于水下工程，由水修法改为干修法，又分别根据河床地质的不同情况采用木桩基础、抛石基础、睡木基础等。其中洛阳桥的抛石桥址、植蛎固基,解决了海口大河的桥基方案。经过这些长期努力和发明创新,于是长桥巨跨,遍布全国,这一丰硕成果的取得,是来自多方面的因素。

二、古桥形式与建筑材料

桥梁的几种主要形式,公元纪年前,在我国已经基本具备。它们是:梁桥(包括伸臂式)、拱桥、索(吊)桥。在以唐、宋为中心上下一千年中间,我国古代的能工巧匠,以精湛的技能,谨严的工艺,大胆的创新,驰骋纵横,贡献出天才智慧,在我国桥梁史上谱写出光辉的篇章。特别值得提出的是:设计方面的首创敞肩拱和筏形基础,施工方面的打桩、建基及架桥等各种方法,包括起吊二百吨重的石梁等,种种技艺奇迹般的出现,更受到古今中外的交口称赞。

桥梁的发展是以生产发展为依据的,生产对桥梁提出要求,也为桥梁提供物质条件。古代的建筑材料,不外乎木、石、竹、铁,在有限的材料品种中,不断地了解掌握材料性能,充分利用,善于筹划,用其所长,补其所短。一方面材料性能决定着桥型结构;另一方面,新的结构又向材料提出新的要求。这种反复结合,深入研究,便形成了一整套经验方法和维修制度,在桥型上也自然形成一派土生土长的民族风格。当然,也并不排除东西方不同民族技术艺术的交流与融合。

上述三种桥型，在不同时期，不同地区，也存在着发展上的不平衡。

梁桥，在历史上较其他桥型出现为早。这是由于它构造比较简单，木材随地皆有，梁柱式木结构使用普遍。但这种木柱木梁结构，很早就出现弱点，不能适应形势发展。起而代之的是石柱木梁。在秦汉时代，建成多跨长桥，如历史上有名的渭桥、灞桥。长桥建于宽广河道，河床地质情况且多变化，这时期，桩基的技术发明了。在文献记载中，出现了铁椎打桩机。石墩的出现，标志着木石组合的桥梁能够跨越较宽大的河道，能够经受汹涌洪流的冲击，但石墩上的木梁，不耐风雨侵蚀。于是建起桥屋，保护桥身，延长木梁使用寿命，还方便往来行旅的憩息。这一桥型在西南、江南一带，相当普遍，而最早见于黄河流域地区。中小型石梁或石板梁，构造方便，材料耐久，维修省力，是民间喜用的一种桥型。南宋绍兴以后，福建泉州商业发达，海船往来，财力殷富，在万安桥成功地建成巨大石梁的影响下，掀起了长大石桥的兴建高潮，做出了不少施工技术上的突破，成为仅见于福建滨海地区的历史盛举。

索桥出现的时代也比较早，古籍中记载很多。这一桥型的出现，首先是由于地势使然。在崇山深谷，急流汹涌的险阻地带，既不能架舟横渡，更无法筑墩架桥，而隔水相望的人民，只有采取凌空悬渡的办法，来解决两地的交通。早期处于试验阶段的"度索寻橦"，是竹索笮桥的先驱。竹索笮桥见于秦李冰七星桥之一，则最原始的藤溜索，时代当然更为古远了。后汉李膺、唐独狐及形容这种溜索，有"人悬半空，度彼绝壑""顷刻不戒，陨无底壑"。唐僧智猛说："窥不见底，影战魂栗。"可见初过悬索心理的紧张，达于极点。由此而并列诸索，又桥面布板，两侧设以扶栏，西藏则有人走在管状布置中的藤网桥，自然安全多了。宋范成大描写索桥："桥于半空，大风过之，掀举幡幡然，大略如渔人晒网，染家晾帛之状，又须舍舆疾走，从容则震悼不可立，同行者失色。"但《元和郡县志》"虽从风摇动而牢固有余，夷人驱牛马来去无惧"，说明经常习惯与偶一通过，其感受是大不相同的。竹索桥比起木桥，更不能经久。著名的四川灌县珠浦索桥，就有一套周密的保管维修制度。铁索桥，在文献上散见于汉晋时代，唐吐蕃建有铁桥。至于建制比较详细的记载，多在晚近明、清时代。《徐霞客游记》对贵州盘江桥曾给以高度评价："望之飘然，践之则屹然不动，日过牛马千百群，皆负重而趋者。"举世闻名的四川泸定铁索桥（公元 1073 年），与欧洲第一座悬桥英国的云溪（Winch，公元 1741 年）桥相比，前者跨长百米，后者仅约二十二米，且只存在很短时间，未可同日而语。

石拱桥，在我国桥梁史上出现较晚，但拱券结构一经形成,便迅猛发展,成为古桥中最富有生命力的一种桥型。由于对各类地基有丰富的经验,使石拱桥遍布南北西东,无地不宜,即在今天,仍有其继续发展的广阔前景。我国历代桥工哲匠,对于石拱桥的造型设计、构造工艺,投入了极大的精力,敢于标新立异,勇于改革创新,反映于石拱桥的多种多样以及因地因时的灵活运用。同时因为石料坚实,经久不坏,遂使历史上一些名桥杰构,留存至今,作为古文化的历史见证。唐张嘉贞称赞赵郡石桥:"制造奇特,人不知其所以为。"对这座独步千载的创造,确是赞无虚誉。而它更重大的意义,还在于大大鼓舞了历代能工巧匠的雄心壮志。除了上文所述的泉州石桥而外,他如河北的卢沟,江苏的宝带,江西的万年,浙江的通济,山东的泗水,安徽的太平等联拱长桥,陆续出现于各地,所可贵的,不在于旧制的模仿复制,而在于自辟途径,花样翻新。

如上所述,古桥在构造形式上,灵活多变而又因地制宜。在梁、拱、吊的基本体系上,又自多种多样,如拱桥有坦拱、陡拱;拱券有尖拱、半圆、圆弧、弓形、多边形等;梁有伸臂、八字撑。从材料上分,拱有石拱、砖拱、木拱、竹拱;梁有木梁、石梁、木石组合梁、石板以及砖、土、石桥面等;索桥有独索、多索、藤网、竹索、铁索;浮桥有用船、皮筏、木排、竹筏等组成的,形式上又分低浮桥、通航浮桥、开合浮桥等。所有这些,它们的设计构思是与力学原理相符合的;它的构造形式是与材料功能相配合的;它的稳定程度是与安全要求相适应的。对于这些工程要素的掌握运用,说明人们已有能力解决多因素的复杂问题,也是技术精湛的体现。

茅以升手绘兰州握桥

三、桥梁艺术与桥梁文学

史学家论我国古代艺术,认为"汉族传统的文化是史官文化"。史官文化的特性,一般地说,就是幻想性少,写实性多;浮华性少,朴厚性多;纤巧性少,宏传性多;静止性少,飞动性多。自东汉建安以后开始发生变革,即由原来寓巧于拙,寓美于朴的作风,渐变为拙朴渐消,巧美渐增的作风。其间佛法东渐,建筑方面在一定程度上受到宗教色调的影响。隋唐时期,中外文化交流甚盛,艺术风格又注入了新的因素。桥梁是水上架空的建筑,除了它所特有的实用功能,和由于实用功能而确定了它的基本形式而外,它不能不受到周围大量建筑群的感染和影响,而表现为某种程度的共性和协调。在另一方面,桥梁又常是置身于山川萦回的林泉胜地,天然风景又要求它以特有的姿态为幽美环境增添风采。而桥梁的本身就是实用与艺术的融合,如桥梁的平直,悬索的凌空,拱券的涵影,它们形象的本身,原来就摇曳着艺术的风姿。英国李约瑟在其专著中说:"没有中国桥欠美的,并且有很多是特出的美。"(《中国科学技术史》四卷三册,145页)

桥梁形式,无论中外,大致相同。采用什么来体现我国古代桥梁所特有的艺术风格呢? 根据文献记载和实物形象,特点有二:一是附属建筑,二是石作雕刻。附属建筑有桥屋、亭阁、栏槛以及牌坊等。桥屋是为保护木梁、铁索以防腐朽锈蚀的,有其实用的必要性。但重阁飞檐,有亭翼然,用之于多跨梁桥,则感到飞动有势;用之于伸臂式桥,则更给人以"飞阁流丹,下临无地"的壮丽景色。又如盘江、泸定等桥,高山急流上的一线悬索,两岸高筑楼台,更增添了雄伟气概,与群山奔湍浑成一体。又如泉州万安、安平等长大石桥,全用厚墩巨梁,偃卧在江海浪潮之上,给人以敦厚质朴之感。采用石凿武士和耸立石塔点缀其间,便觉十分协调。石拱桥的本身形态,原即富有美感,不论圆拱高耸,或者长桥压波,总是能和周围景物相映成趣,更平添了诗情画意。古人常用"长虹饮涧""新月出云"来描绘优美拱桥的形象。比如赵州桥比起欧洲古罗马时代所造的拱桥，就显得前者轻巧，后者笨重。至于风景区和园林建筑群里专为观赏而特建的桥梁，更是独出匠心，纯粹以美术品的要求精工细作，把桥梁所能体现出的美的意境，以实体直觉加以渲染，与实用性的桥梁，又似异中有同。在以实用性为主的石桥中，也有为观赏而设制的装饰，如望柱栏板、桥头石兽的雕刻，这种装饰性的雕刻，一是受周围建筑物的传统影响，更多的原因是石桥的兴起，正当石窟造像风靡一时。据文献记载，郦道元《水经注》："魏太和中，皇都迁洛阳，经构宫极，修理街渠，务穷隐，发石视之，曾无毁坏。又石工细密，非令之所拟，亦奇为精至也。"可见汉晋时代石工严谨不苟，工艺精细。又"建春门石梁不高大，治石工密。旧桥首夹建两石柱，螭矩趺勒甚佳。乘舆南幸，以其制作华妙，致之平城"。这里所说的"工密""华妙"，也如同阳渠桥铭中的"敦勅工匠，尽要妙之巧"。当时对石工的要求是相当严格的，工匠必须通过石作雕刻，才可以表现出他的技能，于是石桥的栏板望柱，便成了石工表现技巧的对象；而当时的风气，建桥石工必须把全桥的一石一拱，进行精细的艺术加工，这一传统便成为后世石工的风范。同时，这类雕刻图式的形成，一方面与同时代的建筑艺术相通；另一方面又与民间神话有密切联系，如治水的龙，分水的犀，降伏水怪的神兽等等，特别是卢沟桥的石刻狮子，名闻世界。

石工建成了桥，还为桥留下了护桥卫士。这些逐渐形成和流传下来的习尚，便形成它的独特风格。

自从建安诗人写桥入诗，首倡了桥梁文学。隋唐以后的诗人词客，都为桥梁写下了大量的清丽动人的诗句，情景交融的描绘，增添了人们对桥梁爱好的情感。一千八百年来，以描述桥梁为主题的诗篇，和以同一主题为内容的铭、赞、颂记，数量之多，几是一个难于统计的巨大数字，它们不只是文人学士一般抒情之作，其中也包含着无比丰富的桥史资料，和对桥梁功用上和艺术上的各种描述与评价。我们研究这些文学、艺术大师的作品，就可能发现我国古代桥梁艺术和风格的精髓所在，这比仅从外形上可见可望的表现要深入得多了。

四、造桥技术与桥工匠师

造桥技术，在我国历史上有卓越的成就。桥梁是技术性相当强的建筑物。尤其是在宽阔河流上，高山峡谷中，要和急流湍溜的水斗，要和卷地怒号的风斗，要和水底冲刷的泥沙斗，要和秋雨冬雪的寒冷天气斗，要和争分夺秒的时间斗，一桥之成，就是一场对大自然的剧烈斗争。因此，建桥之前必须掌握地质情况、水文资料、地震记载等等，桥成之后，又要经历长年的风雨侵蚀、水流冲刷和地震的严峻考验才能屹立无损，安如磐石，达到安全使用的目的。新桥如此，古桥当然也是如此。

我国历代桥工匠师积累了相当丰富的实践经验,创建了多种形式结构。这些成绩的取得,自然是由于诸方面的促进,这与历代匠师的不懈努力、竭精殚思是分不开的。从不少的桥工事例中,说明匠师们不仅仅是善于积累成功经验,取法先进经验,更可贵的是善于吸取失败经验,从失败中吸取教训。特别是在修理旧桥时,他们洞烛要害所在,无异是做了一场试验,整旧如新。一项技术所以成功或失败,必有科学理论上的原因。历代造桥匠师根据实践经验而完成任务,即是遵循理论而不自知。他们的实践,即是应用科学,我们可用理论来评价他们的实践,也可以从他们的实践使我们发现新理论,甚至提供科研新项目。

罗英著《中国石桥》一书中,提出"古经验与新理论"问题,以贵州平越(今福泉县)葛镜桥为例,说明通过石拱辅构件而产生被动压力的重要性。他先以旧理论验算葛镜桥在新的负荷下即不安全,而在新的理论下则并无问题。所谓新理论即是考虑到石拱辅构件的作用。书中对"多铰拱"及"无铰拱"的结构,和"尖拱"及"压拱"的施工,很多独出见解,为研究古桥技术辟一途径。

本书对梁、拱、索三种桥型也分别作了理论分析,与罗著一样,也都提出了不少的问题,需要作进一步的研究。目的是：一方面从理论上发掘古桥所以能"古"而不坏的原因；另一方面则是引导出研究方向，古为今用。

桥梁的文献不少，但技术资料大都散失。我国的绘画技术发达甚早，古代对于大型建筑，都要经过设计制成图式。《说苑》："齐王起九重之台，召敬君图之。"《鲁灵光殿赋》提到殿中的壁画："图画天地，品类群生，杂物奇怪，山海神灵，写载其状，托之丹青。"韩偓《迷楼记》："项升自言能构宫室，后进其图。"《图书见闻志》记："唐朝李涿从高力士故宅保寿寺僧，以缣素换得巨幅桥图，柳公权鉴定为名画家张宣所作石桥图，明皇赐力士，因留寺中。"又记唐明皇的《金桥图》："御容和照夜白，陈闳主之；桥梁、山水、车舆……吴道子主之，狗、马、驴、骡……韦无忝主之。图成，时称三绝。"宋张择端另有一幅名作《金明池夺标图》，画中的"仙桥"，《东京梦华录》记有："桥面三虹，朱漆栏楯，下排雁柱，中央隆起，谓之骆驼虹，若飞虹之状。"关于桥梁工程的设计图式，文献中也有不少的记载。如《宋书·河渠志》："留守向拱重修天津桥，甃巨石为脚，高数丈，锐其前以疏水势，石纵缝以铁鼓络之，其制甚固，四月（建隆三年，公元961年）具图来，降诏褒美。"又为仿赵州桥改建天津桥，令董士锐"彩图作三等样制修砌图本一册"。元虞集《虹桥记》："遂以八月乙丑召工画图，尽撤其旧而新之。"则不仅说明造桥必有图，并且匠师也能制图。

有图必有文字说明，否则工匠将无法据以施工，可惜有些重要文物竟未能保存下来。

关于桥工匠师的社会地位和工作条件，古时有冬官、司空管理百工。《史记·五帝本纪》：“舜……于是以垂为共工。”马融注：“为司空，共理百工之事。”又：“垂主工师，百工致功。”《左传》：“卫文公大布之衣，大帛之冠，务材训农，通商、惠工、敬教、劝学、受方、任能。”“惠工”，有优惠提倡的意思。考唐制，官营手工业，“少府监”（尚方监）总的职掌是管理百工技巧的政务，“将作监”总的职掌是管理土木工匠的政务。类似建筑工程师的高级匠人称为“长上匠”，州出钱雇用，因而也称为“明资匠”，名额二百六十人。柳宗元《梓人传》：“食于官府，吾受禄三倍，作于私家，吾收其值大半焉。”据此则古代匠师是具有专业技能、独立工作和指挥能力的工程技术人员，并且受禄于官府，还允许他作于私家，很像后世的自由职业者。有人以为古代工匠不受重视，没有文化，没有系统的科技知识，是缺乏根据的。

古代造桥技术的传播和继承，和其他手工业者一样采用的是师徒“薪传”的方式，或者子承父业，沿传不绝。这种口传手授的培育方式，不仅造桥技术如此，其他手工业几乎无不如此。师承传习的传统方法，则是历史的产物。我们在大量文献记载中，还看到有不少不是专业匠师而参加了造桥活动,并做出重要成就。其中包括一些佛教和伊斯兰教的宗教徒,如洛阳桥就有僧人做出很大贡献。他们还曾把我国的造桥技术传到国外。如俄国彼得大帝就曾请中国派遣造桥专家西去,传授技术(见李约瑟专著149页)。

五、古代桥梁的社会性与人民性

桥梁,从一开始就以属于人民公用公有的社会性而出现。

封建社会是以私有制为构成杠杆的。即以建筑物而言,宫殿楼阁为皇家贵族所私有,田庄院舍为地主所私有,庙宇坛塔为僧道所私有,甚至渡口船只,也是牟利的私有工具。唯有桥梁,在私有社会里却是社会公有的。无论是官修私建。都有它普遍的公用性,人民造了桥梁,桥梁就为人民服务,成为人民生活中不可缺少的重要建筑物。几千年来,在社会上都有爱桥护路的风尚,都公认"修桥补路"是为人民造福的善举。由于公认造桥是一种公益事业,于是也就被公认为群众性的活动,凡是从事这项活动的事,便受到社会上的支持,从事这种活动的人,便受到社会上的尊敬。所以,"桥梁道路,王政之一端",是为官作宰政绩的体现。士绅们认为造福乡里,可以获得美名,就连方外僧道,也为了"广结善缘",把修桥当作济世渡人的功德。

综观地方志的记载,历来修建桥梁的方式,不外四种。

一是民建,由一家一姓独力成桥;二是募捐集资,士绅僧道,广事劝募,报经官府支持,协力兴工;三是官倡民修,由地方官提倡,商绅附和,并指派官吏士绅主持其事,这类都属于较大工程;四是全由官府拨发公帑施工,也有的以工代赈。其中以募捐集资的一种,最为多见,效果也最好。因为集腋成裘,可以调动社会上一切的积极因素,并且潜力极大。如历史上名桥安济桥、万安桥,都是采用这种方式建成的。建桥如此,修桥也如此。所以我国遍布各地包括穷乡僻壤,桥梁数量之多是惊人的;而千年来这些桥梁的经常维修,同样是十分繁重的工程,它们的数量之大,用工之多,工期之长,可以比拟于万里长城和跨过长江黄河的南北大运河,但修建时民无怨言,乐于输将,这就有别于长城和运河了。

六、从古桥现状看古桥成就

我国古桥遍布全国,是在现代铁路、公路桥梁技术从外国引进以前而建成的。个别古桥已历时千百年以上。近代技术当然大大超过古代,然而古代技术竟能使其作品在某些方面与现代技术的作品争胜,其中奥妙何在? 古桥建成时,并未梦想它有千年寿命,更不料桥上会有今天这样繁重的运输,古桥在结构形式和结构细节上还有近代桥梁所没有的,还可进行大量的科学研究。

研究我国古代桥梁成就的专家学者,国内代不乏人,国外亦有不少，但有一共同情况，即大多系从古书中搜集资料，而很少经过实地观察来立论的，如同李时珍、徐霞客等人那样。古桥亦是古文物，我国自解放后对古代文物特别重视，对其重要者加以保护。三十年来，从地下发掘所得的珍贵文物，数不胜数，其中亦有与桥梁有重大关系的，如刻画汉砖、铁件等。但对造桥工具则未有所闻。它如古代壁画中，间或有古桥图像，过于简略，也无法深入研究。唯有如《清明上河图》中的虹桥，忠实详密，是很重要的资料。足堪欣幸的是，古桥遗迹历来就屹立在地表，都可前往调查、观赏并摄影，然而我国幅员广阔，桥梁又遍地皆是。且或处在穷乡僻壤，深山野谷，或荒凉地带，人迹罕到。要遍察古桥的成就，非一朝一夕之功。

至于在考察所得的结果上，从事理论分析，科学试验，以求推陈出新，应用于近代桥梁，尚有很多工作有待进行。

1983 年

192

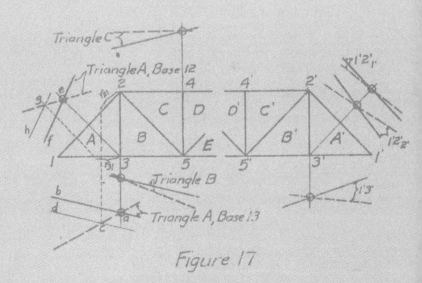

Figure 17

硕士论文

THE DESIGN OF A
TWO-HINGED SPANDREL-BRACED ARCH BRIDGE
WITH AN INVESTIGATION OF ITS
SECONDARY STRESSES

THESIS

Presented to the

Faculty of the Graduate School

Cornell University

for the degree of

Master of Civil Engineering

By

THOMSON MAO

Associate of the Tangshan Engineering College

June, 1917

PREFACE

The object of the present thesis, as the title indicates, is to design a double-track railway arch bridge of the Two-hinged Spandrel-braced type. While the main part of the work is to give numerical computations and designs of the arch-under consideration, equal effort is given to the theory and principles which control the design, and at the same time special attention has been paid to those things which affect the adaptability and utility of the arch. It is the aim of the writer to give critical study to the relative merits of arches and other kinds of bridges having equal chances of adoption; between two-hinged and three-hinged arches; between spandrel-braced trusses and arch ribs either with solid or open webbing and finally between two-hinged spandrel-braced arches with and without cantilever arms. The study is not complete on account of the limited time available, but it disposes the facts and conclusions arrived at in a straightforward manner.

To obtain rigidity in a two-hinged arch and to do away with the uncertainty of stress distribution after erection, an expedient is resorted to which makes the arch a combination type, so called because it possesses the advantages of both two and three-hinged arches. The arrangement is such that the dead load stresses follow the law of the three-hinged arch while those due to live load, impact, wind, temperature and braking forces are controlled by the theory of the two-hinged arch.

Much has been said relating to the strength and stiffness of two and three-hinged arches, both in engineering literature and theses of graduate students in the College of Civil Engineering of Cornell University. As the amount of investigation has been large it seems advisable to collect all the conclusions and results obtained so that it may be in good shape for immediate reference. The present thesis does this summarizing in a very brief form but nothing of importance has been lost sight of.

Perhaps none of the works mentioned above have touched

the problem of secondary stresses in arches. It might be interesting to see how these stresses act in arches of the two-hinged, spandrel-braced type and how they affect the design. For this reason about thirty percent of the present work has been devoted to the calculations of these stresses and to conclusions drawn therefrom. This is a painstaking task as the span is long and influence lines are to be constructed.

As the greater part of the work in calculating stresses in two-hinged arches is required to find the horizontal component of the reaction, a portion of the following pages will be directed to the various methods used in the past, both analytic and graphic, exact and approximate. The relative merits of each will be discussed and the best ones adopted.

The principal dimensions of the arch are proportioned to conform to the best modern practice with respect to aesthetic and economic effects. Since the only argument against the appearance of spandrel-braced arches has been its abrupt termination with a shallow approach span, the present arch will have two cantilever arms supporting two approach simple spans trussed like a portion of the arch.

In regard to the design and detailing of the arch, it may suffice to say that it has been done with great care to follow the most modern practice. Specifications of American Railway Engineering Association were adopted.

In preparation for the writing of this thesis a vast amount of material has been collected from books, thesis and journals. Instead of giving the references separately, each will appear where it is used. Things that are common to all books will not be referenced because only features which differ materially from others merit a reference to its source so that the original article may be consulted for further study.

Attention is called to the following schemes which were developed by the writer: Approximate method of constructing the influence line for horizontal thrust; dead load stresses for unequal panel loads; method of increasing the accuracy of displacement diagrams; and finally the arrangement of table for the solution of equations in calculating secondary stresses.

In conclusion the writer wishes to express his indebtedness to Professor H. S. Jacoby under whose directions the present work is undertaken.

LIST OF TABLES

LIST OF DIAGRAMS

CONTENTS

PART I. INTRODUCTORY CHAPTERS

CHAPTER I. HISTORICAL NOTES

Masonry arch is one of the oldest types of extant bridges and was probably originated in China. [①] Through gradual changes in the art of bridge building and to meet the demand of heavy structures for railway traffic, metallic arches of solid rib type were invented, the first cast iron arch being built at Coalbrookdale, England in 1776. [②] In 1867 there appeared a foot bridge over the Bollat Fall at Hohenschwangau, which was the first arch of the spandrel braced type.

Perhaps the oldest two-hinged spandrel-braced arch is in

---

① A Treatise on Arches. Howe. P. 1.

② History of Bridge Engineering. Tyrrell. P. 309.

America, the one at Corondelet Park, St. Louis, built in 1887. The arch crossing Noce Schluct was built three years later, in 1890. In the year of 1897 the well known suspension bridge at Niagara Falls was replaced by Niagara Falls arch of 550 feet span, then the longest two-hinged spandrel-braced arch in the world. In the same year, Will Street bridge at Watertown, H. Y. was completed. Next on the list comes the one in Germany built in 1899 crossing the Rhine River. The arch of Maine St., Waterport, H. Y. was completed in 1900 and that of Costa Rica, in 1902. In the year of 1905 another monumental arch of the two-hinged spandrel-braced type was built in Africa, the Sambesi bridge crossing Victoria Falls. The first arch of the combination type was completed in the year of 1911, the bridge over Grooked River designed by Modjeski, The arch over St. John River, Canada, was built in 1914; which having a span of 565 feet is the longest two-hinged spandrel-braced arch in existence at the present time.

It is thus seen that the development of two-hinged spandrel-braced arches has been very rapid in recent years. This is partly due to the fact that this structure is the most favorite type among arches for railway traffic,[1] and partly due to the recent perfection of the theory of statically indeterminate structures.

CHAPTER II GENERAL DISCUSSION
OF STEEL ARCHES

Whether an arch bridge is preferable to others has long been a question in engineering profession. Not only there are different opinions between arches and other kinds of bridges, desultory controversies have been kept in arches themselves, among which the problem of the number of hinges is the most serious one. And yet the question is not settled even with the number of hinges known, as for each kind of arch there are different types of trussing and webbing, solid or open, spandrel-braced or braced rib. Undoubtedly to go deep into this problem would be beyond the scope of this thesis. The writer will therefore suggest to give the discussion as brief as possible although it might not be so short as to fail to point out the adaptability of the arch which is going to be designed.

---

[1] Modern Framed Structures. Part III. P. 10.

Article 1. Relative Merits Between Arches and other Kinds of Bridges.

Steel arches have a rational application only where Nature has provided natural abutments[1] In other cases unless the erection difficulties be large other kinds of bridges are generally preferable. It is only where there are deep gorges of shallow streams with rocky bottom that arches are economical,[2] although sometimes they are built for aesthetic purposes. Mr. Grimm[3] considers the arch to be the only type suitable for long spans. He points out the objections of other kinds of bridges as follows: Cantilever bridge is not rigid, great deflection occurs at joint of spans, while its advantage of erection is shared equally well by arches. Continuous bridge has uncertainty of stress distribution due to yielding of support and high secondary stresses. Suspension bridge is influenced by temperature and yielding of anchorage, and further it is not economical.

---

① Engineering Record. Vol. 68. P. 321.

② Bridge Engineering. Waddell. P. 618.

③ Transactions of A. S. C. E.. Vol. 71, P. 233.

Article 2. Relative Merits Between Two-hinged and Three-hinged Arches.

"Three hinged arch is statically determinate while two hinged arch is stiff" is an expression so well known that there is no slightest doubt. The advantages of three hinged arches are: simplicity of computation and adjustment; absence of temperature stresses and adaptability to places with limited head room due to the shallow depth at the crown. [1] The advantages of two hinged arches are stiff, rigidity and absence of kink at centre which is objectionable for railway traffic. It must be acknowledged that the disadvantages of each are shared by the other, although not in the same extent. For instance, the yielding of supports affects greatly two hinged arch, but it also affects three hinged arch in a small extent[2] while the sagging of lower chord of three hinged arch due to temperature changes is also noticeable. [3]

In regard to the stiffness of two hinged and three hinged arches, a general conclusion reached is that the deflection is the

---

[1] Design of Steel Bridge. Kunz. P. 347.

[2] Bridge Engineering. Waddell. P. 618.

[3] Higher Structures. Merriman & Jacoby. P. 218.

212

centre of the three, two, and no hinged arches from a load at centre is about as 6 to 2 to 1. [1] The following investigations are made by the graduate students of Cornell University:

(1) Two hinged arch deflects more at quarter point but less at centre than the three hinged arch for loads near the centre. [2]

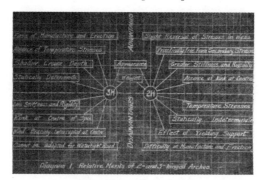

(2) The influence lines for deflections have the same general form for the two arches, the maximum deflection of any panel point due to a single load occurs when the load is over that point. The loading which produces the maximum deflection is not the same as the loading which produces the maximum stress in any part of the members. [3]

(3) The distribution of deflection has been such that a sag

---

① Design of Steel Bridges. Kunz. P. 119.

② Thesis No. 728[#], P. 67. Thesis No. 42[#], P. 59.

③ Thesis No. 42[#]. P. 67.

will be formed in the two hinged arch between the quarter point and the centre while the three hinged arch will give a more uniform grade. [1]

There has been no decisive conclusion arrived as to the relative weights of two and three hinged arches. The analysis of Mr. Hudson showed that two hinged is lighter than three hinged[2] while that of Mr. Davis showed two hinged is heavier than the three hinged arch. [3]

Another advantage in favor of two hinged arches is found in the fact that the removal of centre hinge reduces the reversal of stresses in the web members. [4]

Article 3. Relative Merits Between Two-hinged Spandrel-braced Trusses and Arch Ribs.

In this article the comparison will be made under the following headings: Stiffness, weight and appearance.

(a) Stiffness.   Professors Merriman and Jacoby made the following conclusion: The spandrel braced arch has a much

---

[1]   Thesis No. 42#. P. 68.

[2]   Transactions of A. S. C. E.. Vol. 43. P. 30.

[3]   Thesis No. 59, P. 42.

[4]   Transactions of A. S. C. E.. Vol. 40. P. 135.

smaller range of deflection at every point than the arch rib... so that the arch rib is subject to greater vibration under traffic than the spandrel-braced arch. [1] Mr. Saph has shown that for a full live load the spandrel braced arch has larger deflection than the arch rib but that for partial loads it is smaller. "... We judge that for stiffness we would choose the spandrel braced arch. "[2] Mr. Davis prefers spandrel braced arch by saying that as it seldom happens that the bridge has full load it is more important that there should have a stiffer structure under partial loads, i. e. , the two hinged spandrel braced. [3]

(b) Weight.　　Again there has been much arguments between weights of spandrel braced trusses and arch ribs. Mr. Saph says economy of materials is obtained by using the arch rib[4] while Mr. Davis disagrees with him by showing that the spandrel braced is lighter than the arch rib. [5]

(c) Appearance.　　Almost all authorities on bridges agree

---

[1]　Higher Structure. Merriman & Jacoby. P. 290.

[2]　Thesis No. 47#. P. 39.

[3]　Thesis No. 59#. P. 48.　Also Engineering Record. Vol. 68.　P. 522.

[4]　Thesis No. 47#. P. 37.

[5]　Thesis No. 59#. P. 47.

with the expression that spandrel braced arches are more pleasing to the eye than do the arch ribs, Mr. Grimm, however, advocates braced rib instead of spandrel braced but his conclusion is questioned by most of the prominent engineers. [1]

From the above discussions the following conclusions may be drawn: For railroad bridges the stiffness is of vital importance especially under partial live load and this condition is fulfilled only by the two hinged spandrel braced arches. Where lengths of span or condition of erection precludes the use of simple span, it is the most economic type for railroad traffic. From the standpoint of theoretic design , it is the most logical among arch bridge, it conforms more closely to the demands of bending moment for which truss depth is required. [2]

CHAPTER Ⅲ. COMBINATION TYPE

It is a well known fact that the accuracy of stress distribution in a two hinged arch is dependent upon the conditions of supports. A slight yielding of abutment or pier will produce

---

[1] Transactions of A. S. C. E. . Vol. 71. P. 233.

[2] Transactions of A. S. C. E. . Vol. 71. P. 261.

tremendous effects on the structure and overturns all the stress distribution which was used in the design. Further, considerable difficulties have been met in erection in having the centre chord to be stressed artificially to the exact amount that has allotted to it in the design.[1] As these troubles all arise with the dead load stress only it would seem advisable to provide for the arch a means of adjustment after erection and this is to be found in the addition of a centre hinge at the crown, thus transforming the arch into a three-hinged type. It will be assumed the arch is changed by riveting up the central chord at standard temperatune so that the result will be a three hinged arch for dead load stress and two hinged arch for the other kinds of stresses, live load, impact, wind, temperature etc.

It will be seen that this expedient is very reasonable as the equilibrium polygon for dead load being nearly a parabola will conform to the outline of the bottom chord. No disadvantages of both two hinged and three hinged arches are to be found in this type, but it possesses the merit of both. Moreover, the shop work

---

[1] Engineering Record. Vol. 68. P. 321.

will be made easier as there is no extreme care necessary to secure the exact lengths and sections of members.

Investigations relating the merits of combination type are summarised as follows:

Dr. Waddell says that this type does away about one half of the objectionable ambiguity of stress distribution in the two hinged arch. [1] For spandrel braced arch there is an increase in weight if the temporary hinge be placed in the bottom chord, and a slight saving if it be placed in the top chord. [2]

Mr. Smith prefers the combination type for having the greatest stiffness among the three types of arches, two, three-hinged and combination. [3]

In regard to the stress distribution in combination type, Mr. Mohr found that the maximun and minimum stresses for the combination type lie between those for the pure two, and three hinged types of the same dimensions. [4] The sectional areas for members of combination type agree more closely with the two

---

① Bridge Engineering. Waddell. P. 619.
② Bridge Engineering. Waddell. P. 626.
③ Thesis No. 42#. P. 69.
④ Thesis No. 41#. P. 132.

hinged type. They are greater than the areas for two hinged arches in the upper chords and diagonals and a trifle smaller in some of the verticals. [1]

Another form of combination type is found in the three hinged arches by incorporating the idea of two hinged arches by means of friction joints. This was adopted in the design of the St. Croix River Bridge. As described by Mr. Turner, the scheme is as follows: [2] " in order to combine the advantages of two and three-hinged arches so far as practicable, the connection of the central top chord was made in the form of a friction joint of five plates, two of bronze and three of steel, giving six friction surfaces. As soon as the live load comes on the central chord, these joints lock by friction as firmly as though riveted and the frame then deports itself as a two hinged arch. "

CHAPTER Ⅳ. BIBLIOGRAPHY OF TWO-HINGED SPANDREL-BRACED ARCH BRIDGES

Ⅰ. Corondelet Park, Span 90', built 1887.

---

[1] Thesis No. 41[#]. P. 132.

[2] Transactions of A. S. C. E.. Vol. 75. P. 10.

Ⅱ. St. Guistina crossing Noce Schluct,

Span 196'10″, built 1890.

Engineering News. Vol. 23, p. 103.

Engineering Record. Vol. 21, p. 233.

Engineering Record. Vol. 47, p. 401.

Railroad Gazette. Vol. 22, p. 220.

Ⅲ. Niagara Falls. Span 550', built 1897.

Transactions of A. S. C. E.. Vol. 40, p. 125.

Engineering News. Vol. 36, p. 82.

Engineering News. Vol. 37, p. 252.

Engineering News. Vol. 39, p. 330.

Engineering Record. Vol. 35, p. 447.

Railroad Gazette. Vol. 28, p. 281.

Ⅳ. Mill St. Watertown. Span 165', built 1897.

Engineering Record. Vol. 37, p. 294.

Engineering Record. Vol. 48, p. 277.

Ⅴ. Boun, Germany. Span 307', built 1899.

Engineering News. Vol. 41, p. 242.

Engineering Record. Vol. 47, p. 629.

London Engineering. Vol. 39, p. 153.

VI. Main St. , Waterport. Span, 355', built 1900.

VII. Costa Rica, Span 448'8, built 1902.

Engineering News. Vol. 48, p. 326.

Engineering Record. Vol. 46, p. 390.

VIII. Victoria Falls, Africa. Span 500', built 1905,

Engineering News. Vol. 54, p. 339.

Engineering News. Vol. 53, p. 253.

Engineering Record. Vol. 51, p. 483.

Engineering Record. Vol. 52, p. 346.

IX. Crooked River. Span 340, built 1911.

Engineering News. Vol. 69, p. 550.

Engineering Record. Vol. 68. p. 321.

X. St. John River, Canada. Span 565', built 1914.

Design of Steal Bridges. Kunz, p. 359.

CHAPTER V. RECORDS OF TWO-HINGED SPANDREL-BRACED ARCH BRIDGES

Properties.

Bridge	St. John	Niagara Falls	Victoria Falls	Costa Rica
Traffic	H	H. and 2R	2R	2R
Material	Steel	Steel	Steel	Steel
Curve of L. Chord		Parabola	Parabola	Parabola
Span	565'	550'	500'	$448'8\frac{1}{2}''$
Rise	$61\frac{1}{4}'$	114'	90'	55. 92''
No. of Panels	24	16	20	18
Panel Length	23'4''	$24'4\frac{1}{2}''$	25'	$24'11\frac{3}{8}''$
No. of Ribs	2	2	2	2
Distance Apart	41'	30'	27. 5'	16'
Batt	Vertical	1 – 10	1 – 8	1 – 8
Road	36'	$25\frac{1}{2}'$	–	–
Walk	7'	2 – 11'	–	–
No. of Spans	1	1	1	1

Note. H = Highway. R = S. T. Railway. 2R. = D. T. Railway.

CHAPTER Ⅴ. RECORDS OF TWO-HINGED

SPANDREL-BRACED ARCH BRIDGES(Continued)

Properties.

Bridge	Water-Port	Crooked River	Bonn	St. Guistina	Watertown
Traffic	H	R	R	H	H
Material	Steel	Steel	Steel	W. I.	W. I.
Curve of L. Chord	Circle	Parabola	Circle	Circle	Circle
Span	355'	340'	307'	196'10"	165'
Rise	35.5'	60'	31'	33.6'	11'7"
No. of Panels	20	12	12	14	10
Panel Length	$17\frac{1}{2}'$	28.3'	25.5'	14'	16.5'
No. of Ribs	2	2	4	2	4
Distance Apart	18'	18'	–	15'	9'2"
Batt	1 – 12	1 – 12	Vertical	Vertical	Vertical
Road	–	–	23'5"	$19\frac{1}{2}'$	26'
Walk	–	–	$11\frac{1}{2}'$	–	2 – 6'
No. of Spans	1	1	2	1	1

Proportions.

Bridge	St. John	Niagara Falls	Victoria Falls	Costa Rica
Span	565'	550'	500'	$448'8\frac{1}{2}''$
Rise	$61\frac{1}{2}'$	114'	90'	55.92'
Crown Depth	$8\frac{1}{2}'$	20'	15'	15.57'
Depth at Springing	$60\frac{1}{2}'$	134'	105'	66.5'
Distance Apart	41'	30'	$27\frac{1}{2}'$	16'
$\frac{\text{Rise}}{\text{Span}}\%$	10.8	20.7	18	12.4
$\frac{\text{Crown Depth}}{\text{Span}}\%$	1.5	3.0	3.0	2.8
$\frac{\text{Spacing}}{\text{Span}}\%$	7.25	6.5	5.5	3.57

Proportions. (Continued)

Bridge	Fater-Port	Crooked River	Bonn	St. Guistina	Watertown
Span	355'	340'	307'	196'10"	165'
Rise	35.5'	60'	31'	33.6'	11'7"
Crown Depth	8.5'	12'	4'	6.5'	$3\frac{2}{3}'$
Depth at Springing	44'	72'	33'	40'	16'
Distance Apart	16'	18'	–	16'	9'2"

$\dfrac{\text{Rise}}{\text{Span}}\%$	10	17.6	10	17	7.1
$\dfrac{\text{Crown Depth}}{\text{Span}}\%$	2.4	3.54	9.9	3.3	2.1
$\dfrac{\text{Spacing}}{\text{Span}}\%$	5.1	5.3	–	7.6	5.4

CHAPTER VI. GENERAL DISCUSSION ON THE PROPERTIES OF TWO-HINGED SPANDREL-BRACED ARCHES

Before giving any rule governing the proportions of two hinged arches it is necessary to know the influence of each on the economic and constructional details. The most important among those which deserve considerations are: Rise at bottom chord, crown depth, form of bottom chord, spacing and batter of trusses.

Article 4. Rise at Bottom Chord.

Two hinged spandrel braced arches differ from simple truss in only two points and no more: (1) both ends are hinged, there being no horizontal or vertical movements although rotation is possible. (2) The lower chord is curved or practically polygonal while the top chord is a straight line. On virtue of the first effect there is existed a horizontal thrust acting on the arch from the abutments or piers. The value of the thrust is largely dependent upon the geometrical form of the structure especially the ratio of rise to span. The flatter the arch the greater the thrust for the same position of loading. For economic reasons the rise should be such as to make the thrust approximately equal to the value of the load. Consideration should also be given in respect to appearance so that the rise will satisfy the aesthetic requirements.

Article 5. Depth at Crown.

The economic depth of an arch truss at the crown depends on the span, rise, load and the type of truss. [1] The greater the moment which has to be transmitted from one portion of the arch to the other, the greater should be the depth to give economic section in the chord members. To reduce secondary stresses greater depth has also decisive advantages. As a whole, it may be said that the crown depth is of not much importance on the design and the past practice will furnish enough examples to follow in any special case.

Article 6. Form of Bottom Chord.

The forms of bottom chords that have been used in the past

---

[1] Design of Steel Bridges. Kunz. P. 350.

are of four kinds: (1) straight lines from crown to springing. (2) segmental are of a circle. (3) parabolic and (4) hyperbolic arcs with vertices at crown and axes perpendicular to the span.

From aesthetic points of view the straight chord is out of question although is possesses several advantages which require notice: (1) the straight bottom chords carry the wind stresses directly to the supports without transmitting any appreciable components to the web system of the truss, [1] (2) less shopwork in saving splices thus conducing economy, and (3) ease of erection.

The difference among the other three curvilinear forms has not been great except, perchance on the relative influence on the weight of the truss. For three hinged arches Mr. Schein has shown that the circular arc gives the heaviest truss while the hyperbolic the lightest. [2] For two hinged arches the circular form is undoubtedly heavier but the preference is hardly placed in either parabolic or hyperbolic forms. Probably, hyperbolic is better as it approaches the form of a straight chord but it is not as

---

[1]   Engineering Record. Vol. 68. P. 324.
[2]   Thesis No. 679#. P. 75.

sightly as the parabolic curve.

Article 7. Spacing of Trusses.

The spacing of trusses depends upon two factors: (1) the type of the bridge, and (2) the wind stresses. There is no question about the first consideration as the spandrel braced arch can only be adopted for deck bridges. In respect to the second requirement, the trusses should be spaced as far as possible so that the overturning moment of wind might be more easily resisted. The theoretical spacing of arch trusses would not be uniform through the span, being smaller near the crown and wider at the springing on account of the fact that the wind load is not distributed uniformly. The cost of manufacture, however, prevents this to be carried into practice. The spacing should not be made excessive as it will give a heavy floor which is uneconomical both for the floor itself and the trusses as it increases the dead load.

Article 8. Batter of Trusses.

To keep the floor system from excessively heavy weight and to give sufficient stability of trusses, against wind overturning, arch trusses are generally battered; especially is this obligatory in very long spans. In summary the advantages of giving trusses a batter are as follows: (1) it adds rigidity to the structure, (2) it reduces the total weight,[1](3) it lessens the disagreeable lateral swaying so noticeable in bridges in which the arches are not inclined,[2](4) it improves the appearance of the structure if the batter is small. Practically, however, it must not be forgotten that those are not the only determining factors because the cost of manufacture might offset to some extent the advantages of battering the trusses.

CHAPTER VII. PRINCIPAL DIMENSIONS

Since the present work is not an actual design for any special locality, the principal dimensions of the arch truss will be assumed so as to satisfy the aesthetic and economic requirements of two-hinged spandrel-braced arches. The principal dimensions to be considered are: span, rise of bottom chord, depth at crown, curve of lower chord and spacing between trusses. These will be discussed separately.

---

① Bridge Engineering. Waddell. P. 656.
② Transactions of A. S. C. E.. Vol. 40, P. 153.

Article 9. Span.

Since an arch is not as stiff as a simple truss, it is evidently not suitable for short spans unless the conditions offer difficulty in erection. Spandrel-braced arches, however, are not adopted for very long spans because the web members near the ends will be too long thus causing a waste of material. [1] The investigations made by Mr. Mallison in his thesis [2] shows that only 21% of arches built prior to 1907 have a span greater than 300 feet while there are 59% with spans under 200 feet. However, as arches are only properly and economically used for deep gorges and for locations which offer natural rock supports the span length will probably not be very short, and hence 300 feet may represent a good span. Further, as a two-hinged arch with a long span is better adapted for railway traffic than a three hinged arch, [3] a shorter span might not be so economical. Three hundred feet will therefore be adopted as the span of the arch and it will be assumed to have twelve panels each twenty-five feet long. The

[1] Design of Steel Bridges. Kunz. P. 348.
[2] Thesis No. 728[#]. P. 49.
[3] Higher Structures. Merriman & Jacoby. P. 226. Thesis No. 728[#]. P. 64.

cantilever arms will be made fifty feet long, and the two approach spans, one hundred feet long.

Article 10. Rise of Bottom Chord.

According to Mr. Mallison's investigations for two hinged arches with open webbing built before 1907, the ratio of rise to span varies from 27.8% to 7.1%, the average being 17.6%.[1] Mr. Kunz gives the ratio ranging from $\frac{1}{4}$ to $\frac{1}{12}$ and $\frac{1}{7} = 14.3\%$ as an average economic value. For four two hinged spandrel braced arches built after 1907, the ratio varies from 10.8% to 20%. Since the flatter the arch the greater the effect of horizontal thrust[2] and the greater the rise the greater the effect of full load stresses[3] the value of 18% will be used. The rise of the bottom chord is therefore 0.18 times 300 or 54 feet.

Article 11. Crown Depth.

The average value for the ratio of crown depth to span in two hinged arches with open webbing ranges from 2.3% to 6.7%,

---

1   Thesis No. 728[#]. P. 50.
2   Bridge Engineering. Waddell. P. 633.
3   Thesis No. 728[#]. P. 50.

the average being 3.7% for all bridges built up to 1907. [1] In the four recent arches the value ranges from 1.5% to 3.6%. Mr. Kunz gives $\frac{1}{25}$ to $\frac{1}{20}$, [2] i.e., 4% to 5%. Since the greater the depth the stiffer the structure, [3] 4% will be adopted. Hence the depth at crown is 0.04 times 300 or 12 feet.

Article 12. Curve of Bottom Chord.

Since the bottom chord carries the main portion of the load and the dead load stresses are to follow the law of three hinged arches, the curve is best made parabolic for economic and aesthetic reasons.

Article 13. Spacing Between Trusses.

An examination of all the two hinged open-webbed arches up to the present time shows the ratio between spacing and span to range from 3.57% to 10% for spans under 100 feet and from 3.57% to 7.6% for larger spans. Mr. Kunz gives the ratio to be not less than $\frac{1}{15}$ and at the same time not less than $\frac{1}{3}$ preferably

---

[1] Thesis No. 728#. P. 50.
[2] Design of Steel Bridges. Kunz. P. 350.
[3] Thesis No. 728#. P. 50.

the depth at springing. [1] Dr. Waddell gives $\dfrac{1}{3}$ the depth at the springing as the approximate rule. [2] Twenty-five feet will be adopted in this case.

The sketch diagram is given on the page. It will be seen that it is very pleasing in appearance, the diagonal at the quarter point being inclined at nearly 45° from the vertical. [3]

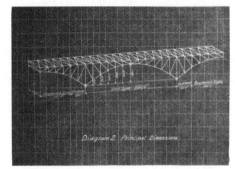

CHAPTER VIII. NOTATIONS

It has been the aim of the writer to cultivate the habit of using same notations throughout the whole work. Unless otherwise stated the following list of important notations will be adhered to in the following pages.

---

[1] Design of Steel Bridges. Kunz. P. 350.
[2] Bridge Engineering. Waddell. P. 633.
[3] Design of Steel Bridges. Kunz. P. 350.

Article 14. Members of Arch.

For the sake of convenience there are three systems of notations used for designating the members of the arch. For most part of the work the letters U, L, D and V with proper subscripts will be used for the upper chords, lower chords, diagonals and verticals. This is shown in the following diagram. Panel points are numbered by Roman capitals except for the lower chord points in which case a subscript of l is attached. For constructing displacement diagrams capital letters will be used for the upper chord panel points and small letters for the lower ones. In calculating secondary stresses, arabic numbers will be used to designate the joints.

In both of the last two cases each member is known by writing in combination the the symbols at its ends, for instance, Hh, 18 – 17.

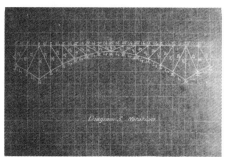

Article 15. Notations Used in the Analysis of Stresses.

V = Vertical component of reaction.

H = Horizontal component of reaction.

L = Length of members. This is used in preference to "l" so that is might not be confused with the number 1.

A = Cross sectional area of members.

I = Moment of inertia of sections of members.

E = Modulus of elasticity of steel taken as 29000000 pounds/in.^{2}.

l = Length of span.

K = Ratio of the distance of a concentrated load from the left hinge to the span of the bridge.

Stress Coefficient = Stress in any member for a unit concentrated load.

Moment Point = centre of moment of any member of the truss.

S' = Stress in members for $H = 0$.

T = Stress in members for $H = 1$.

S = Stress in members for $H = E$.

x, y = coordinates of moment point from the left hinge.

t = lever arms of members.

PART II THEORY OF TWO-HINGED SPARDREL-BRACED ARCHES

CHAPTER IX. INTRODUCTION

Statically indeterminate structures may be classified under two heads: one with redundancy of external forces and the other with redundancy of internal forces. Under the former group are the two hinged and no hinged arches and continuous girders while the latter group embraces those structures where the number of members constituting the truss is greater than the quantity $2j - 3$, where j is the number of joints in the truss.

To analyze stresses in a structure it is necessary to know beforehand the moments and shears at different sections along the span, but before those moments and shears can be found the reactions of supports must be known. It is viol, from the statical indeterminateness of the reactions that the two hinged arches are classified under the head of higher structures. For internal stresses there is little or no difference between the arch and an ordinary simple span.

It is a well-known fact that the moment about pine must be zero, in consequence there of the points of application of the reactions are known because the latter must pass through the hinges. To find the values of the reactions a usual practice is to decompose them into vertical and horizontal components. The vertical reactions are the same as if the arch were a simple truss and are therefore determinable. There remains one statical condition to determine the horizontal components or horizontal thrusts as usually called. By virtue of this condition it is found that the horizontal thrusts must be equal and opposite and in the same line of action for arches with level supports. The value of each of the thrusts is however still unknown and is not determinate by ordinary statics. Therefore it becomes evident that the theory of two-hinged arches is no more or less than that of the horizontal thrust. This will be discussed in full in the next chapter.

CHAPTER X. HORIZONTAL THRUST

Since the redundancy of two hinged arches comes from the fact that the span is immovable due to fixed hinges and the material of which the arch is built is elastic, the determination of the horizontal thrust, H, must be based on these two conditions. The first application of this elastic theory to two hinged spandrel braced arches was made by Maxwell in 1864, from those on a number of methods have been evolved which are more or less similar in character but arrived the some results by entirely different considerations. As a whole two divisions may be made among these methods, the one deals with the displacement of end hinges when one of them is supposed to be removed while the other deals with the principle of least work. Ten methods are given in the following articles which are arranged according to logical sequence rather than chronologic order.

Article 16. Method of Maxwell. [1]

To the eminent processer, Clerk Maxwell, is due the method of calculating horizontal thrusts based on the exact theory

---

[1] Encyclopedia Britannies. Sixth Addition, Vol. 4. P. 319.

of elasticity. His method is the first of its kind to appear in publications and most of the other analytical methods are more or less dependent upon it for their theoretical derivations. Maxwell's formula for horizontal thrust may be obtained from any other formulas derived by different writers by simple modifications and transformations. This formula is therefore of the standard form. It is condensed in the following pages, with its derivations.

Consider member M, and let S be its internal stress. The change of length due to the stress S is

$$\Delta L = \frac{SL}{AE} \quad (1)$$

Proper signs should be given to the arithmetical values of the force S and alteration of length.

If M be the only member under alteration in length and all the other members were absolutely rigid, the span, l, of the truss will undergo a corresponding alteration of Δl such that the ratio $\frac{\Delta l}{\Delta L}$ will depend merely on the geometrical form of the truss. Let this ratio be denoted by T, so that $\Delta l = T\Delta L = T\frac{SL}{AE}$.

Let & be the force produced on the member M by a horizontal force H seting between the springings; then, by principle of virtual velocities.

$$Q:H = \Delta l : \Delta L.$$

Then, $\quad Q = T \cdot H$ or $T = \dfrac{Q}{H}$.

Similarly let S' be the stress which would be produced on M by a vertical force V applied at one springing, while the other springing was held rigidly so that the whole truss could not turn, the ratio between H and V is also constant, depending merely on the form or the frame, so that $\quad H = S'V$.

Where S' is a constant to be found in the same manner as T.

Maxwell pointed out that S' may be defined as equal to the whole stress reduced on the given member by a unit vertical force at the springing, and T as the whole stress produced on the given member by a unit horizontal force at the springing. The truss being held rigidly at the opposite abutment.

The reaction at stringing may be resolved into V and H, and stress S in M will be the sum of stresses of and M or

$$S = HT + VS'.$$

Substituting the value of F in equation(1), we have

$$l = (T^2H + S'TV)\left(\frac{L}{AE}\right).$$

Now let the values of T, S' and $\dfrac{L}{AE}$ be calculated for every member of the frame, then, calling the elongation $\Sigma \Delta l \cdot k$, we have

$$k = \sum \left(T^2 H \frac{L}{AE} + S'TV \frac{L}{AE} \right) .$$

If the abutments do not yield or $k = 0$, we have

$$H = \frac{\sum \dfrac{S'TL}{AE}}{\sum \dfrac{T^2 L}{AE}} V.$$

Article 17. Method of Greene. [1]

The method suggested by Greene is a practical application of the formula for H derived by Clerk Maxwell. It has been shown that the change in span due to change in length of any particular member $= T(TH + S'V) \dfrac{L}{AE}$. Replacing T by $\dfrac{y}{t}$ and S' by $\dfrac{x}{t}$.

$$\Delta l = \frac{y}{t} \frac{Hy + Vx}{t} \frac{L}{AE}.$$

Where Δl = change in span due to change in length of one member.

$$\therefore \quad \Delta l = \frac{Hy^2 + Vxy}{t^2} \frac{L}{AE}.$$

---

[1] Trusses and Arches. C. H. Greene. Part IV.

This same quantity can be calculated for the extensibility due to each member of the frame. The sum of all the changes of span will be

$$H\sum \frac{y^2}{t^2}\frac{L}{AE} + \sum V\frac{xy}{t^2}\frac{L}{AE}.$$

For two hinged arch, assuming the foundation does not yield, this expression should be zero, or

$$H\sum \frac{y^2}{t^2}\frac{L}{AE} + \sum V\frac{xy}{t^2}\frac{L}{AE} = 0,$$

$$H = \frac{\sum V\frac{xy}{t^2}\frac{L}{AE}}{\sum \frac{y^2}{t^2}\frac{L}{AE}}.$$

For practical application: (1) Construct tables of the values $\frac{x}{t}$ and $\frac{y}{t}$ for each member of the frame. Make tables of $\frac{xy}{t^2}$, $\frac{L}{AE}$ and $\frac{y^2}{t^2}\frac{L}{AE}$. Make the summations as indicated. In summing $V\frac{xy}{t^2}$ $\cdot \frac{L}{AE}$ the value of V_1 for left hinge must be used for all members to the left of the loaded joint, and V_2 for right reaction for all the members to the right of the load.

Method of summation. The quantity $\sum \frac{y^2}{t^2}L$ is found in the ordinary way as this is the same for all loading on the frame. In calculating $\dfrac{xy}{t^2}L$ these are divided into four groups for upper chords, lower chords, diagonal and vertical respectively. In each group the value of $\dfrac{xy}{t^2}L$ is arranged in the same order of occurence as the member of which it belongs, starting from left support to the right. Then the value of $\dfrac{xy}{t^2}L$ for the second one is added to the first, that of the third one is added to the sum of the first and second and so on until the last member when all the values will be summed up. Arrange the partial summation in a form of table.

In finding H, proceed as follows: For any position of load note the members on the left and right of that position. From the preceding table for each group of members find that value of $\sum \dfrac{xy}{t^2}L$ opposite the member which is on the left side and is adjacent to the load and add up all these summations for chord and web members. Multiply this by V_1 which will give the summation of $\sum \dfrac{xy}{t^2}L$ for all the members from left support to the load. For the portion of the frame on the right of the load, find from the preceding table that value of $\sum \dfrac{xy}{t^2}L$ which is symmetrically opposite the first one of its group on the right side of the load. Add all these up and multiply by V_2, and the result will be the summation of $\frac{xy}{t^2}L$ for the portion of frame on the right or load. (Adding these two summations and dividing by $\sum \frac{y^2}{t^2}L$, H is found).

Note: For members below the load, as it is acted on by V_1 on one side and V_2 on the other, it is immaterial whether it is considered to lie to the left or the right of the loaded points.

Article 18. Method of Merriman and Jacoby. [1]

The method that gives accurate solutions of horizontal thrust and at the same time involves least amount of work is perhaps the one due to Professors Merriman and Jacoby. It is so easy in application, both in preliminary and final designs that it is by far the most widely adopted method that has ever been given in the history of two hinged spandrel braced arches. A comparison of this method with the method of Greene and Du Bois made by Mr. Muller in his Thesis No. 40[2] shows that "the amount of labor

---

[1]   Higher Structures. Merriman and Jacoby. P. 266.

[2]   Thesis 40#. P. 106.

required to get the stress is much less by the method of M. and J. and not only is that shorter for preliminary stresses, but very much shorter for successive approximations as the only change to be made in the influence diagram is in the influence line for H ... The horizontal displacement of hinges ... renders the determination of temperature stresses an easy matter. "

Before giving the method that properly belongs to Professors Merriman and Jacoby, their analytic derivation of Maxwell's formula will be given first as the proof thereof is the most comprehensible of all and there are not many mathematical operations.

The name reasoning will be made as that of Greene. i. e. , the hinge b is supposed to move horizontally an amount equal to Δ under the action of P and then brought back under the action of H.

Apply a force unity at b, the internal work in any member is equal $\frac{1}{2}$ to $S'(e + e' + e'')$ where e is its change in length due to force unity. e' that due to the force of H after the unit force is deducted, and e'' that due to P. Consider the work due to e only, we have $\frac{1}{2}S'e = \frac{1}{2}S'\frac{TL}{AE}$ because $\frac{T'L}{AE} = e'$. Since the external work is $\frac{1}{2} \cdot \Delta \cdot \perp$, $\Delta = \sum \frac{S'TL}{AE}$.

Let Q be the stress in any member produced by H, $Q = HT$, and by the same reasoning.

$$\Delta = \frac{1}{H} \sum \frac{Q^2 L}{AE} = H \sum \frac{T^2 L}{AE}.$$

Equating the values of Δ, $\quad H = \dfrac{\sum \dfrac{S'TL}{AE}}{\sum \dfrac{T^2 L}{AE}}$,

which is the same as the formula developed by Maxwell.

The method of Merriman and Jacoby is a graphical solution of the formula as developed above but is transformed into the form of $H = \dfrac{\delta}{\delta' P}$ (δ is the vertical deflection of load point and δ' the horizontal deflection of the hinge due to a unit force of H), by the reasoning that in the above formula it is immaterial whether S' be the stress due to the vertical load P and T that due to a horizontal force unity at the hinge, or whether S' be the stress due to a horizontal force P at the hinge and T that due to vertical force of unity applied at the load panel point.

The method makes use of Willot-Molir diagram in finding δ and δ' after the changes in length of all the members of the truss have been obtained. The construction of the displacement will be omitted here; only the following remarks are inserted.

The hinge a is assumed to fixed in place and hinge b is assumed to move freely on a horizontal plane. By constructing onehalf of the displacement diagram of the truss which is symmetrical about the center line, the amount of displacement of b is found and this is equal to δ'. It will be seen that no rotation of the truss about "a" is necessary so that the values of δ are just the vertical components of the line joining the initial and final panel points in the diagram.

It is to be noted that only one set of stresses is needed to determine the values of H because the same diagram will give the deflections at all the panel points of the truss, and hence the different values of δ corresponding to different positions of l ads. This is the principle advantage of the method and it is that which has made the method so popular.

The merits of the method may be summarized as follows:

(1) Celerity and accuracy;

(2) Gross errors easily detected;

(3) Least amount of numerical calculation.

It will be seen that the accuracy of results by use of this method is not very great as compared with that of the analytical method owing to the fact that a slight increase in scale of displacement diagram will require a very big piece of paper which is not convenient to handle. A method evolved by the writer to increase the accuracy of those diagrams will be found in Chapter XVI of part Ⅲ.

Article 19. Method of Cooper. [1]

Cooper's method is based on the principle set forth by Maxwell but has been modified to such an extent that it can be quickly applied in practice without serious drawbacks on account of the labor involved in its numerical calculations. The method of Cooper is as follows:

Stress in member $M = S$. If AB is shortened an amount ΔL, the abutment moves through l due to rotation around E. Then consider E as a fulcrum.

$$\Delta l : \Delta L = y : t, \quad \Delta l = \frac{y}{t}\Delta L,$$

$$\text{But } \Delta L = \frac{SL}{AE}, \quad \therefore \Delta l = \frac{y}{t}\frac{L}{AE}S.$$

---

① Engineering Record. Vol. 46, P. 424.

Now S is the stress in member due to vertical force V and horizontal thrust h.

$$\therefore S = VS' + hT,$$

$$\text{or } \Delta L = \frac{y}{t}\frac{L}{AE}(VS' + hT),$$

$$\text{But } \frac{y}{t} = T,$$

$$\therefore \Delta l = \frac{VS'T\dfrac{L}{AE}}{hT^2\dfrac{L}{AE}}.$$

Since abutment is fixed. $\Delta l = 0$,

$$h = \frac{VS'TL}{AE} \div \frac{T^2L}{AE} = \frac{VS'}{T}.$$

The horizontal thrust due to stresses in all members $= \sum h$,

$$\therefore \ H = \sum h,$$

Let a load P be placed at kl from left and support. The stress in any member on the left half of arch $(l - k)PS'$, and that in a member symmetrically located on the right half of the arch is kPS'. The total stresses in each pair of such symmetrically located members is therefore $(l - k)PS' + kPS' =$

$$PS' = \frac{Py}{t}.$$

Where x = abscissa of moment point. Also

$$T = \frac{y_0 - y}{t}.$$

Where y is the length of vertical at the moment point and y, that at springing.

$$h = \frac{P\dfrac{x}{t}}{2\dfrac{y_0 - y}{t}} = \frac{px}{2(y_0 - y)},$$

Where " 2 " is introduced because the stresses of two members are summed.

Article 20. Method of Turnesure. [1]

This method together with that of Merriman and Jacoby are the two methods followed by the writer in the following calculations. It is a modified form of Cooper's method but allows a more clear understanding and easy application. This method as described by Turnesure in " Modern Framed Structures " is as follows:

Assume the truss loaded with two unit loads P, placed symmetrically. The value of H determined for these two loads will be twice that due to a single load. The stress due to the symmetrical loads are readily calculated. The vertical reactions

---

[1] Modern Framed Structures. Part II. P. 150.

250

are each equal to P. Then for members to the left of P_1, the stress to the same for all positions of the loads, and is equal to $P_1 \dfrac{x}{y}$. For all members between the loads the moment of the external forces is constant and equal to Pkl. A single graphical diagram will give all the stresses $P \dfrac{x}{y}$. The stresses $\dfrac{Pkl}{t}$ are obtained by calculating the values of $\dfrac{l}{t}$ and then multiplying by the various values of k for the different positions of the loads. These will give S' in the formula for H. As P is the same for all the positions of the load, the horizontal thrust may now be readily obtained.

Article 21. Method of Mohr. [1]

This method is derived from the principle of virtual velocities and gives results same as that of Maxwell. The formula has been developed in a more handy form by Winkle and thus allows of easy and rapid application.

Consider the structure as without loads and without weight and free to move at the right hand support, left hand support being fixed in position. As before, let a horizontal outward unit

[1] Engineering News. Vol. 45. P. 114.

force be placed at the right support, there will be and equal outward reaction at the left support and certain stress T in the members.

Now consider the structure as having no horizontal force but bearing a vertical load P at kl from left support. The vertical reactions for the left and right support will be respectively

$$P(l - k) \text{ and } Pk. \qquad (1)$$

This force will produce certain stress in the members which are equal to S'.

The length of span l must be increased somewhat by the action of load, the change being l. Now place an inward horizontal force H at the right support of each quantity as to diminish the span to its length l before P was applied. The stress in any member would then be

$$S = S' - HT, \qquad (2)$$

$$\therefore \frac{\mathrm{d}s}{\mathrm{d}T} = -T. \qquad (3)$$

The internal work upon a homogeneous member of sectional area A for a gradually applied load $S = \dfrac{1}{2}\dfrac{S^2 L}{AE}$, and for the whole system,

$$W = \sum \frac{S^2 L}{2AE}. \qquad (4)$$

Also, if a represent the work due to the change of form of structure in equilibrium under the action of forces applied at its joints and expressed as a function of these forces, and if the point of application of one of the forces P moves by the action of the system of forces a distance δ relative to the line of action of P. then, by Castigliano's Theorem.

$$\delta = \frac{\mathrm{d}W}{\mathrm{d}P}. \qquad (5)$$

If δ is displacement at right support due to H, it because Δl, hence $-\Delta l = \frac{\mathrm{d}W}{\mathrm{d}H}$, $\qquad (6)$

differentiate (4) with respect to H,

$$\frac{\mathrm{d}W}{\mathrm{d}H} = \sum \frac{SL}{AE} \frac{\mathrm{d}S}{\mathrm{d}H}$$

Replacing $\frac{\mathrm{d}W}{\mathrm{d}H}$ by $-\Delta l$ from (6) and $\frac{\mathrm{d}S}{\mathrm{d}H}$ by $-T$ from (3).

$$-\Delta l = \sum \frac{SL}{AE}(-T),$$

or $\Delta l = \sum \frac{TL}{AE} S.$

Substitutes by $S' - HT$ from (2).

$$\Delta l = \sum \frac{S'TL}{AE} - \sum \frac{T^2 L}{AE} H,$$

which solving for H, gives,

$$H = \frac{\sum \dfrac{S'TL}{AE} - \Delta L}{\sum \dfrac{T^2L}{AE}}.$$

If the supports are fixed. $\Delta l = 0$.

$$H = \frac{\sum \dfrac{S'TL}{AE}}{\sum \dfrac{T^2L}{AE}}. \qquad (7)$$

This formula can be simplified as follows:

Take the left hand hinge as the origin of rectangular axes, and let the axis of x pass through right hinge. Let x and y be the coordinates of the moment point of any member, t the lever arm of member, M, the moment of the forces about any moment point, and M' the moment about the same point but neglecting H or consider the frame to be a simple span.

Then $S = \dfrac{M}{t}, M = M' - Hy,$

or $S = \left[\dfrac{M'}{t} - H \dfrac{y}{t} \right].$

Compare with (2) $S' = \dfrac{M'}{t}$, and $H = \dfrac{y}{t}$,

Hence $H = \dfrac{\sum M' \dfrac{y}{t^2} \dfrac{L}{AE}}{\sum \dfrac{y^2}{t^2} \dfrac{L}{AE}}. \qquad (8)$

If we substitute for M' in (8) the values of (1) for the vertical reaction with a load at kl from left support.

$$H = \frac{P\left\{(l - k)\sum \frac{y}{t^2} \frac{L}{AE}x + k\sum \frac{y}{t^2} \frac{L}{AE}x'\right\}}{\sum \frac{y}{t^2} \frac{L}{AE}}, \qquad (9)$$

where the summation of the first term in brackets of the numerator is for members to the left of the load P, and where the summation of the second term is for piece to the right of the load, x' being the distance of each moment point from the right support.

Consider the quantity $\frac{y}{t^2} \frac{L}{AE}$ as vertical force each acting at its respective moment point on a simple beam supported at distance l apart. For chord members these forces will be applied at the same side of the load as the member because the moment points are always lying on the said side of the load as the member. The moment of these forces about any loaded joint, distant kl from left, for all of the members lying on the left of the loaded joint is $kl\left(\frac{y}{t^2} \frac{L}{AE} \frac{x'}{l}\right)$.

The moment of the force $\frac{y}{t^2} \frac{L}{AE}$ about the same loaded joint of all the members lying on the right of the loaded joint is

$$\left(\frac{y}{t^2}\frac{L}{AE}\frac{x'}{l}\right)(l-kl).$$

Total moment of these forces about of chord members about the loaded point is therefore

$$m = \sum\left[\left\{\left(\frac{y}{t^2}\frac{L}{AE}\right)x\right\}(l-k)\right] + \sum\left[\left\{\frac{y}{t^2}\frac{L}{AE}\right\}xk\right]. \quad (10)$$

For web members, the moment point of members generally lies on the opposite side of the loaded joint as the member excepting only few members lying between loaded joint and the centre. In such cases, the moment will be divided into two parts first for those on the same side and the second for those on the other.

For moment centres lying on the same side of the loaded joint as the member, the moment of those arbitrary forces $\frac{y}{t^2}\frac{L}{AE}$ will be the same as before.

$$M = \sum\left[\left\{\left(\frac{y}{t^2}\frac{L}{AE}\right)x\right\}(l-k)\right] + \sum\left[\left\{\frac{y}{t^2}\frac{L}{AE}\right\}xk\right].$$

For members having moment points at opposite sides of loaded joint as the members themselves the moment may be obtained by considering the members lying symmetrically on the opposite side of the center of truss as the member in consideration

256

and load the force $\dfrac{y}{t^2}\dfrac{L}{AE}$ on the moment point of that symmetrical member which being at $(l-x)$ or x' from the left support.

Another method of finding moments for web members is as follows, which being more direct and applying to all web members of the frame thus avoids confusion. Let the force $\dfrac{y}{t^2}\dfrac{L}{AE}$ at moment point be replaced by two equivalent vertical forces on the two upper panel points of the panel whose lower chord has been cut by the moment section for the member under consideration. Let p be the panel length, and d the distance of the nearer lower panel point of the panel from the moment centre, then the equivalent force on the farther side from the moment point is $+\dfrac{d}{p}\dfrac{y}{t^2}\dfrac{L}{AE}$ and that on the nearer side from the moment point is $-\dfrac{d+p}{p}\dfrac{y}{t^2}\dfrac{L}{AE}$.

These equivalent forces may now be treated in the same way as those for chords as they are always on the same side of the loaded joint as the member.

Summarizing, for all members of the truss.

$$m = (l-k)\sum\left(\dfrac{y}{t^2}\dfrac{L}{AE}x\right) + k\sum\left(\dfrac{y}{t^2}\dfrac{L}{AE}x'\right).$$

As this is equivalent to the quantity in the bracket of the numerator of equation (9), we can therefore express the latter as

$$H = P \frac{m}{\sum \frac{y}{t^2} \frac{L}{AE}}.$$

For a practical application of the method, see *Engineering News*. Vol, 42. P. 114 and *Kunz's Design of Steel Bridges*, P. 336.

Article 22. Method of Muller-Breslau.

This method is based on the construction of deflection polygons which are polygons similar in form as equilibrium polygons but with intercepts proportional to the deflection of certain panel points in which a system of load has been placed. First of all let the process of finding H from a deflection polygon be briefly developed and then the method of construction of the polygon which is the principal achievement accompliched by the author.

From the equation $H = \dfrac{\delta}{\delta'}$ for a conventional loading of unity, it is obvious that H is proportional to δ and may be represented by it if a diagram for δ had been constructed. Note that δ' is constant. The diagram for δ has been obtained in a number of ways. Muller-Breslau's method being to construct a deflection polygon which is drawn immediately under the truss in the form of an equilibrium polygon with intercepts representing values of δ for the corresponding as panel points. The same deflection polygon is therefore also the influence line for H, as the polygon shows δ at different panel points, on which the unit load is placed.

The problem of finding H is hence reduced to the construction of deflection polygon. The following is the method of Muller-Breslau as developed by Molitor. [1]

When a homogeneous beam is loaded with any system of forces, the bending produced may be represented by an equilibrium polygon, the moment area being considered as loads. This shows the possibility of obtaining deflection polygons from equilibrium polygons, the only question involved being how to find the arbitrary load. For a homegeneous beam we know it is the moment area but for a graced structure we are confronting a different problem. It is the finding of this arbitrary of fictitious load that constitutes the main part of the operations involved in the method and where the merit is to be found.

---

[1] Kinetic Theory of Engineering Structures. Molitor. P. 53.

Since the deflection of truss is produced by individual changes in lengths of members of the structure, this arbitrary load must in come way by conducted with the stress in the members and the elasticity of the material. From of truss and condition of loading are other controlling factors. Obviously every member of the structure contributes a part of that load and the final result would be a combination of these individual effects. For algebraic convenience the part of load due to stress in a member will be called elastic load of that member and the elastic loads for chord and web members will be treated separately in their logical sequence.

(a) Elastic Loads for Chord Members.

Let any chord member be shortened or lengthed an amount of ΔL due to stress S in the member caused by any system of loading. Let ζ_m be the deflection of the moment point m directly opposite that member due to ΔL. Then putting a unit load acting downward from m the external and internal work must be equal and $l \cdot \zeta_m = S \cdot \Delta L$.

Let M_{ma} represent the static moment about the point m when only the unit load at m is acting. This is easily found from either one of reactions, resulting from the unit load, multiplied by the distance from m to that reaction. Then the stress S_a due to that unit load $= \dfrac{M_{ma}}{t}$. Now if $l \cdot \zeta_m$ be regarded as a moment M_{mw} due to a fictitious load w hung at the point m, then, since moments are proportional to loads,

$$\frac{W}{l} = \frac{M_{mw}}{M_{ma}}, \quad \text{or } W = \frac{l \cdot \zeta_m}{M_{ma}} = \frac{S \cdot \Delta L}{S - t} \quad \text{or } W = \frac{\Delta L}{t}.$$

This fictitious w will be known as the elastic load.

Hence a moment diagram drawn for the elastic load w at m, will give the ordinate $M_{mw} = \zeta_m$ under the point m.

Similarly a moment diagram drawn for an elastic load wn acting at n due to change in length in the chord member opposite n will give ζ_n under the point n.

By Marwell's law these moment diagrams are also the influence lines of deflections for the point when the elastic load moves over the span. Hence, by drawing these diagrams successively for all the panel points of the top and bottom chords, the several partial effects of all the elastic loads w on any particular point m may be found and the summation of these effect $\Sigma \zeta$ will be the total deflection δ_m for the point m resulting from the given case of actual loading. The summation, however, is more readily performed in one operation by laying off the several loads w into a load line ΣW and with a pole distance h, drawing the force and equilibrium polygons. The ordinate δ_m of the resulting polygon, known as deflection polygon, will represent the actual deflection of the corresponding panel point m for that particular case of loading and is equal to the length to the scale used multiplex by h.

WR may now give a more clear understanding of the nature of the elastic load. As can be seen from the above it is a ratio, depending only on the unit stress in the member and the geometric shape of the trues. From this and the real value $W = \dfrac{\Delta L}{t}$ it follows that this elastic load may also be defined as the tangent of the singular change in the angle included by two web members and subtended by the chord whose change in length is ΔL.

It should be noted that while ΔL may be positive or negtive, the elastic load (for chord members) is always positive because the deflection is in any case downward due to the fact that moment M changes sign with L.

For top chords point.

$$\Delta L = \frac{SL}{AE} = \frac{M}{t}\frac{L}{AE} \quad \text{or} \quad W = \frac{\Delta L}{t} = \frac{M}{t^2}\frac{L}{AE}.$$

For bottom chord point

262

$$-\Delta L = \frac{SL}{AE} = \left(-\frac{M}{t}\right)\frac{L}{AE} \quad \text{or} \quad W = \frac{M}{t^2}\frac{L}{AE}.$$

(b) Elastic Loads for Web Members.

Firstly let the moment point of the web member be on the opposite side of load as the member. Let mn be the member is question and the upper chord carries the load. The deflection ζ_0 at the moment point may be found similarly as for chords as follows:

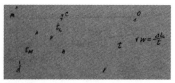

Put a unit load at O, the work produced

$$= l \cdot \zeta_0 = S_{mn} \cdot \Delta L$$

Let M_{mo} be the static moment about O due to that unit load,

$$S_{mn} = \frac{M_{mo}}{t}.$$

Hence $\zeta_0 = \dfrac{M_{mo}}{t} \cdot \Delta L.$

Now consider $l \cdot \zeta_0$ as a moment Mmw produced by an elastic load w, then $\dfrac{W}{l} = \dfrac{M_{mw}}{M_{mo}} = \dfrac{l \cdot \zeta_0}{M_{mo}} = \dfrac{M_m}{M_{mo}} \cdot \dfrac{\Delta L}{t} = \dfrac{\Delta L}{t}.$

This is the elastic load for the web member if the deflection at the moment point of the member be desired. Since it is the deflections at the panel points at the ends of the web member that we are interested, we have to find the equivalent elastic loads at those panel points. We may proceed as follows:

Draw oc from c perpendicular to mn. Taking moments about m,

$$W \cdot \overline{mo} = W_c \cdot \overline{mc}, \quad W_c = \frac{\overline{mo}}{\overline{mc}}W$$

$$= \frac{t}{t_c}W = \frac{t}{t_c}\frac{\Delta L}{t} = \frac{\Delta L}{t_c}.$$

Similarly, taking moments about c,

$$W \cdot \overline{oc} = W_m \cdot \overline{mc}, \quad W_m = \frac{\overline{oc}}{\overline{mc}}W$$

$$= \frac{t}{t_m}W = \frac{t}{t_m}\frac{\Delta L}{t} = \frac{\Delta L}{t_m}.$$

Hence the elastic loads required are

$$W_c = \frac{\Delta L}{t_c} \text{ and } W_m = \frac{\Delta L}{t_m}.$$

Regarding the signs of these two loads, it is seen from the above that the signs must be positive and it is only necessary to find out which is positive. Professor Mohr gives the following simple rule: Calling all top chord members negative and all bottom chord members positive and giving the proper signs to ΔL for the web member in question, then the positive w is found on that side of the section or panel where the sign of ΔL coincides with that of the adjacent chord.

Mr. Molitor gives the following rule: Considering the

quadrilateral mend, assume the bottom chord $\overline{nd}$ to be immovable, then $-\Delta Ll_n \overline{mn}$ will cause the point m to drop, hence $W_m l_M^S$ positive when the top chord is the loaded chord.

(c) Klastic Loads for the Truss.

The elastic loads for all the chord and web members are now found by following the above descriptions and all those acting at the same panel point are algebraically added for total effects. It will be found that for each intermediate panel point there are six such loads, two from the two adjacent chords (loads on lower chord point transferred to the top), two from diagonal and two from verticals.

(d) Horizontal Displacement of Hinges.

It is seen that the elastic loads w are functions of t and ΔL depending on the geometric shape of the truss and the change in length of members. They are independent of the direction of the deflections so that they may be empolyed to find the deflections in any direction. To find the horizontal displacement of hinges, if one of them were free, the force polygon used for the deflection polygon for vertical deflections is to revolve through an angle of $90°$ and an equilibrium polygon constructed on the horizontal elastic loads applied at the panel points will give the horizontal deflections referres to a line which passes through the intersection of the first ray and the force from left hinge, and perpendicular to the closing line for the deflection polygon for vertical deflections.

It is obvious from the construction that the intercept on the prolongation of the chord joining the hinges represents the summation of moments of elastic loads about that chord or is equal to Σwy. Now the intercept is ζ. then $Hx\zeta$ must be equal to Σwy. When H is equal to $l \cdot \zeta$ becomes δ' the displacement of hinge, hence $\delta' = \Sigma wy$.

(e) Horizontal Thrust.

Let now the elastic loads for a conventional load of unity applied at one of the hinges horizontally outward be found for every member of the truss and all these loads come on one joint be algebraically added together. Lay off these elastic loads on a load line, with pole distance h and construct a deflection polygon immediately under the truss.

Then the intercept between any ray of the polygon and the closing line multiped by h is the vertical deflection of the top point due to $H = l$. Horizontal thrust $= H = \dfrac{\delta}{\delta'} = \dfrac{\gamma h}{\delta'}$ where $\gamma =$ the intercept.

If now h be made equal to δ' which, as has been seen, being equal to Σwy, we have $H = \dfrac{\gamma \delta'}{\delta'} = \gamma$.

The whole process may now be summarised true:

(1) Apply a force unity horizontally outward at Z.

(2) Calculate stress in members due to this H.

(3) Find change in length in all members.

(4) Find elastic loads for chords.

(5) Find elastic loads for diagonals.

(6) Find elastic loads for verticals.

(7) Add up all the elastic loads on any panel point algebraically.

(8) Lay off these panel elastic loads on a load line, (vertical).

(9) Find either graphically or analytically Σwy.

(10) Construct an equilibrium polygon with pole distance equal to Σwy.

(11) The resulting deflection polygon is the influence line for H, the horizontal thrust.

Articles 23. Method of Land. [1]

This method, same as the previous one, makes use of the elastic loads for the construction of deflection polygons. The only distinguishable feature of this method lies in the finding of the elastic loads, all the rest being exactly the same as that of Muller-Breslau. The elastic loads are now made dependent upon the stresses in members and the change in angles at the apex where the deflection is required. To show clearly the method of procedure let the deflection of lower chord points be considered. Only the members that constitute the chord will be used forming what is known as a "kinematic chain", the web members connected thereto are not considered, as they have no effect.

In the figure, such a kinematic chain of lower chords members is shown, and for the present, the deflection polygon has been drawn.

Let λ = the angle which any member makes with a line

---

① Kinetic Theory of Engineering Structures. Molitor. P. 102.

parallel to x-axis and through the right hand end of the member. The signs are indicated, as shown.

δ_A, δ_1, δ_2, etc. are the displacements of A, 1, 2, etc., measured from the closing line $A'B'$, and parallel to the y-axis. They are assumed to be known for the present.

W_1, W_2, W_3, are elastic loads required to construct the deflection polygon. They are applied in the same direction as the deflection, i. e., parallel to y-axis.

Draw $l'S$ parallel to $A'S'$ or $X'S$. Hence

$$\because \quad \overline{ce} = (\delta_2 - \delta_1)\frac{d_3}{d_2} \text{ and } \overline{ac} = \delta_2 - \delta_3,$$

$$\therefore \quad \overline{ae} = (\delta_2 - \delta_1)\frac{d_3}{d_2} + (\delta_2 - \delta_3). \quad (1)$$

Since the sides of the deflection polygon are respectively parallel to the rays of the force polygon, then

$$\frac{\overline{ae}}{d_3} = \frac{W_2}{h}, \quad W_2 = = \frac{\overline{ae}}{d_3}h,$$

Substitute (1) for $\overline{ae}$. $W_2 = \left(\frac{\delta_2 - \delta_3}{d_2} - \frac{\delta_3 - \delta_2}{d_3}\right)h,$ (2)

from the figure. $y_1 - y_2 = L_2 \sin \lambda_2$,

the differential of which gives $\Delta y_1 - \Delta y_2 = \Delta L_2 \sin \lambda_2 + L_2 \cos \lambda_2 \, \Delta\lambda_2$.

This expression must also be equal to $\delta_1 - \delta_2$, because they both represent the deflection of point 2 with respect to point 1.

Hence $\delta_1 - \delta_2 = \Delta L_2 \sin \lambda_2 + L_2 \cos \lambda_2 \, \Delta\lambda_2$

Divide this expression by $d_2 = L_2 \cos \lambda_2$,

$$-\frac{\delta_2 - \delta_1}{d_2} = \frac{\Delta L_2}{L_2} \tan \lambda_2 + \Delta\lambda_2. \qquad (3)$$

In similar manner, there is found

$$\frac{\delta_3 - \delta_2}{d_3} = \frac{\Delta L_3}{L_3} \tan \lambda_3 + \Delta\lambda_3. \qquad (4)$$

Making $H \cdot \lambda$ and substituting (3) and (4) in (2).

$$W_2 = -\Delta\lambda_2 + \Delta\lambda_3 - \frac{\Delta L_2}{L_2} \tan \lambda_2 + \frac{\Delta L_3}{L_3} \tan \lambda_3. \qquad (5)$$

Let ψ be the angle included between any two members of the kinematic chain, measured in a clock-wise direction starting from the member on the right of the joint.

Also let α, β and γ be the angles of the triangles meeting at the joint or the kinematic chain under consideration, all angles at verities being denoted by α. Then $\psi = 360° - \Sigma\alpha$.

If $\Delta\alpha$, $\Delta\beta$ and $\Delta\gamma$ are changes in angles α, β and γ due to stresses S_a, S_b, and S_c in the members opposite the angles

α, β and γ respectively, then

$$\left.\begin{array}{l} E\Delta\alpha = (S_a - S_b)\cot\gamma - (S_c - S_a)\cot\beta \\ E\Delta\beta = (S_b - S_c)\cot\alpha - (S_a - S_b)\cot\gamma \\ E\Delta\lambda = (S_c - S_a)\cot\beta - (S_b - S_c)\cot\alpha \end{array}\right\} \qquad (6)$$

Since the sum of the three angles of a triangle must be constant,

$$E\Delta\alpha + E\Delta\beta + E\Delta\gamma = 0.$$

From this relation any one becomes known as soon as the other two are found.

Now
$$\psi = 360° - \Sigma\alpha,$$
$$\Delta\psi = -\Sigma\Delta\alpha, \text{ or}$$
$$E\Delta\psi = -E\Sigma\Delta\alpha.$$

Hence $\Delta\psi$ is known, after $\Delta\alpha'S$ are known,

Again, $180° - \lambda_2 + \lambda_3 = \psi_2$

$\therefore -\Delta\lambda_2 + \Delta\lambda_3 = \Delta\psi_2$

or $-E\Delta\lambda_2 + E\Delta\lambda_3 = E\Delta\psi_2$

$$= -E\Sigma\Delta\alpha. \qquad (7)$$

Also, $\Delta L_2 = \dfrac{S_2 L_2}{E}$, $\Delta L_3 = \dfrac{S_3 L_3}{E}$,

$$\therefore \frac{\Delta L_2}{L_2} = \frac{S_2}{E}, \frac{\Delta L_3}{L_3} = \frac{S_3}{E}. \qquad (8)$$

Substitute (7) and (8) in (3) ×8.

$$EW_2 = -E\Delta\psi_2 - S_2\tan\lambda_2 + S_3\tan\lambda_3.$$

Generally,

$$EW_m = -E\Delta\psi_n - S_n\tan\lambda_n + S_{n+1} + \tan\lambda_{n+1}.$$

For top chord, $\lambda = 0$, and $E\Delta\psi = +E\Sigma\Delta\alpha$, so that $EW_m = E\Delta\psi_n = E\sum\Delta\alpha$.

By this equation the elastic loads can be written for any panel point and the deflection polygon can be drawn in the same way as for method of Muller-Breslau.

Horizontal displacement

$$\delta_{ab} = \sum l^{n-1} y\Delta\psi + \sum_1^n \Delta l \cos(\lambda - \alpha)$$

where α is the angle AB makes with the horizontal.

Article 24. Method of Du Bois. [①]

This method is derived directly from the principle of least work, considering the chord members only.

Let t be the lever arm for any chord member and M the moment at the moment point for that member.

$$\text{Then } S = \frac{M}{t}, \quad \text{Work} = \frac{S^2 L}{2AE} = \frac{M^2}{t^2}\frac{L}{2AE}.$$

The total work of straining all the chord members under the system of loading is

$$\text{Work} = \sum \frac{M^2}{t^2}\frac{L}{2AE}, \qquad (1)$$

which is to be made a minimum.

Now for any point distant x from left support A, between A

---

① Stresses in Framed Structures. Du Bois. P. 196.

and the load.

$$M = Hy - \mathrm{V}_1 x = Hy - P(l - k)x.$$

For any point between P and S.

$$M = Hy - \mathrm{V}_1 x + P(x - kl) = Hy - Pk(l - x).$$

We have then for the work of straining all the chord members, from (1)

$$\mathrm{Work} = \sum\nolimits_{0}^{kl} [P(l - k)x - Hy]^2 \frac{L}{2AEt^2} + \sum\nolimits_{kl}^{\mathrm{L}} [Pk(l - x) -$$

$$Hy]^2 \frac{L}{2AEt^2},$$

$$\frac{\mathrm{d}(\mathrm{Work})}{\mathrm{d}H} = 0 = \sum\nolimits_{0}^{kl} \Big[\frac{Hy^2 l^2 - Ply(l - kl)x}{l^2}\Big] \frac{L}{AEt^2} +$$

$$\sum\nolimits_{kl}^{l} \Big[\frac{Hy^2 t^2 - Pkl^2 y(l - x)}{l^2}\Big] \frac{L}{AEt^2},$$

$$\mathrm{or} \sum\nolimits_{0}^{kl} [Hy^2 - Py(l - k)x] \frac{L}{AEt^2} + \sum\nolimits_{kl}^{l} [Hy^2 - Pky(l - x)]$$

$$\frac{L}{AEt^2} = 0.$$

Hence by reduction.

$$H\sum\nolimits_{0}^{l} \frac{y^2}{t^2} \frac{L}{A} = Pkl \sum\nolimits_{kl}^{l} \frac{y}{t^2} \frac{L}{A} - Pk\Big[\sum\nolimits_{0}^{l} \frac{xy}{t^2} \frac{L}{A} - \frac{l}{k} \sum\nolimits_{0}^{kl} \frac{xy}{t^2} \frac{L}{A}\Big].$$

Article 25. Method of Mirei. [1]

The formula derived for horizontal thrust by Mirei is also the direct application of the principle of least work. The stresses in each member of the truss are found in terms of H and other external forces and substituted in the formula for internal work. By summarizing the work in all the members of the structure and working it a minimum, the value of H is obtained by differentiating the work equation with respect to H.

Assume the arch to be loaded with two equal loads of P each, distance up from each end (P = panel length), find the stresses in members by taking moments at successive sections.

Introduce these in the expression for total internal work $\sum \dfrac{S^2 L}{2AE}$ we get a relation between H and the work. Differentiating this with respect to H, and setting the differential coefficient equal to zero, there is at once obtained an expression for the values of H for $\sum p$. The required H will be one half the amount found above.

---

[1]   Statically Indeterminate Stresses. Mirei. P. 86.

$$H = P \frac{\sum_0^{np}\left[\dfrac{Vp^2(V_0-V)}{r^2A_n} + \dfrac{VpV_0L}{n^2A_l} + \dfrac{lV_0V}{(l-Vp)^2AV}\right] + \dfrac{(V_0+V_1)V_0}{pAV_0}}{2\sum_0^{\frac{l}{2}}\left[\dfrac{(V_0-V)^2P}{t^2A_n} + \dfrac{V_0^2L}{h^2A_l} +\right.}$$

$$\frac{+\sum_{np}^{\frac{l}{2}}\left[\dfrac{np^2(V_0-V)}{VA_n} + \dfrac{npV_0L}{l^2A_l} + \dfrac{npV_0V}{(l-Vp)^2A_V} + \dfrac{npV_0d}{t^2A_d}\right]}{\left.\dfrac{V_0^2V}{(l-np)^2A_L} + \dfrac{V_0^2d}{t^2Ad}\right] + 2\dfrac{(V_0-V_1)^2V_0}{p^2AV_0}}$$

Where r represents, in case of chord members, the distance – in number of panels – of the panel point opposite the member under consideration, and in case of web members the ordinal number – from nearest support-of the upper chord opposite the web member in question, n = lever arm of lower chord, v = length of verticals, A_n = area of upper chord. A_l, A_d, A_V = areas of lower chord, diagonal & vertical. h = lever arm of lower chord, t = lever arm of diagonal.

Article 26. Comparison.

There is no doubt that the method of Du Bois and Mirei are not suitable for practical purpose as the formulas are complex and do not allow of easy application. The method of Greene is the same to that of Maxwell only the formula is expressed in a more explicit manner. Its application is shortened by the use of sysmetrical members by Cooper and sysmetrical loads by

Turnesure. The method of Turnesure may therefore be taken as the representative of the methods of Maxwell, Greene and Cooper. The method of Muller-Breslau is a graphical treatment of that of Mohr as the elastic loads therein termed are no more or less than the loads $\dfrac{y}{t^2}\dfrac{L}{AE}$ used in the latter method. Either one of them may be used in comparison, the method of Mohr will be taken for convenience. The method of Land is not practical as it requires the same amount of work in constructing the deflection polygon while the finding of the elastic loads is more troublesome than that of Muller-Breslau. It now requires to compare the method of Merriman and Jacoby with those of Turnesure and Mohr. As for as the amount of calculation is concerned the method of Merriman and Jacoby is the best while it is inferior to the other two methods in that it is not as accurate in result. There is not much differences between the methods of Turnesure and Mohr although the former seems to have the advantage of being more really comprehensible.

The method of Merriman and Jacoby and that of Turnesure are the two methods used in this thesis. They are so entirely different in character that they serve good checks to each other, thus avoiding mistakes in the computations.

CHAPTER XI. REACTION LOCUS

When an arch is loaded by a single load the structure is under the influence of three external forces, two of which are supplied by the reactions at the springings. The points of application of all these forces are known and also the direction of the load. If there can be found a point on the line of action of the load through which the reactions have to pass, the reactions may be defined completely by joining this point to the springings and thereby the directions and consequently the magnitudes of the reactions may be obtained. This gives H for one position of load.

When the load moves along the span, the point of intersection moves in a similar manner but besides having a horizontal motion, it also has some sort of vertical motion, the exact nature of which depends on the geometrical form of the truss. The path along which this point moves is called a reaction locus or intersecting locus. If this locus be known, the reactions can be easily determined for loads occupying different points along the span. The following are two methods of constructing the locus after the values of H are known.

Article 27. Method of Ritter.

The method of Ritter has been demonstrated by Du Bois by

a graphical application of his formula given in Article 24. It is there claimed that his formula is tedious in application and it is better to adopt a graphical treatment by constructing the reaction locus.

Perform the summations denoted by $\sum_0^l \dfrac{y}{t^2} \dfrac{L}{A}$ and $\sum_0^l \dfrac{y^2}{t^2} \dfrac{L}{A}$ and determine the distance $\bar{y}$ giving

$$\bar{y} = \frac{\sum_0^l \dfrac{y^2}{t^2} \dfrac{L}{A}}{\sum_0^l \dfrac{y}{t^2} \dfrac{L}{A}}.$$

At this distance $\bar{y}$ above AB draw a line $A'S'$ parallel to AB.

Consider the quantity $\dfrac{y}{t^2} \dfrac{L}{A}$ for each chord panel as a weight acting at the moment point for that chord. Lay off these fictitious loads on a vertical AP through A to any convenient scale. Take a pole distance $h = AB$ or $h = \sum_0^l \dfrac{y}{t^2} \dfrac{L}{A}$ and construct an equilibrium polygon. AB will be its closing line.

For load P at any apex let A be the corresponding intercept of the equilibrium polygon.

The equation for H can be written in the form

$$H\sum\nolimits_0^l \frac{y}{t^2}\frac{L}{A} = Pkl\sum\nolimits_0^l \frac{y}{t^2}\frac{L}{A} - Pk\sum\nolimits_0^l \frac{xy}{t^2}\frac{L}{A}$$

$$- P\sum\nolimits_0^{kl} \frac{(kl-x)yL}{t^2 A}.$$

Let $\sum\nolimits_0^l \frac{y}{t^2}\frac{L}{A}$ be denoted by w. Then, since

$$W\overline{y} = \sum\nolimits_0^l \frac{y}{t^2}\frac{L}{A} \text{ and } \sum\nolimits_0^l \frac{y}{t^2}\frac{L}{A}x = W\frac{L}{L},$$

we have, $HW\,\overline{y} = P\left[\frac{wkl}{2} - \sum\nolimits_0^{kl} \frac{(kl-x)yL}{t^2 A}\right]$

Now $h\,x\,z$ is equal to the moment under P and is equal to the quantity in the parentheses, for $\frac{w}{z}$ is the reaction for all the fictitious loads and the Σ is the loads between reaction and section.

$\therefore Hw\overline{y} = Phs$,

Now h is made equal to $AB = w$,

$\therefore H\overline{y} = Ps$.

Since $V_1 = P(l-k)$, $\dfrac{H}{V_1} = \dfrac{Z}{(l-k)\overline{y}}$

If, now, we draw the line BE through the foot of the ordinate z, we have $Ae = \dfrac{z}{(l-k)}$.

Hence, $\dfrac{H}{V_1} = \dfrac{Ae}{\overline{y}}$.

If then we lay off the distance Ae horizontally from A and draw vertical of cutting $A'B'$ at P, AF produced will meet P at one point of the reaction locus.

Article 28. Method of Balet. [①]

Let $A3$, $A5$, $A7$, and $A9$ be equal to the values of δ at 3, 5, 7 and 9 obtained from a displacement diagram. Lay off δ' on AD from A to D, and

$\frac{1}{2}\delta'$ from C to X on CK, C being at centre of the span. The line DK cuts off ordinates on the verticals I, II, III, etc. , which are equal to vertical reactions at A when the load δ' is placed respectively at I, II, III, etc. , These reactions are now transferred to the verticals created at 3, 5, 7, etc. , giving the points a', b', c', etc. , The line Aa' is then the reaction for a load placed at I, and I is a point of the intersection locus. The line PC which passes through all the points similarly determined is therefore the intersecting locus.

---

①   Elastic Arches. Balet. P. 56.

CHAPTER XII. APPROXIMATE METHODS

Since the exact formulas for horizontal thrust involves the unknown cross sectional area, it is necessary to make a preliminary design before the correct value of H can be obtained. To shorten the work, approximate methods of finding reactions have been proposed from time to time. They may all be included under three heads:

(1) By certain approximations in the exact formulas.

(2) By using a mathematical curve for the reaction locus.

(3) By finding an equation for the influence line of the horizontal thrust.

These will be considered separately.

Article 29. Approximations in Exact Formulas.

This is by far the most common method used in the preliminary design when the sectional areas are yet unknown. The approximations that have been used in the formulas by the various writers may be summarized as follows:

(a) The areas of all members are assumed equal, thus doing away with the unknown term λ in the formula for H. This is most commonly used in practice and has been found to involve but a small error. With constant E, the formula of H becomes

$$H = \frac{\sum S'TL}{\sum T^2 L}.$$

This is the formula used in the preliminary design of the present work.

(b) The effect of web members may be neglected. This has been advocated by the majority of authorities on engineering structures, but in a letter to the Engineering Record, Mr. Godfrey opposed it very strongly by working out an actual example and showing that the discrepancy was too large to be neglected. [1]

(c) Neglecting the effect of web members, the length of all the chord members may be assumed to be constant. [2]

$$H = \frac{\sum S'T}{\sum T^2}.$$

Article 30. Mathematical Curves of Reaction Locus.

By the use of a reaction locus the stresses may be found

---

[1] Engineering Record. Vol. 46. P. 604. A series of letters discussing this matter may be found in the same journal. Vol. 47. P. 61, 180, 227, 396, 442 and 702.

[2] Design of Steel Bridges. Kunz. P. 328.

more quickly than with the use of vertical and horizontal components of the reactions. Thus arises the use of mathematical curves for such locus so that it may be drawn as soon as the geometrical proportions of the truss are known. The following are some of the methods that have been proposed.

(a) Method of Johnson.

Two equations have been suggested by Johnson, the first assumes the reaction locus to be of parabolic form while the second assumes it to be of hyperbolic form, the latter being claimed to be more accurate than the first. [1]

The equation of the parabolic locus referred to the center of the line joining the two hinges as origin, is

$$y = \frac{2.5(r - D_c)x^2}{l^2} + r + 2.2D_c,$$

where y = ordinate to the locus curve, x = distance out to right or left from the center, l = span of arch, r = rise of arch and D_c = depth of arch at crown.

The hyperbolic locus has the following equation

$$y = \sqrt{\frac{8}{3}\left(\frac{x}{2}\right)^2 (Q^2 - K^2) + K^2}.$$

---

[1] Transactions of A. S. C. E.. Vol. 40. P. 165.

Where $Q = 1.35r + 1.35D_c$,

$K = r + 2.2D$.

(b) Method of Treeman.

This is a sine locus and is given in Proc.

Inst. C. X. Vol. 147.

$$y = 1.3(r + D_c) - 0.2(r + D_c)\sin\left[\overline{90°^2} - \frac{X}{\frac{1}{2}L}\overline{180°^2}\right].$$

(c) Method of Balet.

This method consists in finding the intersection locus from what is known as a standard locus which is drawn for an arch rib with parabolic axis, the rise of rib axis $= 1$, and the span of the axis $= 2$. The following are the coordinates for the standard diagram referred to the left springing as origin.

x	0.1	0.2	0.3	0.4	0.5	0.6	0.7	0.8	0.9	1.0
y for rib axis	0.19	0.36	0.51	0.64	0.75	0.84	0.91	0.96	0.99	1.00
y for-locus	1.526	1.468	1.420	1.378	1.347	1.322	1.305	1.290	1.283	1.280

To correct the standard diagram for an arch axis which is not a parabola, compute the area enclosed by the axis of the arch and the axis. Then compute the rise of parabolic arch having the same area and same span. Plot the intersection locus for the resulting equivalent parabola. Let r = rise of the arch in *consideration*, D_c = crown depth, D_s = depth at springing and c = a factor depending on the form of the structure, being equal to 1. 3 for a Pratt bracing.

Plot $\dfrac{D_c}{D_s} \times 1. 3r$ as an ordinate from the left end of a straight line whose length is equal to the span of the arch. Join the upper end of the ordinate to the right end of the span, giving the ordinates of the line at different panel points of the truss.

Construct a reaction locus from the standard diagram by using a rise equal to that of the equivalent parabola. At the different panel points along the span, add the ordinates of line just obtained to those of the standard locus. The resulting curve will be about the same as the true locus.

Article 31. Equation of Influence Line for Horizontal Thrust.

This method is the most expedient among others in that it gives directly the stresses in members if the method of Merriman and Jacoby given in Article 32 be used in the analysis. Two methods will be given, one of Mr. Kunz and the other of the writer.

(a) Method of Kunz. [1]

---

[1] Design of Steel Bridges. Kunz. P. 335.

Two equations are proposed for unit loads both being obtained from a parabolic braced arch with approximately parallel chords. The first is

$$y = \frac{5L}{8r}(k - 2k^3 + k^4)\frac{1}{c},$$

where $\quad c = 1 + \frac{15}{32}\frac{D_c^2}{r^2}.$

The second equation is $\quad y = \frac{3}{4}\frac{L}{r}(k - k^2)\frac{1}{c}.$

Where c has the same significance as in the first.

(b) Method Proposed by the Writer.

By an analysis of a large number of two hinged spandrel-braced arches with economic proportions in regard to rise, crown depth and form of lower chord, it is found that the value of H for a concentrated load at the centre of span is very nearly equal to the value of load. By virtue of this fact the writer has found another approximate method for constructing the influence line of the horizontal thrust. At the center of the span erect an ordinate with value equal to that of the load or unity for influence lines. Draw a hyperbola through the ends of the span with this ordinate as the middle ordinate. The resulting curve will represent very closely the influence line of H. For the arch under design, this method is so exact that it is far better than the influence line obtained from the formula

$$H = \frac{\sum S'TL}{\sum T^2 L},$$

as it is very close to the final H curve. See plate b. The writer

does not claim that this result will be obtained in every case as it is entirely over-thrown by an uneconomic proportion of the arch dimensions.

The following ordinates of the hyperbola were calculated for the construction.

CHAPTER XIII. ANALYSIS OF STRESSES

As soon as the external forces are known the stresses in members of the arch may be found by any convenient method just as easy as for a simple truss. For full loads, the process is simple because the positions of the loads are fixed. To inquire into the effect of live loads considerable thought has to be given to the arrangement of the loads such that the stress in a given member might be a maximum. To facilitate the work there are four different methods proposed, each of which has its own merit in a particular case.

(1) Graphic method of Merriman and Jacoby.

(2) Analytic method.

(3) By the use of a reaction locus.

(4) By a correction of corresponding stresses in a three hinged arch of the same proportions.

Article 32. Live Load Stresses by the Method of Merriman and Jacoby. [1]

Let M' be the bending moment about the moment point of any member for $H = O$, i, e., as if the arch acted as a simple truss. Then the required bending moment is $M = M' - Hy$, and the stress is $\quad S = \dfrac{M}{t} = \dfrac{1}{t}(M' - Hy) \quad$ or $\quad S = \dfrac{y}{t}\left(\dfrac{M'}{y} - H\right)$.

The greatest tension and compression in the member due to live load may therefore be found by constructing the influence line of $\dfrac{M'}{y}$ in such a relation to that of H that the ordinate between them will represent the difference between their respective ordinates and after adding the positive and negative

---

① Higher Structures. Merriman and Jacoby. P. 271.

ordinates separately, multiplying the respective sums by the quantity $\frac{y}{t}$ ($= T$) which is known as the multiplier of the influence diagram or influence coefficient. [1]

The influence lines of $\frac{M'}{y}$ are straight lines and can be constructed most expeditiously in the following manner: Let the moment point of a certain member be at a distance x from the left support. For a load on the right of the moment point. $M' = \frac{x'x}{l}$ where x' is the distance of the load from the right support. If this relation were also true when the load is on the left of the moment point the value or $\frac{M'}{y}$ at the left end would be $\frac{x}{y}$, because $x' = l$, Let this value be laid off as an ordinate at the left end and draw a right line from right and to the top of the ordinate. This line will cut the line of action of the load. That part of the line which is intercepted between the right end and the load is one part of the influence line. The other part may be obtained in a similar manner.

---

[1] Design of Steel Bridges. Kunz. P. 61.

Article 33. Live Load Stresses by Analytical Method.

This method consists in finding the stresses in every member for all the positions of load and summing up the stresses that are of same sign. It is best to find the stresses separately, those due to vertical forces, ($H = 0$) are found first and then the stresses due to H for the corresponding position of the load are added to it algebraically.

Article 34. Live Load Stresses by Reaction Locus.

This method is very much the same as that used for the calculation of maximum stresses in three-hinged spandrel-braced arches. [1]

After the reaction locus is drawn, to find the maximum live load stresses in any member draw a right line through the moment point of that member and the left hinge of truss if the member is on the left side and right hinge if it is on the right side of the truss. The point at which this line cuts the reaction locus is therefore the position of the load which gives zero stress in the member. For all loads on one side of this point of division there will be only one kind of stress in the member (except for web

---

[1]   Higher Structures. Merriman and Jacoby. P. 182.

members which may have another point of division on one side of the previous one depending on whether or not the chords cut by the section through the web member intersect beyond the line of action of the reaction), the summation of which will give the greatest stresses of that kind. Similarly the summation of stresses produced by the loads on the other side of the division point will give the greatest stress of the other kind. This method is useful in being capable of showing the kind of stress just by inspection of the diagram.

Article 35. Live Load Stresses by a Correction of the Corresponding Stresses in a Three-hinged Arch of the Same Proportions.

This method is best suitable in case of alternate designs of two and three-hinged arches of the same proportions. It consists in finding the live load stresses by a correction of the stresses resulting from the three-hinged arch. It is very useful for combination type.

For a load P[1] at any point consider the members

---

① Statically Indeterminate Stresses. Hodson. P. 104.

U_6 and U_7 out at point Ⅵ and compute the stresses in all members as for a three-hinged arch. Then compute the horizontal motion Δ of the two cut ends with reference to each other under the load P. Then compute the horizontal motion δ of the two cut ends with reference to each other for a horizontal force of unit at the cut ends. Then the stress S_a in U_6 for the two cut ends brought together is $S_a = \dfrac{\Delta}{\delta}$. If S_2 is the stress in any member of the truss when considered as a two-hinged arch and S_3 the stress in the same member when the structure is considered to have three hinges and S_c = stress in the member when the structure is considered to have three hinges. due to S_2, then $S_2 = S_3 S_c$.

Article 36. Comparison.

It will be seen that each method has its superiority in certain cases. To say as a whole the method of influence lines of Merriman and Jacoby is the best of the four for the following reasons:

(1) Same diagram may be used for both preliminary and final stresses by changing the H curve.

(2) It gives loaded length in impact formula.

(3) It gives a method to find the maximum stresses due to wheel concentrations instead of uniform panel loads.

(4) It shows clearly the effect on the stresses for any slight variation of H.

(5) It is the most easy in application. As a consequence of this method of influence lines is used in the present thesis in calculating the live load stresses. To check the result analytic methods were also used.

PART Ⅲ. PRELIMINARY STRESSES

CHAPTER ⅩⅣ. INTRODUCTION

To keep pace with the most modern practice, stresses arising from the following items will be adequately considered:

(1) Dead Load;

(2) Live Load;

(3) Impact;

(4) Wind and Lateral Forces; and

(5) Temperature.

As stated in the preface, the present work endeavors to make some investigations in regard to the introduction of cantilever arms at the sides of the truss and the effect of erecting the arch by the insertion of a crown hinge at the centre of the span. To obtain a reliable result it is necessary to make a design of each of the cases considered and compare the estimated weights. It is a good plan but on account of lack of time, the writer will not attempt such a refined process but instead will make the comparisons based on the total stresses produced in the various types. The errors thus resulted are small because the weight of the main shapes is proportional to the stress the only variation being in the details which do not differ much in the three types under consideration. The following comparisons will be considered.

$$(a) \begin{cases} \text{Combination type without cantilever.} \\ \text{Combination type with cantilever.} \end{cases}$$

$$(b) \begin{cases} \text{Two hinged type with cantilever.} \\ \text{Combination type with cantilever.} \end{cases}$$

In the first case, since the stresses produced from the various loads are entirely different from each other, the comparison must be made on the basis of total stresses, while in the second case, only the dead stresses will be sufficient, because all the other stresses remain the same in both types. Again, in the first case the stresses in both types are statically indeterminate (except D. L) , thus making it permissible to compare the preliminary stresses only as the sectional areas of the corresponding members will not very much. For the second case, however, the dead load stresses, for the combination type is exact and therefore in making comparisons the revised stresses of the two-hinged arch must be used. To meet these variations the comparisons will be carried out in the following manner.

(a) First Case. In calculating the live load stresses for the two hinged arch the truss is at first assumed to have no cantilever arms. After the stresses are obtained the effect of the cantilever arms is considered by computing the stresses in the arch span produced by loads on the cantilever arms and also the stresses in the arms themselves. This is possible because for arches having cantilever arms, the latter have no effect upon the truss, so long as the loads are placed on the inside of the arch span.

(b) Second Case. After the revised stresses are obtained the dead load stresses for the two-hinged arch may be found by multiplying the stress coefficient by the proper panel loads and summing up the stresses for any member. These stresses may now be compared with those obtained by considering the arch as having a crown hinge.

On the following page is shown a table of elements of the members of the truss giving lengths, coordinates of moment points and lever arms. This table will be used extensively in the calculations of all kinds of stresses.

Table 1. Elemens of Members.

Member	L	x	y	t
U_1	25.	25.	16.5	49.5
U_2	25.	50.	30.	36.
U_3	25.	75.	40.5	25.5
U_4	25.	100.	48.	18.
U_5	25.	125.	52.5	13.5
U_6	25.	150.	54.	12.
L_0	29.954	0.	66.	55.084
L_1	28.413	25.	66.	43.554
L_2	27.116	50.	66.	33.192
L_3	26.101	75.	66.	24.424
L_4	25.402	100.	66.	17.715
L_5	25.045	125.	60.	13.476
D_1	55.456	100.000	66.	89.260
D_2	43.830	116.668	66.	75.291
D_3	35.711	135.714	66.	61.206
D_4	30.806	160.000	66.	49.665
D_5	28.413	200.000	66.	47.512
D_6	27.731	350.000	66.	97.362
V_0	66.	–	–	–
V_1	49.5	116.668	66.	91.668

V_2	36.	135.714	66.	85.714
V_3	25.5	160.000	66.	85.000
V_4	18.	200.000	66.	100.000
V_5	13.5	350.000	66.	225.000
V_6	12.	–	–	–

CHAPTER XV. DEAD LOAD STRESSES

Article 37. Design of floor Systems.

E – 50 Loading.

(a) Timber Ties.

Spacing of stringers 6'6".

Load from one wheel = 25 kips

Add 100% impact = 25

50 kips

Load on one tie $\dfrac{50}{3}$ = 16670 lbs.

Moment = 16670 × 10.75 = 179200 in. lbs.

Take unit stress of timber = 2000 lbs./in.^{2} width = 8".

$$d = \left(\frac{6\,M}{2000\,b}\right)^{\frac{1}{2}}$$

$$= 0.0548\ (M/b)^{\frac{1}{2}}$$

$$= 0.0548\ \left(\frac{1.79200}{8}\right)^{\frac{1}{2}} = 8.2 \text{ inches.}$$

The timber ties used are therefore made 10″ deep, 8″ wide, 10′ long and spaced at 5″ clear or 13″ centres.

Take weight of timber to be 4.5 lbs. per F. B. K.

Wt. of l tie = 300 lbs.

St. of ties per foot of bridge = $\dfrac{300}{13} \times 12 = 277$ lbs.

St. of track and fastenings = $\underline{150}$

$\underline{427}$ lbs./ft.

= say 450 lbs. To include guard rails. etc.

(b) Stringers. (span 25′)

St. of track = 450. lbs. per ft. of truss.

Assume weight of 2 stringers = $\underline{450}$ lbs. per ft.

$\underline{900}$

D. L. per ft. per stringer = 450 lbs.

D. L. Moment = $450 \times 25^2 \times \dfrac{12}{8} = 422000$ in. lbs.

D. L. Shear = $450 \times \dfrac{25}{2}$

$ = 5620$ lbs.

For Cooper′s E 50 loading, it is found for 25′ span

maximun L. L. centre moment = 381300 ft. lbs.

$ = 4570000$ in. lbs.

maximun L. L. end shear $ = 71000$ lbs.

impact ratio $= \dfrac{300}{325}$ $\qquad = .923$

impact moment $\qquad = .923 \times 4570000$ in. lbs.

$\qquad = 4220000$ in. lbs.

Impact shear $\qquad = .923 \times 71000$ lbs.

$\qquad = 65600$ lbs.

Moment		Shear	
D. L.	422000	D. L.	5620
L. L.	4570000	L. L.	71000
Impact	4220000	Impact	65600
	9212000 in. lbs.		142220 lbs.

For economic reasons the depth of stringer should be from $\dfrac{1}{6}$ to $\dfrac{1}{7}$ th the span or 50 to 42.8 inches.

Take the depth of web plate to be 44″.

Gross area required for web $= \dfrac{142220}{10000}$

$\qquad = 14.22$ sq. in.

thickness of web $= \dfrac{14.22}{44} = .323$ inches,

Take $\dfrac{3}{8}''$ $\qquad = .375$ in.

therefore, web plate is made $44'' \times \dfrac{3}{8}$.

Distance b. to b. of flange angles = $44 \frac{1}{4}''$ (top of angles flush with web).

Assume affective depth to be $40''$. Stress in flanges

$$= \frac{9212000}{40} = 230500 \text{ lbs.}$$

Net area required $= \dfrac{230500}{16000} = 14.4$ sq. in.

$2 \underline{|S} \ 6 \times 6 \times \frac{5}{8} - 2 - 1'' \text{ holes} = 14.22 - 1.25 = 12.97$

$\dfrac{1}{8}\text{web} = \dfrac{375 \times 44}{8} \qquad = \underline{2.06}$

Net area supplied $\qquad = \underline{15.03}$ sq. in.

Therefore flange angles will be made $6'' \times 6'' \times \dfrac{5}{8}''$.

Kivet Pitch.

L. L. Shear at $\dfrac{1}{2}$ point $\qquad = \dfrac{5}{8}$ end shear.

L. L. Shear at centre $\qquad = \dfrac{2}{7}$ end shear.

Point	D. L.	L. L.	Impact	Total
End	5620	71000	65600	142200
Quarter	2810	44300	41000	88110
Center	0	20300	18800	39100

Total vertical load on stringer $= \dfrac{5000}{3 \times 13} + \dfrac{225}{12}$

$= 1280 + 18.7 = 1300$ lbs. per in.

Point	Shear	Longitudinal Shear		Vert. Shear	Resultant Shear
		On Flange	On Rivet		
End	142220	3490	3010	1300	3280
Quarter	88110	2170	1870	–	2280
Center	39100	957	673		1460

Value of $\dfrac{7}{8}$ rivet in bearing on $\dfrac{7}{8}''$ plate

$= 7880$ lbs. $/\text{in.}^2$

Pitch required:

End $-2.47''$;   quarter $-3.41''$;   Center $-5.32''$

Estimate of Weight.

1 web $44 \times \dfrac{3}{8}$ 25'lg   56.1 $= 1400$

$4 \lfloor S\ 6 \times 6 \times \dfrac{5}{8}$ 25'lg   24.2 $= 2420$

10 stiff, $\lfloor S\ 4 \times 3 \times \dfrac{3}{8}$ $3'5\dfrac{3}{4}''$ lg   8.5 $= 305$

4conn. $\lfloor S\ 5 \times 3\dfrac{1}{2} \times \dfrac{1}{2}$ $3'5\dfrac{3}{4}''$ lg   13.6 $= 195$

4fills $\dfrac{5}{8} \times 7$ $2'8\dfrac{1}{4}''$ lg   14.88 $= 160$

Half lateral systems

$2\frac{1}{2}\ \lfloor S\ 4\times4\times\frac{3}{8}\quad 8.2'\quad$ lg   0.8 $\qquad\qquad =\underline{201}$

$\qquad\qquad\qquad\qquad\qquad\qquad\qquad\qquad 4681$

$\qquad\qquad\qquad$ Add 20% $\qquad\qquad\quad =\underline{936}$

$\qquad\qquad\qquad\qquad\qquad\qquad\qquad\qquad \underline{5617}$

$\qquad\qquad\qquad\qquad\qquad =225\quad$ lbs. per ft.

For 2 stingers, $\qquad\qquad\qquad =450\quad$ lbs. per ft.

(c) Floor beam(span 25′)

$\qquad\qquad$ Wt. of track $=450$

$\qquad\qquad$ Wt. of stringer $=\underline{450}$

$\qquad\qquad\qquad\qquad\quad \underline{900}$ lbs. per ft. of truss.

D. L. Concentration from two pairs of stringers

$\qquad\qquad\qquad =900\times25 =22500$ lbs.

For one pair of stringers. $=11250$ lbs.

For live loads ($x-50$). Maximum floor beam reaction is found when wheel 4 is placed at the beam and the reaction = 94540 lbs. From each stringer.

Impact ratio $=\dfrac{200}{350} =.857$

Impact reaction $=.857\times94500 =81000.$

Stringer Concentration:

D. L. 11250

L. L. 94540

Imp. <u>81000</u>

186790 lbs.

say 186800 lbs.

Assume wt. of floor beam itself as 9000 lbs, or 360 lbs./ft.

Moment at center due to it

$$=\frac{1}{8} \times 9000 \times 25 \times 12 = 338000 \text{ in. lbs.}$$

End shear due to it

$$=\frac{1}{2} \times 9000 = 4500 \text{ lbs.}$$

End reactions due to

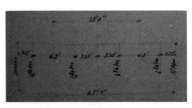

stringer concentrations

$$= 2 \times 186800 = 373600 \text{ lbs.}$$

Moment at center $= 373600 \times 12.5 - 186800(3.25 + 9.75)$

$$= 4675000 - 2430000$$

$$= 2245000 \text{ ft. lbs.} = 27000000 \text{ in. lbs.}$$

Moment due to weight of floor beam $= \underline{338000}$

<u>27340000</u> in. lbs.

Maximun end shear $= 373600$

End shear due to wt. of floor beam $= \underline{4500}$

378100 lbs.

Gross area of web plate required = 37. 81 sq. in.

Depth of web to be greater then $\frac{1}{6}$ th the span to give

Stiffness – assume 62″.

Thickness of web required = $\frac{37.81}{62}$ = . 61″.

Thickness used = $\frac{5}{8}″$ = . 625″.

Therefore, web plate will be 62″ × $\frac{5}{8}″$ (no splice required).

Assume effective depth to be 60″. Stress in flange

= $\frac{27340000}{60}$ = – 455000 lbs.

Area required = 28. 4 sq. in.

2 $\underline{\mathsf{S}}$ 6 × 6 × $\frac{3}{4}$ – 2 × 8. 44 – 2 × 1 × $\frac{3}{4}$

(Deduct 2 – 1″ holes) = 15. 38

1 cover plate 14 × $\frac{3}{4}$ = 10. 5 – 2 × 1 × $\frac{3}{4}$ = 9. 00

$\frac{1}{8}$ × 62 × $\frac{5}{8}$ = $\underline{4.84}$

Net area supplied = 29. 55 sq. in.

Effective depth = 60. 35″, Therefore flange will be made of

$$2 \underline{\lfloor S} \ 6 \times 6 \times \frac{3}{4} \left.\right\} \ \ 25' \ \text{lg}$$
$$1 \ \text{pl} \ 14 \times \frac{3}{4}$$

Plange Rivets.

(a) Between first stringer connection and end of beam.

Moment at first stringer connection

$$= 373600 \times 2.76 \text{lbs.}$$

$$= (4600 \times 2.75 - 360 \times 2.75 \times \frac{12}{2})$$

$$= 12480000 \ \text{in. lbs.}$$

Stress in flange

$$\frac{12480000}{60.35} = 206500 \ \text{lbs.}$$

Stress on rivet $= 206500 \times \frac{24.38}{29.22} = 172000$ lbs.

value of $\frac{7}{5}''$ rivet in bearing on $\frac{5}{8}''$ plate. $= 13130$ lbs.

No. of rivets required $= \frac{172000}{13130} = 14$.

Shear at end $= 378100$ lbs.

$$\frac{378100}{60.35} = 6270 \ \text{lbs. per lin. in.}$$

$$= 6270 \times \frac{24.38}{29.72} = 5220 \ \text{lbs.} / \text{in.}$$

Spacing required $= \dfrac{13130}{5220} = 2.5''$.

(b) Between stringer connections under each track, Shear at first stringer connection

$$= 373600 - 186800(4500 - 2.75 \times 360)$$

$$= 19000 \text{ lbs.}$$

$\dfrac{190000}{60.35} = 3150 \text{ lbs. per lin. in.}$

$3150 \times \dfrac{24.38}{29.22} = 2620 \text{ lbs. per lin. in. on rivets.}$

$\text{Spacing} = \dfrac{13130}{2620} = 5.01 \text{ in.}$

(c) Between inner stringer connections:

Let one track be loaded.

$B_1 = \dfrac{186800(15.75 + 22.36)}{25} = 264000 \text{ lbs.}$

Shear in second

stringer connection

$$= 186800 + 186800 - 28400 = 89600 \text{ lbs.}$$

Shear per lin. in. $= \dfrac{89600}{60.35} = 1485 \text{ lbs.}$

Shear per lin. in. on rivet $= 1485 \times \dfrac{24.28}{29.22} = 1237 \text{ lbs.}$

Spacing required $= \dfrac{13130}{1237} = 10.6''$, $6''$ used.

Estimate of Weight.

$$1 \text{ web } 62 \times \frac{5}{8} \quad 25' \text{ lg } \quad 131.8 \qquad\qquad = 3300$$

$$4 \; \lfloor S \quad 6 \times 6 \times \frac{3}{4} \quad 25' \text{ lg } \quad 28.7 \qquad\qquad = 2870$$

$$2 \quad \text{cover plates } 14 \times \frac{3}{4} \quad 25' \text{ lg } \quad 35.7 \qquad\qquad = 1790$$

$$6 \quad \text{stiff. } \lfloor S \; 4 \times 3 \times \frac{3}{8} \quad 5'\frac{3}{4}'' \text{ lg } \quad 8.5 \qquad\qquad = 255$$

$$4 \quad \text{connecting } \lfloor S \quad 6 \times 4 \times \frac{1}{2} \quad 5'\frac{3}{4}'' \text{ lg } \quad 16.2 = 324$$

$$4 \quad \text{fillers } 12 \times \frac{3}{4} \quad 4'2\frac{1}{2}'' \text{ lg } \quad 30.6 \qquad\qquad = \underline{512}$$

$$\underline{9051}$$

Article 38. Weight of Two-hinged Spandrel-braced Arch Bridges.

It is still a mooted question as to the relative weights of arch and simple trusses both of the same span. An analysis made by Dr. Waddell shows results in favor of arch bridges, the ratio of weight of metal in trusses, laterals and floor system of the riveted arch span to that of same in the corresponding simple truss, riveted span being about 0.75. Another spandrel-braced arch of

290' span and rise 18% of span gives the ratio of 0.83.[1] Mr. Kunz states "the steel weight of arch bridges up to 300 feet span and of economic rise (about $\frac{1}{5}$ to $\frac{1}{7}$ of span) may be assumed the same as for simple spans. Arches of greater span and economical rise weigh from 10 to 15% less than simple spans."[2] To be on the safe side for preliminary stresses and on account of lack of data in regard to weight of existing arches the rule of Kunz will be adopted in spite of the decisive advantages of arches as advocated by Waddell. The following is a collection of weights of double track simple spans of 300' span deducted from the formulas and diagrams given by various authorities.

(a) Merriman and Jacoby's Roofs and Bridges, part I, Art. 20:

$$W = 1.85 \times 8.63(l + 1.3w - 140)$$

$$= 4400 \text{ lbs. per ft. of span.}$$

part III. Art. 27:

$$w = 1070 + 10.7l = 4880 \text{ lbs. per ft. of span.}$$

(b) Waddell's Bridge Engineering, p. 1231. From diagram

---

[1] Bridge Engineering. Waddell. P. 637.

[2] Design of Steel Bridges. Kunz. P. 354.

for Waddell's Class 50 loading which is about 1% heavier than Cooper's Corresponding class, the weight in trusses is given as 4850 lbs. per ft. of span. On page 1240, the weight is given as 3900 lbs. , on page 1243 as 4800 lbs. per linear foot of span.

(c) Design of Steel Bridges, Kunz, page 224: weight in truss equals 3540 lbs, per ft. of span. (Transformed from 200' span by Waddell's formula given on page 1286 of his Bridge Engineering) Page 225: Total weight in bridge = $w - 14l + 1200$ = 5400 lbs. per ft. of span.

(d) Modern Framed Structures, Part I, p. 116, $w = 1.85$ $(8l + 700) = 5550$ lbs. per. ft. of span.

From the above data it may be safely said that for simple trusses of 300' span, the weight of truss only is somewhere between 4000 to 5000 lbs. per ft. of span. For arches it might be lighter but as this is unreliable for comparatively short spans, it may be taken as equal to 5000 lbs. per ft. of span as the value to be used in the preliminary stresses. Hence the weight of one truss is 2500 lbs. per ft. and for arch span proper it is 300 × 2500 = 750000 lbs.

Article 39. Dead Panel Loads.

In most of the books on bridges it will be found that the dead load for spandrel braced arches is assumed to uniform throughout the whole span. Professors Merriman and Jacoby, however, opposed it strongly by saying that "in spandrel braced arch bridges, the differences between the actual panel dead loads are so large in many cases that it is desirable... To assume panel loads of unequal magnitudes. "[1]

The same statement is also endorsed by Dr. Waddell. [2] To follow this method it is necessary to know the law of variation of the weights of trusses. While it is impossible to give a rigid rule of such a variation, the general nature of which may be mastered by the following considerations.

(a) The weight is greater at the springings than at the crown.

(b) The weight of any member is carried equally to its ends, acting as a beam to give equal and reactions.

(c)The details are concentrated about the joints of the truss.

From the above facts the following method may be used in allocating the weights of the truss to the different panel points along the span. Consider the weight concentrated to each panel point to be directly proportional to the sum of the lengths of the

---

[1]　Higher Structures. P. 282.
[2]　Bridge Engineering. P. 94.

members assumed carried to the same point. This assures the constant cross sectional areas of the members and is therefore in the same degree of accuracy as the preliminary live load stresses. The aggregated lengths of the members of the arch proper (excluding the cantilever arms) is 1496. 956 feet, the weight represented by each foot of length is therefore $\frac{750000}{1496.956} = 500$ lbs. By finding the lengths of members carried to any panel load of the truss the dead panel loads at that point may be found by multiplying the sum by 500. The loads for the whole truss are given in the following table.

Table 2. Dead Panel Loads Due to Weight of Truss. (in kips)

Panel Point	Length of Members carried to point	Panel load in kips
II′	30. 5	15. 2
I′	71. 2	35. 6
0	113. 5	56. 3
I	71. 2	35. 6
II	60. 9	30. 4
III	53. 2	26. 6
IV	48. 2	24. 1
V	45. 7	22. 8
VI	31. 0	15. 5

II_1'	54.1	27.1
I_1'	81.7	40.8
0_1	62.0	31.7
I_1	81.7	40.8
II_1	67.6	33.8
III_1	57.3	28.7
IV_1	50.1	25.0
V_1	46.2	23.1
VI_1	58.8	29.5

Weight of floor system:

Weight of stringer = 900 lbs. /ft.

Weight of track   = <u>900</u> lbs. /ft.

1800 lbs. /ft.

Weight of floor beam = 9500 lbs.

Panel Load = 25 × 1800 + 9500

= 54500 lbs.

For one truss = 27250 lbs.

= 27.3 kips applied at upper panel points.

Weight of lateral system assumed at 600 lbs. per foot of span or 300 lbs. per foot of truss:

Panel load = 300 × 25

= 7500 lbs.

For upper panel = 3.8 kips

Lower panel　　$= 3.8$ kips

Weight of suspended span ($100'$) long assumed at 3000 lbs./ft. of span.

Total weight of suspended span $= 300000$ lbs.

For one truss $= 150000$ lbs.

End reaction $= 75$ kips applied at upper end panel point of cantilever arm.

Table 3. Dead Panel Loads. (in kips)

Panel Point	Truss	Floor	Lateral System	Total
II'	15.2	27.3	3.8	121.3
I'	35.6	27.3	3.8	66.7
0	56.3	27.3	3.8	87.4
I	35.6	27.3	3.8	66.7
II	30.4	27.3	3.8	61.5
III	26.6	27.3	3.8	57.7
IV	24.1	27.3	3.8	65.2
V	22.8	27.3	3.8	53.9
VI	15.5	27.3	3.8	46.6
II_1'	27.1	0.0	3.8	30.9
I_1'	40.8	0	3.8	44.6
0_1	31.7	0	3.8	35.5
I_1	40.8	0	3.8	44.6
II_1	33.8	0	3.8	37.6
III_1	28.7	0	3.8	32.5

IV_1	25.0	0	3.8	28.8
V_1	23.1	0	3.8	25.9
VI_1	29.5	0	3.8	33.3

Article 40. Dead Load Stresses.

The dead load stresses in the members of the truss will be calculated according to the law of three hinged arches. After the vertical and horizontal components of the reactions are obtained by taking moments about the three hinges, the work reduces to the simple operation of finding moments of the stresses in the members about their moment points. This, however, involves a great amount of labor because of the irregular panel loads. To overcome this difficulty, the writer has evolved a method which is believed to be simple in application and accurate in result. It should be noted that this method is applicable to any structure where the panel loads are not uniform but symmetrical about the centre. Further, the panel lengths must be uniform.

The calculations are divided into two parts, the first for the chord members and the other for the webs.

(a) Stresses in chord members.

In Table 4a there are written in the first line the different panel loads in their proper order of occurrence, starting from the center which corresponds to the second last column in the table. On the extreme right column, the panel lengths are put down with their Partial summations. For instance, the second line represents the length of two panels, third line of three panels, and so on. There will be as many these lines as there are panel loads on one half of the truss.

Draw the step lines starting from the right first cell and continue on to the left as shown. Now fill the spaces on the left of these step lines as follows:

Multiply all the panel loads by the panel length, 25. And write the products in the proper place in the first line of the table. Add to them the values equal to themselves and write the sums in the next line below, omitting the number on the right of the step lines. Add the values of the first line to those of the second line, which gives the sum to be filled in the third line. Repeat the process until all the cells will be filled by numbers obtained from successive additions of the values of the first line. Check the last number in each vertical column, by multiplying load by number on right column. Now add up the numbers in the different cells occurring in the diagonal that runs parallel to the step lines. Put the summations under the column H. This will give the sum of moments of different panel loads about the moment point of the member which is marked in the column to the left of M. As a check, add up all the numbers in different columns, and find the sum of these summations. This should equal the sum of numbers in colum M.

Next find the moment of reactions about the moment points of different members and arrange as shown in Table 4b.

For all chord members, the moment of vertical reactions is always positive, of H negative and of panel loads also negative.

In Tables 4d and 4e the positive and negative moments of stresses in all chord members are tabulated, Add them algebraically give total M, and dividing by lever arm gives the stresses.

(b) Stresses in web members.

The most important part of this method of finding stresses lies in the finding of the moments in web members. This will be described in detail.

Let D be the web member whose stress is to be found. Let

m be its moment point at distance x from A and o from P_1.
Then, moment of loads about m

$$= P_1 C + P_2 (C + p) + P_3 (C + 2p) + P_4 (C + 3p) + P_5 (C + 4p) + P_6 (C + 5p) + \text{etc.}$$

$$= P_1 C + P_2 C + P_3 C + P_4 C + P_5 C + P_6 C + \text{etc.}$$

$$+ P_2 p + P_3 \times 2p + P_4 \times 3p + P_5 \times 4p + P_6 \times 5p + \text{etc.}$$

$$= (P_1 + P_2 + P_3 + P_4 + P_5 + P_6 + \text{etc}) C +$$

$$P_2 p + P_3 \times 2p + P_4 \times 3p + P_5 \times 4p + P_6 \times 5p + \text{etc.}$$

Now the expression

$$P_2 p + P_3 \times 2p + P_4 \times 3p + P_5 \times 4p + P_6 \times 5p + \text{etc.}$$

is the same as the sum of moments of loads about the point n and is equal to the value of M in Table 4a—corresponding to members in this particular case.

Hence the moment of loads is composed of two parts: first, the product of the distance from moment point to the nearest load on the left of the section passed through the panel and the summation of loads on the left of moment section; and second, the moment M of Table 4a.

The construction of the table 4f is as follows:

Write the members in the first column as shown.

In the second column the panel loads are written in the order from left to right until the one at the center is reached. On the third column are given the partial summations of those loads; for instance, in second column, $263.5 = 111.3 + 152.2$, in the third column, $386.5 = 123 + 263.5$, in the fourth column $497.8 = 111.3 + 386.5$, and so on.

In the fourth column are given the distances of moment point of the member from the load on left of section. They are equal to the lever arm of verticals in the panel immediately on the left of the panel in which the diagonal is located.

The products of the third and fourth columns are written in the fifth column marked "1st term". In the sixth column marked "End term" are copied the numbers under column M of Table 4a, opposite the lower chord members in the same panel as the diagonals.

The sum of the fifth and sixth columns gives the moments of the loads required.

After the moments are obtained the stresses in webs are found in the same manner as for chords and are arranged in Table 4b.

It should be noted that for verticals, there is an additional column under the moment of load marked a plus sign (Table 4i). This is the moment of the upper chord panel load that will be on the left of the section for diagonals but will be on the right of the section for verticals.

The dead load stresses this obtained are checked by graphic method which is shown in plate I. In the construction of the stress diagrams, the loads were all assumed on the top chord, the stresses in verticals being afterwards corrected by the amount of lower panel loads. The stresses thus obtained are given in Table 5.

In Table 6 the dead load stresses for the same arch without cantilever arms are given. The panel loads inside the arch span are the same as for the arch with cantilevers, the only difference being the loads over the support, which is now reduced to one half its former value.

Table 4. Dead Load Stresses for the Arch With Cantilever Arms.

(a) Table of Negative Moments for Chord Members.

Member	M	152.2	111.3	123.0	111.3	99.1	90.2	84.0	80.8	
L_1U_1'	3805~3805	2782.5	3075	2782.5	2477.5	2255	2100	2020		25
$P_1L_1U_1$	10392.5-7610	5565	6150	5565	4955	4510	4200			50
$V_1P_2L_2U_1$	20055-11415	8347.5	9225	8347.5	7432.5	6765				75
$L_3L_4U_1$	32500-15220	11130	12300	11130	9910					100
$P_4L_4U_1$	47742.5-19025	13912.5	15375	13912.5						125
$V_5L_5U_1$	64600-22830	16695	18450							150
$P_6L_6U_1$	78875-26635	19477.5								175
U_6	105175-30440									200

367812.5 136980 77910 64575 41737.5 24775 13530 6300 2020 367847.5

(b) Table of Positive Moments of V for Chord Members.

($V = 891.8$ kips)

Member	U_1, L_1	U_2, L_2	U_3, L_3	U_4, L_4	U_5, L_5	U_6
Distance	25	50	75	100	125	150
Moment	22295	44590	66885	89180	111475	133770

(c) Table of Negative Moments of K for Upper Chords.

($N = 530$ kips)

Member	U_1	U_2	U_3	U_4	U_5	U_6
Ordinate	15.5	30	40.5	45	52.5	54
Moment	8745	15900	21466	25440	27825	28620

For lever chords, negative moment $= 530 \times 66$

$$= 34980 \text{ kip feet.}$$

(d)Stresses in Upper Chords.

Member	M of Load ($-$)	M of Reaction +	M of Reaction $-$	Total M ($-$)	Lever Arm	Stress
U_2'	0	0	0	0	36	0
U_1'	3805	0	0	3805	49.5	+76.9
U_1	20055	22295	6745	6605	49.5	+131.5
U_2	32500	44590	15900	3810	36	+105.6
U_3	47422.5	66885	21466	2003.5	25.5	+78.4
U_4	64500	64180	25440	960	18	+47.7
U_5	83677.5	111473	27825	237.5	13.5	+16.8
U_6	105175	133770	28620		12	0

(e)Stresses in Lever Chords.

Member	M of Load ($-$)	M of Reaction +	M of Reaction $-$	Total M ($-$)	Lever Arm	Stress
L_1'	3805	0	0	3805	43.554	-87.3
L_0'	10398.5	0	0	10392.5	55.084	-188.5
L_0	10392.5	0	34980	46372.5	66.084	-823.0
L_1	800055	22295	34980	32740	43.554	-750.0
L_2	32600	44390	34880	22800	33.192	-690.0
L_3	47422.6	66886	34980	155175	24.424	-635.0

L_4	54600	88180	34980	10400	17.716	-587.0
L_5	92877.5	111475	34980	7382.5	12.476	-548.0

(f) Table of Negative Moments for web Members.

Member	Load	Σ Loads	Distance	1st Term	End Term	M
V_1', D_2'	152.2	152.2	66.008	10160	0	10160.0
D_1'	111.3	263.5	15.000	13175	3605.5	16980.0
D_1	123.0	386.5	100.000	38680	10392.5	49042.5
V_1, D_2	111.3	497.8	91.063	45632	20055.0	65687.0
V_2, D_3	99.1	596.9	85.714	51665	22300.0	83665.0
V_3, D_4	90.2	687.01	85.000	56404	47422.5	105826.5
V_4, D_5	84.0	771.1	100.000	77110	54600.0	141710.0
V_5, D_6	85.8	861.9	225.000	191680	83877.6	275557.5

(g) Table of Positive Moments of V for Web Members. ($V = 891.8$ kips)

Member	D_1	D_2, V_1	D_3, V_2	D_4, V_3	D_5, V_4	D_6, V_5
Distance	100	116.668	136.714	160	200	360
Moment	89180	104057	121029	142690	178360	312136

Negative Moment of H for all web members

$= 530 \times 66 = 34980$ kip feet.

(h) Stresses in Diagonals.

Member	M of Load	M of Reactions		Total M (+)	Lever Arm	Stress
		+	−			
D_2'	+ 10150.0	0	0	10150	75.291	+ 134.5
D_1'	+ 15967.6	0	0	16980	89.260	+ 189.9
D_1	− 40042.3	89180	34980	8157.5	89.260	+ 57.8
D_2	− 85687.0	104057	34980	3390	75.291	+ 45.0
D_3	− 83665	121089	34980	2284	61.206	+ 38.9
D_4	− 105826	142690	34980	1884	49.665	+ 37.9
D_5	− 141710	178360	34980	1670	47.512	+ 35.1
D_6	− 275557	312136	34980	1599	97.368	+ 16.5

(i) Stresses in Verticals.

Member	M of Load		M of Reaction		Total M (+)	Lever Arm	Stress
	−	+	+	−			
V_2'	−	−	0	0	−	85.714	− 121.3
V_1'	+ 10150	− 6100	0	0	4050	91.668	− 44.2
V_0	−	−	−	−	−	−	− 299.0
V_1	65687	6110	104057	34980	9500	91.668	− 103.8
V_2	83665	5270	121029	34980	7684	95.714	− 89.5
V_3	108826.5	4900	142890	34980	6783	85.000	− 79.6
V_4	141710	5620	178360	34980	7190	100.000	− 71.9
V_5	275557.6	12130	312130	34980	13729	223.000	− 61.0
V_6	−	49.6	−	−	−	−	− 46.6

Table 5. Dead Load Stresses for the Arch With Cantilever Arms by Graphic Method. (in kips)

Graphic Method

Member	Stress	Member	Stress
U_2'	0	D_2'	+133
U_1'	+78	D_1'	+177
U_1	+130	D_1	+65
U_2	+100	D_2	+45
U_3	+80	D_3	+36
U_4	+62	D_4	+35
U_5	+28	D_5	+29
U_6	0	D_6	+30
L_1'	−89	V_2'	−121.3
L_0'	−190	V_1'	−44.2
L_0	−981	L_0	−300.0
L_1	−761	V_1	−102.4
L_2	−690	V_2	90.4
L_3	−640	V_3	77.5
L_4	−594	V_4	71.2
L_5	−560	V_5	65.1
		V_6	46.5

Table 6. Dead Load Stresses for the Arch Without Cantilever

Arms. (in kips)

(a) Table of Negative Moments for Chord Members.

Member	M	61.5	111.3	99.1	90.2	84	80.8	
V_1, D_2, L_1, U_1	1538.0	-1538	2782.5	2477.5	2255	2100	2020	25
V_2, D_3, L_2, U_2	5857.5	-3075	5565.0	4955.0	4510	4200		50
V_3, D_4, L_3, U_3	12655.5	-4613	8347.5	7432.5	6765			75
V_4, D_5, L_4, U_4	21707.5	-6150	11130.0	9910.0				100
V_5, D_6, L_6, U_5	32860.5	-7688	13912.5					125
U_6	46032.5	-9225						150
	120651.5	32288	41737.5	24775.0	13530	6300	2020	120651.

(b) Table of Positive Moments of V for Chord Members.

$V = 566.9$ kips

Member	U_1, L_1	U_2, L_2	U_3, L_3	U_4, L_4	U_5, L_5	U_6
Distance	25.0	60.0	75.0	100.0	125.0	150.0
Moment	14172.6	28345.0	42517.5	56690.0	70862.5	86035.0

(c) Table of Negative Moments of H for Upper Chords.

$H = 721$ kips

Member	U_1	U_2	U_3	U_4	U_5	U_6	All $L'S$
Distance	16.5	30	40.5	48	52.5	54	66
Moment	11900	21630	29200.5	34608	37865	38964	47566

(d) D. L. Stress in Upper Chords.

Member	M of Load (−)	M of Reactions		Total M (+)	Lever Arm	Stress
		+	−			
U_1	1538.0	14172.5	11900	734.5	49.5	+41.8
U_2	5857.5	28345.0	21630	857.5	36.0	+23.8
U_3	12655.5	42517.5	29200	662.0	25.5	+26.1
U_4	21707.5	56690.0	34608	374.5	18.0	+20.8
U_5	32860.5	70862.5	37835	167.0	13.5	+12.4
U_6	46032.5	86035.0	38964	0	12.0	0

(e) Stresses in Lower Chords.

Member	M of Load (−)	M of Reactions		Total M	Lever Arm	Stress
		+	−			
L_0	0	0	47566	−47566.0	55.084	−863.3
L_1	1538.0	14172.5	"	−34932.0	43.554	−801.4
L_2	58575.0	28345.0	"	−25078.5	33.192	−757.6
L_3	12655.5	42517.5	"	−177.4.0	24.424	−783.7
L_4	21707.5	56690.0	"	−12583.5	17.715	−710.2
L_5	328605.0	70862.5	"	−9564.0	13.476	−710.1

(f) Table of Negative Moments for Web Members.

Member	Load	ΣLoads	Distance	1st Term	2nd Term	M
D_1	61.5	61.5	100.00	6150	0.	6150.0

V_1, D_2	111.3	172.8	91.67	15820	1538.0	17358.0
V_2, D_3	99.1	271.9	85.71	23310	5857.5	29167.5
V_3, U_4	90.2	362.1	85.00	20805	12655.5	43460.5
V_4, D_5	84.0	446.1	100.00	44610	21707.5	66317.5
V_5, D_6	80.8	526.9	225.00	118450	32860.5	151310.0

(g) Table of Positive Moments of V for Web Members.

Member	D_1	D_2, V_1	D_3, V_2	D_4, V_3	D_5, V_4	D_6, V_5
Distance	100	116.67	135.71	160	200	350
Moment	56690	66132.00	76620.00	90733	113380	198200

Negative moment of H for all web members

$$= 721 \times 66 = 47566 \text{ kip feet.}$$

(h) Stresses in Diagonals.

Member	M of Load (−)	M of Reactions		Total M	Lever Arm	Stress
		+	−			
D_1	6150	56690	47566	+ 2974	89.26	− 332.2
D_2	17358	66132	47566	+ 1208	75.29	+ 16.1
D_3	29167.5	76620	47566	− 113.5	61.21	− 1.8
D_4	43460.5	90733	47566	− 293.	49.67	− 6.9
D_5	66317.5	113380	47566	− 503.5	47.51	− 10.6
D_6	151310.	198200	47566	− 676.	97.36	− 6.9

(i) Stress in Verticals.

Member	M of Load		M of Reactions		Total M (+)	Lever Arm	Stress
	−	+	+	−			
V_0	−	−	−	−	−	−	− 72. 1
V_1	17358. 0	6100	66132	47566	7208. 0	91. 67	− 78. 8
V_2	29167. 5	5270	76620	47566	5156. 5	85. 71	− 60. 2
V_3	43460. 5	4900	90733	47566	4606. 5	85. 00	− 54. 2
V_4	66317. 5	5520	113380	47566	5016. 5	100. 00	− 50. 2
V_5	151310. 0	12130	1985200	47566	11454. 0	225. 00	− 50. 5
V_6	−	−	−	−	−	−	− 46. 5

CHAPTER XVI. LIVE LOAD STRESSES

Article 41. Introduction.

The live load stresses will be calculated by the method of Merriman and Jacoby and that of Turnesure. The stress coefficients (stresses in members for a unit load) for the arch without and with cantilever arms are first computed for all the positions of loads and then multiplied by the proper live panel loads to get the actual stresses. In each case the analytic method is first considered and then checked by the graphic result.

Article 42. Stress Coefficients in the Arch without Cantilever Arms.

(I) Stresses Due to Vertical and Horizontal Forces.

The values of T for a unit horizontal thrust (acting outward) of all the members of one half the truss are given in Table 7. The

Table 7. Stresses for $H = 1$.

Member	y	t	T	Member	y	t	T
U_1	16.5	49.5	$-.3333$	D_1	66	89.260	$+.7394$
U_2	30	36	$-.8334$	D_2	66	75.291	$+.8766$
U_3	40.5	25.5	-1.5883	D_3	66	61.206	$+1.0783$
U_4	48	18	-2.6667	D_4	66	49.665	$+1.3289$
U_5	52.5	13.5	-3.8889	D_5	66	47.512	$+1.3891$
U_6	54	12	-4.5000	D_6	66	97.362	$+.6779$
L_0	66	55.084	$+1.1982$	V_0	–	–	$-.6600$
L_1	66	43.554	$+1.5153$	V_1	66	91.668	$-.7200$
L_2	66	33.192	$+1.9885$	V_2	66	85.714	$-.7700$
L_3	66	24.424	$+2.7022$	V_3	66	85.000	$-.7765$
L_4	66	17.715	$+3.7255$	V_4	66	100.000	$-.6600$
L_5	66	13.476	$+4.8977$	V_5	66	225.000	$-.2933$
				V_6	–	–	0

Table 8. Stresses for $H = 0$.

(a) For Unit Left-hand Reaction.

Member	x	t	S'	Member	x	t	S'
U_1	25	49.5	−.5050	D_1	100.000	89.260	+1.1203
U_2	50	36.0	−1.3889	D_2	116.668	75.291	+1.5497
U_3	75	25.5	−2.9411	D_3	133.714	61.206	+2.2.73
U_4	100	16.0	−5.5556	D_4	160.000	47.665	+3.2216
U_5	125	13.5	−9.2594	D_5	200.000	47.512	+4.2094
U_6	150	12.0	−12.500	D_6	350.000	97.362	+3.5948
L_0	0	55.084	0	V_0	−	−	−1.0000
L_1	25	43.554	+.5740	V_1	116.668	91.668	−1.2729
L_2	50	33.192	+1.5062	V_2	135.714	85.714	−1.5833
L_3	75	24.424	+3.0707	V_3	160.000	85.000	−1.8823
L_4	100	17.715	+45.6449	V_4	200.000	100.000	−2.0000
L_5	125	13.476	+9.2760	V_5	350.000	225.000	−1.5556
				V_6	−	−	0

(b) For Unit Right-hand Reaction.

Member	x'	t	S'_r	Member	x'	t	S'_r
U_1	275	49.5	−5.5555	D_1	200.000	89.260	+2.2406
U_2	250	36.0	−6.9445	D_2	183.332	75.291	+2.4349
U_3	225	25.5	−8.8234	D_3	164.286	61.206	+2.5842
U_4	200	18.0	−11.1111	D_4	140.000	49.665	+2.8189

U_5	175	13.5	-12.9630	D_5	100.000	47.512	$+2.1047$
U_6	150	12.0	-12.5000	D_6	-50.000	97.362	$-.5120$
L_0	300	55.084	$+5.4463$	V_0	–	–	0
L_1	275	43.554	$+6.3140$	V_1	183.332	91.668	-2.0000
L_2	250	33.192	$+7.5323$	V_2	164.286	85.714	-1.9167
L_3	225	24.414	$+9.2122$	V_3	140.000	85.000	-1.6471
L_4	200	17.715	$+11.2897$	V_4	1400.000	100.000	-1.0000
L_5	175	13.475	$+12.9865$	V_5	-50.000	225.000	$+.2222$
				V_6	–	–	0

values of S' and S'_r for $H = 0$ and unit left and right reactions are given in Table 8. On Plate 2 is drawn a truss diagram of the arch with the principal elements of the members marked thereon. Stresses for $H = 1$ and $V = 1$ are also given as a check of the above mentioned tables.

The values of S' thus obtained are the stresses in members which are located outside two symmetrical unit loads. They are the same for whatever the positions of the loads so long as the latter are placed at equal distances from the center of the span. For members which lie between the loads the stresses are found from the expression $S' = kl \div t$. These values are given in Table

9. The values of $\frac{1}{t}$ or $\frac{300}{t}$ are first found for every member of one half the truss. Dividing those numbers by 12 gives the value of S' for the loads at points I and XI of the span. Call these values C. Now add C to S' for $k = \frac{1}{12}$ or for loads at I and XI, the stresses S' for $k = \frac{2}{12}$ or loads at II and X are obtained, add C to S' for $k = \frac{2}{12}$, S' for $k = \frac{3}{12}$ are obtained, and so on until the points V and VII are reached. Now multiply C by 5. If they are the same as the stresses thus obtained for $k = \frac{5}{12}$ the correctness of the whole work is ensured.

Table 9. Stresses in Members Lying Between the Loads Placed Symmetrically About Center of Span.

Member	t	$\frac{1}{t}$	S' for $k =$				
			$\frac{1}{12}$	$\frac{2}{12}$	$\frac{3}{12}$	$\frac{4}{12}$	$\frac{5}{12}$
U_1	49.5	6.0606	.50505	1.01010	1.51515	2.02020	2.52525
U_2	36.0	8.3334	.69445	1.38890	2.08335	2.76781	3.47225
U_3	25.5	11.765	.98040	1.35080	2.94120	3.92160	4.90200
U_4	18.0	16.667	1.3889	2.7778	4.1667	6.5556	6.9445

U_5	13.5	22.222	1.3519	3.7038	5.5557	7.4076	9.2595
U_6	12.0	25.000	2.0833	4.1666	6.2499	8.3332	10.4165
L_0	55.084	5.4462	.46386	.90772	1.36158	1.81544	2.26930
L_1	43.554	6.8880	.57400	1.14800	1.72200	2.29600	2.87000
L_2	33.192	9.0388	.75323	1.50646	2.26969	3.01292	3.76615
L_3	24.424	12.283	1.0236	2.0472	3.0708	4.0944	5.1180
L_4	17.715	16.934	1.4112	2.8224	4.2336	5.6448	7.0560
L_5	13.476	22.262	1.8552	3.7104	5.5656	7.4208	9.2760
D_1	89.260	3.3609	.28008	.56016	.84024	1.12032	1.40040
D_2	75.291	3.9845	.33204	.66408	.99612	1.32816	1.6602
D_3	61.206	4.9016	.40846	.81692	1.22538	1.63384	2.04230
D_4	49.665	5.0404	.50337	1.00674	1.51011	2.01348	2.51685
D_5	47.512	6.5141	.52517	1.05234	1.57851	2.10468	2.63085
D_6	97.362	3.0813	.25677	.51364	.77031	1.02708	1.28385
V_0	–	–	–	–	–	–	–
V_1	91.668	3.2727	.27272	.54544	.81816	1.09088	1.36360
V_2	85.714	3.5000	.29167	.58334	.87501	1.15668	1.45835
V_3	86.000	3.3294	.29411	.58822	.88233	1.17644	1.47055
V_4	100.000	3.0000	.25000	.50000	.75000	.10000	.12500
V_5	325.000	1.3333	.11111	.22222	.33333	.44444	.55555
V_6	–	–	–	–	–	–	–

The stresses in Tables 8 and 9 may now be combined to give the stresses due to vertical forces for two loads occupying every pair of symmetrical panel points along the span. These are given in Table 10 and are obtained in the following manner: After the table is constructed draw in the step lines as shown. Fill in the cells above the step lines by the values of S' given in Table 8a. For the cells below the step lines copy the numbers below the step lines of Table 9, taking the numbers from those cells that are similarly located as the table in construction. It will be seen that there are two good checks for the table thus obtained. For the upper chords the difference between the numbers on each side of the vertical step line must be equal to the number in the column under the point I and XI, along the same horizontal line. For lower chords these two numbers are equal. If careful work be done the complete table for chords may be obtained by calculating the numbers in column I, XI, only, all the rest may be found by successive additions.

In the above table the value of S' for loads at 0 and XII are omitted. Under this condition only V_0 and V_{12} are stressed, the value of each being -1.0000.

Table 10. Stresses in Members for Two Symmetrical Loads ($H = 0$).

($U-,L+,D+,V-$)

Member	Loads at					
	I , XI	II , X	III , IX	IV , VIII	V , VII	VI
U_1	.5050	.5050	.5050	.5050	.5050	.5050
U_2	.6945	1.3889	1.3889	1.3889	1.3889	1.3889
U_3	.9804	1.9608	2.9411	2.9411	2.9411	2.9411
U_4	1.3889	2.7778	4.1667	5.5556	5.5556	5.5556
U_5	1.8519	3.7038	5.5557	7.4076	9.2594	9.2594
U_6	2.0833	4.1666	6.2499	8.3332	10.4165	12.5000
L_0	0	0	0	0	0	0
L_1	.5740	0.5740	.5740	.5740	.5740	.5740
L_2	.7532	1.5065	1.5065	1.5065	1.5065	1.5065
L_3	1.0236	2.0472	3.0707	3.0707	3.0707	3.0707
L_4	1.4112	2.8224	4.2336	5.6449	5.6449	5.6449
L_5	1.8552	3.7104	5.5656	7.4208	9.2760	9.2760

Member	Loads at					
	I , XI	II , X	III , IX	IV , VIII	V , VII	VI
D_1	1.1203	1.1203	1.1203	1.1203	1.1203	1.1203
D_2	.3320	1.5497	1.5497	1.5497	1.5497	1.5497

D_3	.4085	.8169	2.2173	2.2173	2.2173	2.2173
D_4	.5034	1.0067	1.5101	3.2216	3.2216	3.2216
D_5	.5262	1.0523	1.5785	2.1047	4.2094	4.2094
D_6	.2568	.5135	.7703	1.0271	1.2839	3.5948
V_0	1.0000	1.0000	1.0000	1.0000	1.0000	1.0000
V_1	1.2729	1.2729	1.2729	1.2729	1.2729	1.2729
V_2	.2915	1.5833	1.5833	1.5833	1.5833	1.5833
V_3	.2941	.5882	1.8823	1.8823	1.8823	1.8823
V_4	.2500	.5000	.7500	2.0000	2.0000	2.0000
V_5	.1111	.2222	.3333	.4444	1.5556	1.5556
V_6	–	–	–	–	–	2.0000

Table 11. Values of T^2L and STL.

Member	U_1	U_2	U_3	U_4	U_5	U_6	
L	25.000	25.000	25.000	25.000	25.000	25.000	
T	0.3333	0.8334	1.5883	2.6667	3.8889	4.5000	
TL	8.333	20.835	39.708	66.668	97.223	112.500	
T^2L	2.776	17.363	63.067	177.783	378.092	506.244	1145.337
0, XII	0	0	0	0	0	0	0
I, XI	4.208	14.469	38.930	92.594	160.050	234.368	564.619
II, X	4.208	28.938	77.860	185.188	360.160	468.736	1125.090
IV, IX	4.208	28.938	116.789	277.782	540.150	703.104	1670.971

Ⅳ, Ⅷ	4.208	28.938	116.789	370.383	720.020	937.472	2177.810
Ⅴ, Ⅶ	4.208	28.938	116.789	370.383	900.220	1171.840	2592.378
Ⅵ, Ⅵ	4.208	28.938	116.789	370.383	900.220	1406.233	2826.551

Member	D_0	D_1	D_2	D_3	D_4	D_5	
L	29.964	28.413	27.116	26.101	25.402	25.046	
T	1.1982	1.5153	1.9885	2.7022	3.7255	4.8977	
TL	25.801	43.056	53.921	70.531	94.636	122.663	
T^2L	42.003	66.244	107.244	190.591	352.567	600.358	1359.387

0, Ⅻ	0	0	0	0	0	0	0
Ⅰ, Ⅺ	0	24.714	40.614	72.196	133.540	227.563	498.626
Ⅱ, Ⅹ	0	24.714	81.228	144.390	267.080	455.126	972.538
Ⅳ, Ⅸ	0	24.714	81.230	216.585	400.620	682.589	1405.838
Ⅳ, Ⅷ	0	24.714	81.230	216.585	534.160	910.252	1766.941
Ⅴ, Ⅶ	0	24.714	81.230	216.585	534.200	1137.815	1994.544
Ⅵ, Ⅵ	0	24.714	81.230	216.585	534.200	1137.820	1994.544

Member	D_1	D_2	D_3	D_4	D_5	D_6	
L	55.456	43.830	35.711	30.806	28.413	27.731	
T	.7394	.8766	1.0783	1.3289	1.3891	0.6779	
TL	41.005	38.420	38.508	40.938	39.468	18.798	
T^2L	30.318	33.679	41.524	54.403	54.825	12.743	225.492

0, XII	0	0	0	0	0	0	
I , XI	45.938	12.756	15.730	20.608	20.768	4.827	120.627
II , X	45.938	59.540	31.460	41.216	41.536	9.654	229.344
IV , IX	45.938	59.540	85.384	61.824	62.304	14.481	329.471
IV , VIII	45.938	59.540	85.384	151.887	83.072	19.308	425.129
V , VII	45.938	59.540	85.384	131.887	166.138	24.136	513.022
VI , VI	45.938	59.540	85.384	131.887	166.138	67.577	556.464

Member	V_0	V_1	V_2	V_3	V_4	V_5	V_6	
L	66.00	49.5	36.0	25.5	18.0	13.5	12.0	
T	.6600	.7200	.7700	.7765	.6600	.2933	0	
TL	43.559	35.639	27.720	19.800	11.880	3.960	0	
T^2L	29.743	25.659	21.344	15.374	7.841	1.162	0	100.129

0, XII	43.559	0	0	0	0	0	0	43.559
I , XI	43.559	45.364	8.086	5.823	2.970	4.400	0	110.202
II , X	43.559	45.364	43.889	11.646	5.940	8.799	0	150.197
IV , IX	43.559	45.364	43.889	37.270	8.910	13.199	0	192.191
IV , VIII	43.559	45.364	43.889	37.270	23.760	17.599	0	211.441
V , VII	43.559	45.364	43.889	37.270	23.760	6.160	0	200.002
VI , VI	43.559	45.364	43.889	37.270	23.760	6.610	0	200.002

(II) Horizontal Thrust by Analytic Method.

After the values S' and T are obtained, the products T^2L and $S'TL$ may be performed for use in the approximate formula of horizontal thrust, which being $H = \dfrac{S'TL}{T^2L}$.

These products are given in Table 11, in which the signs of S' and T are omitted because the products $S'TL$ and T^2L are always positive. Same checks may be found as for Table 10. The summations of $S'TL$ and T^2L are made for each group of members and are given in the same table. In Table 12 are given the summations for the whole half truss, from which the preliminary values of H may be obtained for all the positions of the loads. The value of $\Sigma S'TL$ is for one half of truss only but as it is produced by two symmetrical loads, it will be the summation for the whole truss. The value of ΣT^2L does not include the effect of the symmetrical loads and therefore should be multiplied by 2 to obtain the result for the whole truss. The horizontal thrust is then obtained from the expression $H = \dfrac{\sum S'TL}{2\sum T^2L}$.

Table 12. Values of Horizontal Thrust.

	Member	U	L	D	V	Σ
	$0, \mathrm{XII}$	0	0	0.	43.559	43.559
	I, II	564.619	498.626	120.627	110.202	1294.074
	II, X	1126.090	972.538	229.344	159.197	2486.169
STL for Loads at	$\mathrm{III}, \mathrm{IX}$	1670.971	1405.838	329.471	192.191	3598.471
	$\mathrm{IV}, \mathrm{VIII}$	477.810	1766.941	425.129	211.441	4581.321
	V, VII	2592.378	1994.544	513.022	200.002	5299.946
	VI	2826.771	1994.544	556.469	200.002	5577.781
	$T^2 L$	1145.327	1359.387	227.492	100.129	2833.335

$$2 \sum T^2 L = 5664.670$$

Member	0	I	II	III	IV	V	VI
H	.0077	.22844	.43889	.63624	.80874	.93500	.98465

The influence lines of $\Sigma S'TL$ for chord and web members are plotted in Diagram 4, the scales used therein are not the same as the aim is rather to show the distribution than the magnitude of these quantities. The values of $2 \sum T^2 L$ are also marked. It will be observed that the lines are almost straight from the panel points 0 to IV showing that the forms are pretty close to hyperbolas.

In Diagram 5 the values of $\Sigma S'TL$ for the members are added together showing the distribution of same among chords and webs. The influence of chord members upon the values of H is thus seen to be very great, the importance increasing toward the center of span. The effect of verticals is very small accounting to less than 10% of the total.

Besides the above, the following statements may be made:

(1) The form of the curve for total $S'TL$ of the truss is very closely represented by an hyperbola.

(2) The value of $\Sigma S'TL$ for a load at center of span is very nearly equal to that of $2\Sigma T^2 L$. This first together with the previous one gives rise to the approximate method of finding H as proposed by the writer. See Article 31.

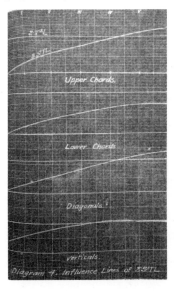

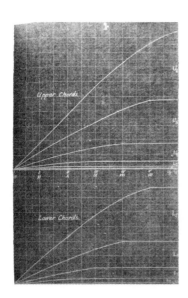
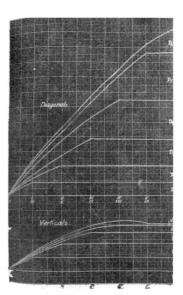

(3) The value of $\Sigma S'TL$ for verticals is very uniform for different positions of loads.

The distribution of $S'TL$ in different members is shown in Diagram 6. The effect of U_6 upon the value of $\Sigma S'TL$ is so great that it contributes almost one half of the total value. This leads to the method of calculating preliminary temperature stresses.

(Ⅲ) Horizontal Thrust by Graphic Method.

As said in the introduction the stress coefficients are obtained by two methods, analytic and graphic. The method of Merriman and Jacoby will therefore next be considered. In Table

11 are given the values of TL of members. These will be the change in lengths in members if the cross sectional area be assumed equal to 1 sq. in. and E equal to 1 unit per sq. in. , the deformations being expressed in fact. Multiply these values of TL by 12 thereby obtaining the deformations in inches as given in Table 13. With these axial deformations a displacement diagram (H and Hh fixed) is constructed on Plate III shown by dotted lines. Measuring the vertical components of the displacement of upper panel points and dividing them by twice the horizontal distance from a' to H', the values of horizontal thrusts are obtained

Table 13. Axial Deformations for $H = 1$.

Member	TL feet	TL inches	Member	TL feet	TL inches
U_1	6.33	$-.100$	D_1	41.01	$+.493$
U_2	20.84	$-.250$	D_2	38.42	$+.461$
U_3	39.71	$-.476$	D_3	38.51	$+.462$
U_4	66.68	$-.800$	D_4	40.94	$+.491$
U_5	97.22	-1.167	D_5	39.47	$+.474$
U_6	112.50	-1.350	D_6	18.80	$+.226$
L_0	35.89	$+.431$	V_0	43.56	$-.523$
L_1	43.06	$+.517$	V_1	35.64	$-.427$

L_2	53.92	+.647	V_2	27.72	-.332
L_3	70.53	+.846	V_3	19.80	-.238
L_4	94.64	+1.135	V_4	11.88	-.143
L_5	122.66	+1.472	V_5	3.96	-.048
			V_6	0	0

for the different positions of the load. The result is as follows:

$$\delta' = 2 \times 33.35 = 66.70.$$

	0	**I**	**II**	**III**	**IV**	**V**	**VI**
$\delta =$	.52	15.16	29.53	42.53	54.08	63.28	66.60
$H =$	.0077	..2275	.442	.638	.810	.948	.998

The values of H check fairly closely with those obtained by the analytic method. The discrepancy is unavoidable due to the limited space available for the drawings.

To increase the accuracy of the displacement diagram the writer has evolved a method which permits of constructing a diagram of twice its size without enlarging the size of paper. Before going to give the method it will be necessary to point out some of the salient points that deserve consideration.

(1) It is the ratio of deflections that is required so that any attempt at increasing the scale of arch is useless unless the

accuracy of the drawing be carried to such a degree as is consistent with the scales used.

(2) As the displacement diagram is based on the construction of parallel and perpendicular lines, the greatest care must be taken to draw the lines truly parallel and perpendicular. For this reason any attempt to draw a line parallel to another one that is shorter than the line itself is unwarrantable.

(3) Since there are many deflections to measure, the process must not be complicated to such an extent as to make errors in reading.

(4) The method must be simple and it must show the whole field so as to make apparent any discrepancies that might result from any source of error.

(5) The method must not require any greater size of paper than that required by the ordinary method. The writer must confess that his method is not applicable to any other kinds of displacement diagrams, it being useful only in two-hinged spandrel-braced arches due to the following reasons:

① There is no rotation required so that the resulting diagram is just a Willet diagram.

②The form of the displacement diagram is more or less similar to the form of the truss.

③The lengths of construction lines, i. e. , parallels and perpendiculars are considerably longer than the elongations and contractions in members. This is the reason why the scale of the diagrams is always small in order to accomodate the drawing in a convenient size of paper.

④The truss is symmetrical about its center line.

On a piece of paper (Refer to Plate IV) draw the four boundary lines which will form the base lines to which all the parallels and perpendiculars will be referred. Draw an outline of the truss using any scale that is advisable and letter up all the panel points. Use a scale as large as possible, draw in the panel of the arch at the springing and locate it on the paper in such a way that it is everywhere perpendicular to that of the small truss diagram, as shown by Aaco in Plate IV. From c drop a perpendicular to Aa meeting at c' and on Cc measure off Cd equal to the length of the right vertical of the second panel and join c' d. In the same way find f, g, h, d', f', g' and draw the diagonals as shown. Join A to each of the points c, d, e, etc. , on *Cc*. Now there is formed a diagram in which all the panels of the half truss have been superposed on each other and each is perpendicular to its corresponding ones on the truss diagram. It is therefore equivalent to enlarging the truss diagram to a larger scale, the largest scale obtainable on the paper. Hereafter all the parallels and perpendiculars will be drawn from this diagram, the truss diagram is only used to show the letters of panel points.

Now construct the Willet diagram in the ordinary way by taking *H* as fixed in position and *Hh* fixed in direction, until some of its constructing lines strike the boundary line on account of lack of space on the paper. Now suppose there is a considerable space in the paper beyond the boundary lines and the diagram is extended beyond the limited space. If the diagram is folded over the paper along the boundary line and the lines of the extended portion are projected on the paper it will be seen that all the new lines are now making an angle with the boundary line equal to that made by them and the boundary line before they pass out the paper. For instance, if, in the figure, *ab* is the boundary line and *cd* is the constructing line that passes out the paper, making an angle α with *ab*, after the upper portion is

folded over the lower portion
along *ab*, *cd* takes the new
position of on and making the
same angle α with the line *ab*.

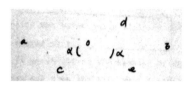

Hence by folding once we can accomodate a diagram on a
paper that will afford enough space only for one half of th diagram
by using the ordinary method. It will be observed that no trouble
need be taken to measure the angles made by constructing lines
with the boundary because in the second operation, we can use
the other half of the truss where every member makes the same
angle with horizontal as do the corresponding members of the half
we are considering.

It will be seen that the method as described has several
objections. They are the following:

(1) As the diagram is nearly of the same shape as that of the
truss diagram, the folding will cause the vertical lines (in the
right of the diagram in this case) of the second part to overlap
those of the first part and thereby causing confusion which
violates one of the principles stated above.

(2) It necessitates the drawing of the second half of truss

and then the big panel diagram corresponding to it.

To get rid of these objections the following expedient has been evolved.

Let AB be the boundary line, ag, ac and dc be the constructing lines that would appear should the paper have enough space beyond the 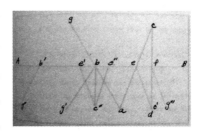 boundary. Now fold the upper portion upon the lower one along AB, c will have the new position of c' on fd, and ec' makes the same angle with AB as does ae. Similarly $g'b$ makes the same angle with AB as ab. It is obvious that fd and fc' overlap each other. Now suppose the portion folded (b,e,f,c',g') be shifted along to the left until f occupies the position of b, terminal point of ab, e at e', c' at c'' and g' at g''. Then $e'c''$ must be parallel to ec' and $b'g''$ parallel to bg'. Again, suppose the left portion $bc''e'g''$ be folded on the right portion along the vertical line bc'', we have the new figure $bc''e''g'''$, $c''e''$ making same angle with bc'' as does $e'c''$.

Then $\qquad \angle bc''e'' = \angle bc''e' = \angle fc'e = \angle fce.$

Hence $c''e''$ is parallel to ac. Similarly $f''g'''$ is parallel to bs. Hence follows the construction:

Continue the diagram until some of its constructing lines as ac, ag, dc, etc. , strike the boundary AB. On b draw a line bc'' parallel to df, on f a line parallel to ab. Find point e'' by making fe'' equal to be. Draw $e''c''$ parallel to ae until it intersects the line bc'' at c''. From c'' and fg''', the rest of the diagram can be constructed by inverting the drawing paper or by changing the signs of alterations in lengths of the different members. The latter method is to be preferred. After the diagram is finished, draw a vertical through H, cutting boundary line at H'''. Measure off $K''H'''$ equal $C''H''$ and draw a vertical through H''' cutting the horizontal line Aa' at v. Then $va' = \dfrac{1}{2}\delta'$ The $\delta's$ may be measured off in the ordinary way.

By the use of this method the following result is obtained:

$$\delta' = 2 \times 33.75 = 67.50$$

0	I	II	III	IV	V	VI
$\delta = .053$	15.40	29.62	42.82	54.63	63.08	66.41
$= .007$	.228	.439	.635	.809	.935	.984

(Ⅳ) Stress Coefficients by Analytic Method.

Having obtained the values of H for different positions of loads, the stress coefficient for H may be obtained by multiplying T for the members by the values of the thrust. Since the thrusts from both springings are equal for any position of load it will be sufficient to consider only one half of the panel points of the truss. On Table 14 are given the stresses in members due to horizontal thrust (acting inward) corresponding to the positions of loads.

Table 14. Stresses Due to Horizontal Thrust. (Acting inward)

Member	U_1	U_2	U_3	U_4	U_5	U_6
T	+.3333	+.8334	+1.5883	+2.6667	+3.8889	+4.5000
0	.0026	.0064	.0123	.0206	.0300	.0347
I	.0762	.1904	.3628	.6092	.8884	1.0280
II	.1463	.3658	.6971	1.1707	1.7068	1.9750
III	.2117	.5294	1.0089	1.6940	2.4704	2.8587
IV	.2696	.6740	1.2845	2.1566	3.1451	3.6393
V	.3119	.7797	1.4860	2.4960	3.6385	4.2102
VI	.3282	.8205	1.5639	2.6258	3.8292	4.4309

Member	L_0	L_1	L_2	L_3	L_4	L_5
T	-1.1982	-1.5153	-1.9885	-2.7022	-3.7255	-4.8977

0	.0092	.0117	.0153	.0208	.0287	.0377
I	.2737	.3462	.4543	.6173	.8511	1.1188
II	.5259	.6651	.8728	1.1860	1.6351	2.1496
III	.7611	.9626	1.2632	1.7166	2.3666	3.1112
IV	.9690	1.2255	1.6082	2.1854	3.0130	3.9610
V	1.1210	1.4177	1.8606	2.5282	3.4856	4.5822
VI	1.1798	1.4921	1.9580	2.6608	3.6683	4.8224

Member	D_1	D_2	D_3	D_4	D_5	D_6
T	$-.7394$	$-.8766$	-1.0783	-1.3289	-1.3891	$-.6779$
0	.0057	.668	.0083	.0102	.0107	.0052
I	.1689	.2002	.2463	.3036	.31473	.1549
II	.3245	.3847	.4733	.5832	.6097	.2975
III	.4697	.5568	.6850	.8442	.8824	.4306
IV	.5980	.7089	.8721	1.0723	1.1234	.5482
V	.6918	.8901	1.0089	1.2433	1.2996	.6342
VI	.7281	.8631	1.0618	1.3085	1.3678	.6675

Member	V_0	V_1	V_2	V_3	V_4	V_5	V_6
T	$+.6600$	$+.7200$	$+.7700$	$+.7765$	$+.6600$	$+.2933$	0
0	.0051	.0055	.0059	.0060	.0051	.0023	0
I	.1508	.1645	.1759	.1774	.1508	.0670	0
II	.2897	.3160	.3379	.3408	.2897	.1287	0

III	.4193	.4574	.4891	.4932	.4193	.1863	0
IV	.5337	.5823	.6227	.6280	.5337	.2372	0
V	.6175	.6736	.7204	.7265	.6175	.2744	0
VI	.0499	.7089	.7582	.7645	.6499	.2888	0

Table 10 gives the values of stresses due to vertical forces ($H = 0$) in the different members of the truss for a concentrated load occupying different panel points along the span. For members on the left of the load the stresses are obtained form S' for unit left hand reaction (Table 6a) and for members on the right of the load they are obtained from $S'r$ for unit right reaction (Table 6b). The details for the calculation of the table are as follows:

After completing the outline of the table, draw in the step lines as shown. Divide the values of S' for different members given on Table 6a by 12 and write the results in the line where the load is at the point XI. Call this value $\frac{S'}{12}$ equal to C. Add C to the stresses obtained for load at point XI, the stresses obtained for load at X is obtained. Adding c to the stresses thus obtained, the stresses for the load at XI are obtained, repeating the process until the step lines are reached. This completes the large part of the table by filling in all the cells below the step lines. Next *divide* $S'r$ for unit right reaction (Table 6b) by 12 and write the results in the air at line of the table. Call this value $\dfrac{S'r}{12} = K$.

Table 15. Stresses Due to Vertical Forces.

(a) Upper Chords. (All negative)

Load at	U_1	U_2	U_3	U_4	U_5	U_6
0	0	0	0	0	0	0
I	.4630	.5787	.7353	.9259	1.0803	1.0417
II	.4209	1.1574	1.4706	1.8518	2.1605	2.0833
III	.3788	1.0417	2.2059	2.7777	3.2408	3.1250
IV	.3367	.9259	1.9608	3.7038	4.3211	4.1667
V	.2946	.8102	1.7157	3.2408	5.4013	6.2084
VI	.2525	.6944	1.4706	2.7778	4.6297	6.2502
VII	.2105	.5787	1.2255	2.3149	3.8581	5.2085
VIII	.1634	.4630	.9804	1.8519	3.0865	4.1668
IX	.1263	.3472	.7352	1.3889	2.3149	3.1251
X	.0842	.2315	.4902	.9259	1.5432	2.0834
XI	.0421	.1157	.2451	.4630	.7716	1.0417
XII	0	0	0	0	0	0

(b) Lower Chords. (All positive)

Load at	L_0	L_1	L_2	L_3	L_4	L_5
0	0	0	0	0	0	0
I	0	.5261	.6277	.7677	.9408	1.0822
II	0	.4783	1.2554	1.5364	1.8816	2.1644
III	0	.4305	1.1299	2.3031	2.8224	3.2467
IV	0	.3826	1.0043	2.0472	3.7632	4.3289
V	0	.3348	.8788	1.7913	3.2928	5.4110
VI	0	.2870	.7532	1.5354	2.8224	4.6380
VII	0	.2392	.6277	1.2795	2.3520	3.8650
VIII	0	.1913	.5022	1.0236	1.8816	3.0920
IX	0	.1435	.3766	.7677	1.4112	2.3190
X	0	.0957	.2511	.5118	.9408	1.5460
XI	0	.0478	.1255	.2559	.4704	.7730
XII	0	0	0	0	0	0

(c) Diagonals. (All positive)

Load at	D_1	D_2	D_3	D_4	D_5	D_6
0	0	0	0	0	0	0
I	1.0270	.2029	.2237	.2349	.1764	−.0426
II	.9336	1.2914	.4474	.4698	.3508	−.0852
III	.8402	1.1623	1.6630	.7047	.5262	−.1278

IV	.7469	1.0331	1.4782	2.1478	.7016	-.1704
V	.6535	.9040	1.2935	1.8793	2.4555	-.2130
VI	.5602	.7748	1.1087	1.6108	2.1047	1.7974
VII	.4668	.6457	.9239	1.3424	1.7539	1.4979
VIII	.3734	.5166	.7391	1.0739	1.4031	1.1983
IX	.2801	.3874	.5543	.8054	1.0523	.8987
X	.1867	.2583	.3696	.5369	.7016	.5991
XI	.0934	.1291	.1848	.2685	.3508	.2996
XII	0	0	0	0	0	0

(d) Verticals. (All negative)

Load at	V_0	V_1	V_2	V_3	V_4	V_5	V_6
0	1	0	0	0	0	0	0
I	.9167	1.1668	.1597	.1373	.08333	.0185	0
II	.8333	1.0607	1.3195	.2745	.16666	.0370	0
III	.7500	.9546	1.1876	1.4117	.2500	.0556	0
IV	.6667	.8486	1.0556	1.2549	1.3333	.0741	0
V	.5933	.7426	.9237	1.0980	1.1667	.9073	0
VI	.5000	.6364	.7917	.9412	1.000	.7777	1.000
VII	.4167	.5304	.6598	.7843	.8333	.6481	0
VIII	.3333	.4243	.5278	.6274	.6667	.5185	0
IX	.2500	.3182	.3959	.4706	.5000	.3889	0

| X | .1667 | .2121 | .2639 | .3137 | .3333 | .2592 | 0 |
| XII | .0833 | .1061 | .1320 | .1569 | .1667 | .1296 | 0 |

Add K to the stress for load at I the stresses for load at II is obtained and similarly obtain the stresses for all the positions of the load by adding successively the value K to the stresses for the previous position of load. This process is repeated until the step lines are reached, and thus complete the table.

To avoid error the following checks may be observed:

(1) In the table the number which is immediately above or below the step line and is the last one obtained by successive additions should be checked up by multiplying the value of S' by the proper value of K.

(2) For chord members, both upper and lower, the difference between the two numbers that are immediately above and below any horizontal step line must be equal to the number on the first line of the table (load at I) under the same vertical column, i. e., for the same number. This check is very valuable as it ensures the correctness of the whole table.

(3) For diagonals and verticals the following check may be useful. For any position of load, add up the stresses in the

members that are symmetrically located in the truss. The result should be the same as the stresses for that member given on Table 10.

Table 16. Stress Coefficients by Analytic Method.

(a) Upper Chords.

Load at	U_1	U_2	U_3	U_4	U_5	U_6
0	.0026	.0064	.0123	.0206	.0300	.0347
I	-.3868	-.3883	-.3726	-.3167	-.1919	-.0137
II	-.2746	-.7916	-.7735	-0.6811	-.4537	-.1083
III	-.1671	-.5123	-1.1970	-1.0837	-.7704	-.2663
IV	-.0671	-.2519	-.6763	-1.5472	-1.1760	-.5274
V	.0173	-.0305	-.2297	-.7458	-1.7628	-.9982
VI	.0757	.1261	.0933	-.1520	-.8005	-1.8193
VII	.1014	.2010	.2605	.1801	-.2196	-.9983
VIII	.1062	.2110	.2041	.3047	.0586	-.5275
IX	.0854	.1822	.2737	.3051	.1555	-.2664
X	.0621	.1343	.2069	.2448	.1636	-.1084
XI	.0341	.0747	.1177	.1462	.1168	-.0137
XII	.0026	.0064	.0123	.0206	.0300	.0347

(b) Lower Chords.

Load at	L_0	L_1	L_2	L_3	L_4	L_5
0	−.0092	−.0117	−.0153	−.0208	−.0287	−.0377
I	−.2737	.1799	.1734	.1504	.0897	−.0366
II	−.5259	−.1868	.3826	.3494	.2465	0148
III	−.7611	−.5321	−.1333	.5865	.4558	.1355
IV	−.9690	−.8429	−.6039	−.1382	.7502	.3679
V	−1.1210	−1.0829	−.9817	−.7369	−.1928	.8288
VI	−1.1798	−1.2051	−1.2048	−1.1254	−.8459	−.1844
VII	−1.1210	−1.1785	−1.2328	−1.2487	−1.1336	−.7122
VIII	−.9690	−1.0342	−1.1060	−1.1618	−1.1314	−.8690
IX	−.7611	−.8191	−.8866	−.9489	−.9554	−.7922
X	−.5259	−.5694	−.6217	−.6742	−.6943	−.6036
X	−.2737	−.2984	−.3288	−.3688	−.3607	−.3458
XII	−.0092	−.0017	−.0153	−.0208	.0287	−.0377

(c) Diagonals.

Load at	D_1	D_2	D_3	D_4	D_5	D_6
0	−.0057	−.0068	−.0083	−.0102	−.0107	−.0052
I	.8581	.0027	−.0226	−.0687	−.1419	−.1819
II	.6091	.9067	−.0259	−.1134	−.2589	−.3515
III	.3705	.6055	.9780	−.1395	−.3562	−.5116
IV	.1489	.3242	.6061	1.0755	−.4218	−.6562
V	−.0383	.0840	.2846	.6360	1.1559	−.7692

VI	$-.1679$	$-.0883$	$.0469$	$.3023$	$.7369$	1.1299
VII	$-.2250$	$-.1744$	$-.0850$	$.0991$	$.4543$	$.8637$
VIII	$-.2246$	$-.1923$	$-.1330$	$.0016$	$.2797$	$.6501$
IX	$-.1896$	$-.1694$	$-.1307$	$-.0388$	$.1699$	$.4681$
X	$-.1378$	$-.1264$	$-.1307$	-0463	$.0921$	$.3016$
XI	$-.0755$	$-.0711$	$-.0615$	$-.0351$	$.0335$	$.1463$
XII	$-.0057$	$-.0068$	$-.0083$	$-.0102$	$-.0107$	$-.0052$

(d) Verticals.

Load at	V_0	V_1	V_2	V_3	V_4	V_5	V_6
0	$-.9949$	$.0055$	$.0059$	$.0060$	$.0051$	$.0023$	0
I	$-.7659$	-1.0023	$.0162$	$.0401$	$.0675$	$.0485$	0
II	$-.5436$	$-.7447$	$-.9816$	$.0663$	$.1230$	$.1657$	0
III	$-.3307$	$-.4972$	$-.6985$	$-.9185$	$.1693$	$.2419$	0
IV	$-.1330$	$+.2663$	$-.4329$	$-.6269$	$-.7996$	$.3113$	0
V	$.0342$	$-.0689$	$-.2033$	$-.3715$	$-.5492$	$-.6329$	0
VI	$.1499$	$.0725$	$-.0335$	$-.1767$	$-.3501$	$-.4889$	-1.000
VII	$.2008$	$.1432$	$.0606$	$-.0578$	$-.2158$	$-.3737$	0
VIII	$.2004$	$.1580$	$.0949$	$.0006$	$-.1330$	$-.2813$	0
IX	$.1643$	$.1392$	$.0932$	$.0226$	$-.0807$	$-.2026$	0
X	$.1230$	$.1039$	$.0740$	$.0271$	$-.0436$	$-.1305$	0
XI	$.0675$	$.0584$	$.0439$	$.0205$	$-.0159$	$-.0626$	0
XII	$.0051$	$.0055$	$.0059$	$.0060$	$.0051$	$.0023$	0

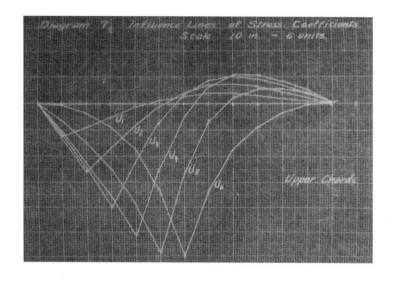

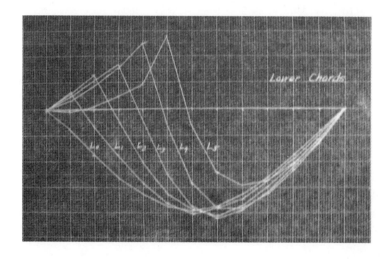

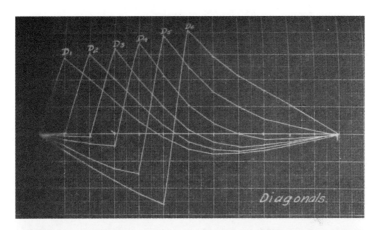

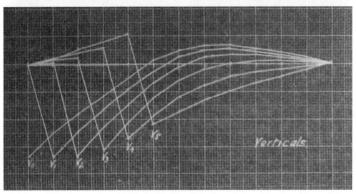

Table 17. Summation of Positive and Negative Stress Coefficients.

Member	Positive	Negative	Member	Positive	Negative
U_1	0. 4874	0. 8956	D_1	1. 9866	1. 0701
U_2	0. 0421	1. 9746	D_2	1. 9231	0. 8355

U_3	1.2808	3.2490	D_3	1.9156	0.5790
U_4	1.2221	4.5255	D_4	2.1145	0.4622
U_5	0.5545	5.3749	D_5	2.9223	1.2002
U_6	0.0694	5.6475	D_6	3.5587	2.4808
L_0	0	8.4996	V_0	0.9502	2.7681
L_1	0.1799	7.7728	V_1	0.6862	2.5794
L_2	0.5560	7.1302	V_2	0.3946	2.3498
L_3	1.0863	6.4371	V_3	0.1892	2.1514
L_4	1.5422	5.3915	V_4	0.3700	2.1879
L_5	1.3470	3.6242	V_5	0.7720	2.1725
			V_6	0	1.0000

After the stresses due to vertical and horizontal forces are known the stresses due to combined effects of them may be obtained by algebraic summation. These are given in Table 16. In Table 17 are given the summations of the positive and negative coefficients.

(V)Stress Coefficients by Graphic Method.

The method of Merriman and Jacoby is now used to find the stress coefficients by means of influence lines. In Table 18 are given the ordinates $\dfrac{x}{y}$ and $\dfrac{1-x}{y}$ to be laid off at the ends of span for influence lines of $\dfrac{M'}{y}$ which being constructed in Plate V. The influence lines of H are there shown by dotted lines. Scaling off the differences in ordinates of the curves for H and $\dfrac{M'}{y}$ and multiplying by the values of T, the stress coefficients are obtained. These are given in Table 19. It will be seen that a close agreement exists between these values and those obtained in Table 16.

Table 18. Values of End Ordinates of $\dfrac{M'}{y}$.

Member	x	y	$\dfrac{x}{y}$	Member	x	$1-x$	$\dfrac{x}{y}$	$1-\dfrac{x}{y}$
U_1	25	16.5	1.515	D_1	100	200	1.515	3.03
U_2	50	30.	1.665	D_2	116.7	183.3	1.752	2.78
U_3	75	40.5	1.85	D_3	135.7	164.3	2.055	2.49
U_4	100	48.	2.08	D_4	160.	140.	2.425	2.12
U_5	125	52.5	2.38	D_5	200.	100	3.03	1.515
U_6	150	54.	2.78	D_6	350.	−50.	5.31	−.757
L_0	0	66	0	V_1	116.7	183.3	1.752	2.78
L_1	25	60	.379	V_2	135.7	164.3	2.055	2.49
L_2	50	66	.758	V_3	160	140	2.425	2.12

L_3	55	66	1.135	V_4	200	100	3.03	1.515
L_4	100	66	1.515	V_5	350	−50	5.31	−.757
L_5	125	66	1.895					

Table 19. Stress Coefficients by Graphic Method.

(a) Upper Chords.

Load at	U_1	U_2	U_3	U_4	U_5	U_6
0	.0036	.006	.012	.021	.030	.034
I	−.390	−.379	−.368	−.320	−.195	−.010
II	−.274	−.784	−.771	−.667	−.427	−.090
III	−.163	−.509	−1.190	−1.050	−.748	−.225
IV	−.064	−.250	−.667	−1.540	−1.170	−.517
V	.017	−.029	−.222	−.733	−1.750	−.990
VI	.076	.125	.095	−.150	−.784	−1.800
VII	−102	.204	.262	.186	−.214	−.990
VIII	.100	.213	.300	.298	.044	−.520
IX	.085	.183	.278	.306	.156	−.247
X	.060	.133	.199	.240	.156	−.135
XI	.035	.075	.118	.147	.148	−.020
XII	.003	.006	.012	.021	.030	.034

(b) Lower Chords.

Load at	L_1	L_2	L_3	L_4	L_5
0	− .012	− .015	− .021	− .029	− .037
I	.182	.179	.162	.093	.020
II	− .182	.377	.338	.242	.015
III	− .532	− .129	.568	.432	.098
IV	− .833	− .597	− .148	.745	.342
V	− .995	− .973	− .730	− .186	.830
VI	− 1.210	− 1.193	− 1.132	− .840	− .195
VII	− 1.180	− 1.232	− 1.242	− 1.112	− .735
VIII	− 1.030	− 1.105	− 1.160	− 1.118	− .882
IX	− .820	− .884	− .936	− .932	− .773
X	− .568	− .616	− .676	− .672	− .588
XI	− 0.296	− 0.318	− 0.370	− 0.380	− 0.342
XII	− 0.012	− 0.015	− 0.021	− 0.029	− 0.037

(c) Diagonals.

Load at	D_1	D_2	D_3	D_4	D_5	D_6
0	− .005	− .007	− .008	− .012	− .011	− .005
I	.860	.010	− .011	− .060	− .132	− .193
II	.610	.897	− .016	− .166	− .255	− .380
III	.370	.600	.972	− .133	− .347	− .557

IV	.151	.320	.610	1.075	-.417	-.710
V	-.044	.082	.285	.630	1.150	-.850
VI	-.166	-088	.054	.380	.736	1.130
VII	-.222	-.175	-.075	.100	.458	.862
VIII	-.222	-.87	-.129	.007	.284	.652
IX	-.189	-.171	-.129	-.034	.167	.468
X	-.136	-.125	-.105	-.047	.097	.302
XI	-.076	-.070	-.059	-.034	.210	.143
XII	-.005	-.007	-.008	-.012	-.011	-.005

(d) Verticals.

Load at	V_1	V_2	V_3	V_4	V_5
0	.006	.006	.006	.005	.002
I	-.993	.020	.035	.063	.050
II	-.744	-.980	.062	.119	.165
III	-.498	-.700	-.920	.165	.240
IV	-.266	-.434	-.629	-.810	.308
V	-.067	-.199	-.370	-.540	-.635
VI	.072	-.038	-.179	-.350	-.490
VII	.144	.060	-.060	-.218	-.372
VIII	.155	.092	.000	-.133	-.281
IX	.140	.092	.019	-.079	-.096

X	.104	.069	.027	−.046	−.131
XI	.058	.042	.019	−.032	−.062
XII	.006	.006	.006	.005	.002

Article 43. Stress Coefficients in the Arch with Cantilever Arms.

As previously stated the reactions and stresses in this case are precisely the same as if there were no cantilever arms provided the loads do not extend outside the springings. For any load placed inside the arch span there is no moment or shear outside the support, which makes zero S' and T in the members of cantilever arms and consequently those members bear no effect upon the values of horizontal thrusts. The stress coefficients for loads placed inside the arch span (including those over the supports) are therefore the same as for the arch without cantilever arms and are those given in Table 10. Under this condition the stresses in members of cantilever arms are equal to zero, the above mentioned tables may therefore be used without addendum for the structure having cantilever arms and loads placed inside the span.

To find the stress coefficients for loads placed outside the supports, it is necessary to carry through the same process as for the arch without cantilever arms and every step prescribed in the previous article has to be followed in full. As before the horizontal thrust will be found first and then the stress coefficients.

(a) Horizontal Thrust.

The same method of symmetrical loads will be used here as in the case of pure arch span. When two loads are placed symmetrically on the cantilever arms, the vertical reactions are each equal to the load. To find the stresses S' three cases will be considered: (1) for members which lie outside the loads; (2) for members between the loads and the springings, and (3) for members inside the arch span.

For the first case the stresses are obviously zero because there are no moments or shears outside the loads. For the second case, the stresses may be obtained by any convenient method, analytic or graphic. For the third case the moment inside the springings is constant and is equal to $P(kl)$ so that the stress in any member is equal to $\dfrac{Pkl}{t}$. This last quantity has been calculated already and is given in Table 6, the only difference being that the signs are reversed in the present case.

The values of T for the members inside the arch span remain the same as before Table 4, there being no such stresses in the cantilever arms.

After the values of S' and T are known the products $S'TL$ may be performed, these being given in Table 20. The values of TL are taken from Table 11. Summing up the values of $S'TL$ and dividing it by $2\sum T^2L$ which remains the same as before, the values of horizontal thrusts are obtained.

$$2\sum T^2L = 5664.670$$

	Load at I′	Load at II′
$S'TL =$	1170.948	2385.454
$K =$	$-.2067$	$-.4211$

The sign of H is minus, meaning that it directs away from the center of span.

After the analytical values of H are obtained it becomes necessary to find the same by the method of Merriman and Jacoby. Since the values of T for members of cantilever arms are equal to zero, there will be no alterations in lengths in those members. The displacement diagram for those members may therefore be constructed by continuing the diagram for the members inside the span without laying off any axial deformations. This is shown at the lower part of Plate 3 in which the points A and a are shifted from the original diagram. Dotted lines refer to preliminary values of H.

Table 20. Values of $S'TL$ for Loads on Cantilever Arms.

Member	TL	Load at I′,XI′		Load at II′, X′	
		S'	$S'TL$	S	$S'TL$
U_2'	0	0	0	0	0
U_1'	0	0	0	.5050	0
U_1	8.333	.5051	4.208	1.0101	8.417
U_2	20.835	.6945	14.469	1.3889	28.936
U_3	39.708	.9804	38.930	1.9068	77.860
U_4	66.668	1.3889	92.596	2.7778	185.192
U_5	97.228	1.8519	180.046	3.7038	360.092
U_6	112.500	2.0833	234.274	4.1666	468.748
			Σ 564.623		1129.246
L_1'	0	0	0	-.5740	0
L_0'	0	-.4539	0	-.9077	0
L_0	35.891	-.4539	16.289	-.9077	32.578
L_1	43.056	-.5740	24.714	-1.1480	49.428
L_2	53.921	-.7532	40.615	-1.5065	81.230

L_3	70.531	-1.0236	72.193	-2.0472	144.386
L_4	94.635	-1.4112	133.521	-2.8224	267.042
L_5	122.663	-1.8552	227.562	-3.7104	455.125
			Σ 514.895		1029.790

Member	TL	Load at I′, XI′		Load at II′, X′	
		S'	$S'TL$	S'	$S'TL$
D_2'	0	0	0	8855	0
D_1'	0	8402	0	5602	0
D_1	41.005	-.2801	11.484	-.5602	22.968
D_2	38.420	-.3320	12.757	-.6641	25.514
D_3	38.608	-.4085	15.729	-.8169	31.458
D_4	40.938	-50.34	20.607	-1.0067	41.214
D_5	39.468	-.5262	20.767	-1.0523	41.534
D_6	18.798	-.2568	4.827	-.5135	9.654
		Σ	86.171		172.342
V_2'	0	0	0	-1.0000	0
V_1'	0	-1.0000	0	-.7273	0
V_0	43.559	-.5000	-21.779	0	0
V_1	35.639	.2727	9.720	.5454	19.439
V_2	27.720	.2917	8.085	.5833	16.170
V_3	19.800	.2941	5.823	.5882	11.647

V_4	11.880	.2500	2.970	.5000	5.940
V_5	3.960	.1110	.440	.2222	.880
V_6	0	0	0	0	0
			Σ 5.259		54.076

The values of H obtained from measurement are as follows:

$$\delta' = 2 \times 233.35 = 66.70$$

	Load at I′	Load at II′
$\delta =$	13.68	28.00
$H =$	$-.205$	$-.420$

It will be seen that were it not for the effect of the member of V_0 the value of H for load at II′ would be just twice that for load at I′. As the influence of V_0 is always inappreciable it might be well to bear the fact that the influence line of H for loads on the cantilever arms is very nearly equal to a straight line.

(b) Stress Coefficients.

After the values of H are obtained the stresses in members produced by them may be obtained from the values of T. These are given in Table 21.

The stresses due to vertical forces considering the truss as a simple span may be found from the values of S' and in Tables 5a and 5b (except members of cantilever arms). For a load on the left cantilever arm the stress in any member is equal to S'_r multiplied by the right hand reaction. For a load on the right cantilever arm, the stress in the same member is equal to S' multiplied by the left hand reaction. A more direct method of obtaining those stresses is to take the values from Table 15. Proceed thus: For a load at I′ copy stresses given in the first lines of Table 15, for a load at II′, add to the stresses thus obtained an amount equal to the stresses themselves. For a load at XI′ copy the stresses given in the last lines of Table 15, for a load at X′ add to the stresses thus obtained an amount equal to stresses themselves. There is a good check to ensure the accuracy of the work by summing up the stresses of symmetrical members under any positions of load. These should be the same as S' given in Table 20.

It should be noted that the above statement applies only to members that are inside the arch span, for cantilever arms the stresses are obtained separately.

The stresses due to vertical forces only are given in Table 13, those due to combined effects of both vertical and horizontal forces are given in Table 14.

Table 21. Stresses Due to Horizontal Thrust for Loads on

Cantilever Arms.

Member	T	Load At		Member	T	Load At	
		I′	II′			I′	II′
U_2'	0	0	0	D_2'	0	0	0
U_1'	0	0	0	D_1'	0	0	0
U_1	−.3333	−.0689	−.1404	D_1	+.7394	+.1529	+.3114
U_2	−.8334	−.1723	−.3510	D_2	+.8766	+.1812	+.3692
U_3	−1.5883	−.3284	−.6688	D_3	+1.0783	+.2229	+.4540
U_4	−2.6667	−.5512	−1.1230	D_4	+1.3289	+.2747	+5596
U_5	−3.8889	−.8039	−1.6376	D_5	+1.3891	+.2871	+.5850
U_6	−4.5000	−.9302	−1.8950	D_6	+.6779	+.1410	+.2854
L_1'	0	0	0	V_2'	0	0	0
L_0'	0	0	0	V_1'	0	0	0
L_0	+1.1982	+.2477	+.5046	V_0	−.6660	−.1364	−.2780
L_1	+1.5153	+.3132	+.6382	V_1	−.7200	−.1488	−.3032
L_2	+1.9886	+.4120	+.8374	V_2	−.7700	−.1592	−.3242
L_3	+2.7022	+.5418	+1.1380	V_3	−.7765	−.1605	−.3270
L_4	+3.7255	+.7695	+1.5688	V_4	−.6600	−.1364	−.2780
L_5	+4.8977	+1.0122	+2.0624	V_5	−.2933	−.0606	−.1236
				V_6	0	0	0

Table 22. Stresses Due to Vertical Forces for Loads on Cantilever Arms.

Member	Load At			
	I′	**II′**	**XI′**	**X′**
U_2'	0	0	0	0
U_1'	0	+ .5060	0	0
U_1	+ 0.4630	+ 0.9260	+ 0.0421	+ 0.0842
U_2	+ 0.5787	+ 1.1574	+ 0.1157	+ 0.2315
U_3	+ 0.7353	+ 1.4706	+ 0.2461	+ 0.4902
U_4	+ 0.9259	+ 1.8518	+ 0.4630	+ 0.9259
U_5	+ 1.0803	+ 2.1605	+ .7716	+ 1.5432
U_6	+ 1.0417	+ 2.0833	+ .10417	+ 2.0834
L_1'	0	− .5740	0	0
L_0'	− .4532	− .9064	0	0
L_0	− ..4532	− .9064	0	0
L_1	− .5261	− 1.0522	− .0478	− .0957
L_2	− .6277	− 1.254	− .1255	− .2511
L_3	− .7677	− 1.5354	− .2559	− .5118
L_4	− .9408	− 1.8616	− .4704	− .9408
L_5	− 1.0822	− 2.1644	− .7730	− 1.9460

Member	Load at			
	I′	II′	XI′	X′
D_2'	0	.8855	0	0
D_1'	.8402	.5602	0	0
D_1	−.1865	−.3730	−.0934	−.1867
D_2	−.2029	−.4058	−.1291	−.2583
D_3	−.2237	−.4474	−.1848	−.3696
D_4	−.2349	−.4698	−.2685	−.5369
D_5	−.1754	−.3508	−.3508	−.7016
D_6	−.0270	−.0540	−.2996	−.5991
V_2'	0	−1.0000	0	0
V_1'	−1.0000	−.7273	0	0
V_0	−.6667	−.1667	.0833	.1667
V_1	.1667	.3334	.1061	.2121
V_2	.1597	.3194	.1320	.2639
V_3	.1373	.2745	.1569	.3137
V_4	.0833	.1667	.1667	.3333
V_5	−.0185	−.0370	.1296	0.2632
V_6	0	0	0	0

Table 23. Stress Coefficients for Loads on Cantilever Arms.

Member	Load At			
	I′	II′	XI′	X′
U_2'	0	0	0	0
U_1'	0	+ .5050	0	0
U_1	+ .3941	+ .7856	− .0268	− .0562
U_2	+ .4064	+ .8064	− .0566	− .1195
U_3	+ .4069	+ .8018	− .0833	− .1786
U_4	+ .3747	+ .7288	− .0882	− .1971
U_5	+ .2764	+ .5229	− .0323	− .0944
U_6	+ .1115	+ .1883	+ .1115	+ .1884
L_1'	0	− .5740	0	0
L_0'	− .4532	− .9064	0	0
L_0	− .2055	− .4008	+ .2477	+ .5046
L_1	− .2129	− .4140	+ .2654	+ .5425
L_2	− .2159	− .4180	+ .2865	+ .5863
L_3	− .2259	− .3974	+ .2859	+ .6262
L_4	− .1713	− .3128	+ .2991	+ .6280
L_5	− .0700	− .1020	+ .2392	+ .5164

Member	Load At			
	I′	**II′**	**XI′**	**X′**
D_2'	0	+ .8855	0	0
D_1'	+ .8402	+ .6602	0	0
D_1	− .0336	− .0616	+ .0595	+ .1247
D_2	− .0217	− .0366	+ .0621	+ .1109
D_3	+ .0008	+ .0066	+ .0381	+ .0844
D_4	+ .0398	+ .0898	+ .0062	+ .0227
D_5	+ .1117	+ .2342	− .0637	− .1166
D_6	+ .1140	+ .3394	− .1586	− .3137
V_2'	0	− 1.000	0	0
V_1'	− 1.0000	− .7273	0	0
V_0	− .8036	− .4447	− .0521	− .1113
V_1	+ .0179	+ .0302	− .0427	− .0911
V_2	+ .0007	− .0048	− .0272	− .0883
V_3	− .0232	− .0525	− .0136	− .0133
V_4	− .0531	− .1113	+ .0303	+ .0552
V_5	− .0791	− .1605	+ .0690	+ .1357
V_6	0	0	0	0

In Table 24 are given the summation of the stress coefficients which being the sum of stresses of Table 17 and those due to loads on I′ and XI′; for the time being, the stresses due to loads at ends of cantilever arms are not included.

To check the results obtained by analytic methods, influence lines of H and $\dfrac{M'}{y}$ are extended beyond the supports as shown in Plate V. The influence lines of $\dfrac{M'}{y}$ are straight lines and are prolongations of the same for loads near the ends of the span. It is found that the graphical values check very closely with those obtained analytically (Table 23).

Article 44. Live Load Stresses.

After the stress coefficients are obtained the live load stresses may be calculated for any value of panel loads. As the typical loading of wheel concentrations is nothing more than arbitrary and is never fully realized in practice there is no gain in calculating the stresses to such hair-splitting exactness as to adopt those various concentrations as the natual loads.[1] For this reason equivalent uniform loads will be used. In Diagram 8 is given the equivalent uniform live loads for Cooper's E 50 loading for

---

[1]   Bridge Engineering, Waddell, P. 161.

bridges of 300′ span at different points along the span. It will be seen that there is a critical change at about the quarter point of the span. This value of equivalent uniform load may therefore be advantageously adopted and assumed to be the same throughout the entire span.

For one rail, equivalent load at quarter point = 2832 lbs. per ft. Assume both tracks loaded in finding the maximum stresses, uniform load = 5664 lbs. per ft. of truss.

∴ Panel live load = 141600 lbs

$$= 141.6 \text{ kips.}$$

(According to the theory of probability, it is permissible to use a lighter load for the truss than for the floor system,[1] but to provide for future increase in traffic, the same loads will be used without reduction.)

Maximun live load reaction carried to the ends of the cantilever arm from loads on suspended span is the same as the maximum shear for the end of the suspended span, which for E 50.

---

① Bridge Engineering. Waddell. P. 107.

Table 24. Summation of Stress Coefficients for the Arch With Cantilever Arms. (Stresses due to loads at ends of cantilever arms not included.)

Member	Positive	Negative	Member	Positive	Negative
U_2'	0	0	D_2'	0	0
U_1'	0	0	D_1'	.8402	0
U_1	.8802	.9237	D_1	2.0489	1.1009
U_2	1.3453	2.0344	D_2	1.9786	.8538
U_3	1.6817	3.3383	D_3	1.9611	.5790
U_4	1.5865	4.6250	D_4	2.1707	.4682
U_5	.8160	5.4231	D_5	3.0394	1.2685
U_6	.2578	5.6475	D_6	3.7284	2.6377
L_1'	0	0	V_2'	0	0
L_0'	0	.4532	V_1'	0	1.0000
L_0	.2523	8.6996	V_0	.9502	3.6295
L_1	.4512	7.9798	V_1	.7013	2.6249
L_2	.8492	7.3392	V_2	.3946	2.3823
L_3	1.3994	5.5902	V_3	.1892	2.1842
L_4	1.8562	5.5479	V_4	.3977	2.3436
L_5	1.6052	3.6752	V_5	.8398	2.2528
			V_6	0	1.0000

loading and 100′ span is 157.5 kips for one rail, of 375 kips for one truss. Therefore, total maximum live load at the end of cantilever are

$$= \frac{141.6}{2} + 375$$

$$= 445.8 \text{ kips.}$$

Having found the live panel loads the live load stresses in the truss may be obtained by multiplying the stress coefficients by the value of the load. On Table 25 are

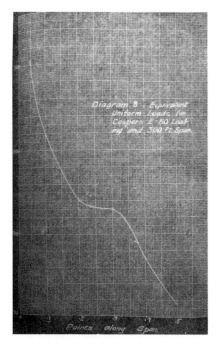

given the live load stresses for the arch without cantilever arms, the values of which are obtained by multiplying the values of Table 17 by 141. 6. On Table 17 are given the same stresses for the arch with cantilever arms. These values are obtained in two terms, the first is the stresses due to all loads except those at the ends of cantilever arms and is the product of 141. 6 and coefficients of Table 15; the second term is the stresses due to load on the ends of cantilever arms and is the product of 445. 8

and the values of coefficients in Table 23 for loads at the point II′ and X′.

Table 25. Live Load Stresses for the Arch Without Cantilever Arms. (in kips)

Member	Positive		Negative	
	Coefficient	Stress	Coefficient	Stress
U_1	0.4874	69.1	0.8956	126.9
U_2	0.9421	133.2	1.9746	279.8
U_3	1.2808	181.7	3.2490	461.5
U_4	1.221	173.3	4.5265	642.5
U_5	0.5545	78.5	5.3749	761.1
U_6	0.0694		5.6475	799.8
L_0	0	0	8.4996	1203.0
L_1	0.1799	25.5	7.7728	1099.1
L_2	0.5560	78.8	7.1302	1011.2
L_3	1.0863	153.9	6.4371	912.3
L_4	1.5422	218.4	5.3915	764.2
L_5	1.3470	190.9	3.6242	513.4

Member	Positive		Negative	
	Coefficient	Stress	Coefficient	Stress
D_1	1.9866	281.5	1.0701	151.7

D_2	1.9231	272.7	0.8355	118.2
D_3	1.9156	271.3	0.5790	82.2
D_4	2.1145	299.1	0.4622	65.4
D_5	2.9223	413.7	1.2002	170.1
D_6	3.5587	503.4	2.4808	351.7
V_0	0.9502	134.5	2.7681	391.5
V_1	0.6862	97.1	2.5794	366.8
V_2	0.3946	55.9	2.3498	333.1
V_3	0.1892	26.8	2.1514	304.8
V_4	0.3700	52.4	2.1879	309.2
V_5	0.7720	109.2	2.1725	207.9
V_6	0	0	1.0000	141.6

Table 26. Live Load Stresses for the Arch With Cantilever Arms. (in kips)

(a) Positive Stresses.

Member	Coefficient	1st Term	Coefficient	2nd term	Stress
U_2'	0	0	0	0	0
U_1'	0	0	0.5050	227.1	227.1
U_1	0.8802	124.7	0.7856	350.4	475.1
U_2	1.3453	190.5	0.8064	359.6	550.1
U_3	1.6817	238.1	0.8018	357.0	597.1

	Coefficient	1st Term	Coefficient	2nd Term	Stress
U_4	1.586	224.6	0.7288	325.0	549.6
U_5	0.8160	115.5	0.5229	233.4	348.9
U_6	0.2578	36.6	0.3767	167.1	204.7
L_1'	0	0	0	0	0
L_0'	0	0	0	0	0
L_0	0.2523	35.8	0.5046	225.1	260.9
L_1	0.4312	639.	0.5426	242.0	305.9
L_2	0.8492	120.3	0.5863	261.3	381.6
L_3	1.3994	198.4	0.6262	289.2	487.6
L_4	1.8562	262.8	0.6280	280.1	542.9
L_5	1.6052	227.3	0.5164	230.0	457.3

Member	Coefficient	1st Term	Coefficient	2nd Term	Stress
D_2'	0	0	0.8855	395.0	395.0
D_1'	0.8402	119.0	0.5602	249.9	368.9
D_1	2.0489	290.6	0.1247	55.6	346.2
D_2	1.9786	280.4	0.1109	49.5	329.9
D_3	1.9611	277.8	0.0910	40.7	318.5
D_4	2.1707	307.7	0.1125	50.2	357.9
D_5	3.0394	430.4	0.2342	104.3	534.7
D_6	3.7284	527.9	0.3394	151.3	679.2
V_2'	0	0	0	0	0

V_1'	0	0	0	0	0
V_0	0.9502	134.4	0	0	134.4
V_1	0.7013	99.4	0.0302	13.5	112.9
V_2	0.3946	56.0	0	0	66.0
V_3	0.1892	26.8	0	0	26.8
V_4	0.3977	56.4	0.0552	24.6	81.0
V_5	0.8398	118.9	0.1357	60.5	179.4
V_6	0	0	0	0	0

(b) Negative Stresses.

Member	Coefficient	1st Term	Coefficient	2nd Term	Stress
U_2'	0	0	0	0	0
U_1'	0	0	0	0	0
U_1	0.9237	130.8	0.0562	25.0	155.8
U_2	2.0344	288.1	0.1195	53.3	341.4
U_3	3.3383	472.7	0.1786	79.7	552.4
U_4	4.6250	654.9	0.1971	87.9	742.8
U_5	5.4221	767.8	0.0944	42.2	810.0
U_6	5.6475	799.7	0	0	799.7
L_1'	0	0	0.5740	256.1	256.1
L_0'	0.4532	64.3	0.9064	404.1	468.4
L_0	8.6996	1231.9	0.4008	178.7	1410.6

L_1	7.9798	1129.9	0.4140	184.6	1314.5
L_2	7.3392	1039.2	0.4180	186.7	1225.9
L_3	5.5902	791.6	0.3974	177.1	968.7
L_4	5.5479	785.6	0.3128	139.6	925.2
L_5	3.6752	520.4	0.1020	45.5	565.9

Member	Coefficient	1st Term	Coefficient	2nd Term	Stress
D_2'	0	0	0	0	0
D_1'	0	0	0	0	0
D_1	1.1009	155.9	0.0616	27.5	183.4
D_2	0.8538	121.9	0.0366	16.3	138.2
D_3	0.5790	82.1	0	0	82.1
D_4	0.4622	65.4	0	0	65.4
D_5	1.2585	178.3	0.1166	52.2	230.5
D_6	2.6377	373.5	0.5137	140.0	513.5
V_2'	0	0	1.0000	445.8	445.8
V_1'	1.0000	141.6	0.7273	324.2	465.8
V_0	3.6295	514.0	0.5560	248.2	762.2
V_1	2.6249	371.7	0.0911	40.7	412.4
V_2	2.3823	345.2	0.0651	29.1	374.3
V_3	2.1842	309.3	0.0658	29.4	338.7
V_4	2.2436	317.7	0.1665	75.9	393.6

| V_5 | 2.2528 | 319.0 | −.1605 | 73.2 | 392.2 |
| V_6 | 1.0000 | 141.6 | 0 | 0 | 141.6 |

CHAPTER XⅧ. IMPACT STRESSES

In accordance with the clauses of specifications the effect of impact is taken care of by the formula suggested by Schneider:

$$I = S' \frac{300}{300 + \tau}$$

where I is the impact stress, S' is the live load stress, and τ is the loaded length of track in feet producing the maximum stress in the member.

The loaded lengths of tracks are obtained by measurements from the influence lines of stresses (Diagram 7) and doubled to care for the effect of double tracks. These lengths may be assumed as those covered by loaded care while the remaining span by empty cars. [1] The impact stresses for the arch without cantilever arms are given in Table 27 and those for the arch with cantilever arms in Table 28. In those tables the column "ratio"

---

[1] Design of Steel Bridges. Kunz. P. 22.

stands for the quotient of $\dfrac{300}{300 + \tau}$ and the column "L. L." for the corresponding live load stresses.

Table 27. Impact Stresses for the Arch Without Cantilever Arms. (in kips)

Member	Positive				Negative			
	I	Ratio	L. L.	Stress	I	Ratio	L. L.	Stress
U_1	360	.455	69.1	31.5	240	.556	126.9	70.5
U_2	340	.470	133.2	62.7	260	.337	279.8	150.3
U_3	316	.487	181.7	88.4	284	.514	461.5	237.0
U_4	280	.317	173.3	89.5	320	.484	642.5	311.4
U_5	210	.588	78.5	46.2	390	.436	761.1	331.5
U_6	10	.970	9.8	9.5	59	.337	799.8	269.1
L_0	–	–	–	–	600	.333	1203.0	401.5
L_1	80	.790	25.5	20.1	520	.366	1099.1	402.7
L_2	140	.682	78.8	53.8	460	.395	1011.2	400.0
L_3	190	.512	153.9	94.1	410	.422	912.3	385.2
L_4	240	.557	218.4	116.1	360	.454	764.2	347.1
L_5	290	.609	190.9	97.1	310	.492	513.4	252.3

茅以升全集②

Member		Positive				Negative		
	I	Ratio	L. L.	Stress	I	Ratio	L. L.	Stress
D_1	240	.555	281.5	156.1	360	.454	151.7	68.8
D_2	270	.527	272.7	143.7	330	.476	118.2	56.2
D_3	220	.577	271.3	156.1	380	.442	82.2	36.2
D_4	260	.536	299.1	161.0	340	.468	65.4	30.7
D_5	380	.517	413.7	213.6	220	.677	170.1	98.1
D_6	330	.567	303.4	265.9	270	.527	351.7	185.2
V_0	600	.333	134.5	44.7	600	.333	391.5	129.3
V_1	330	.454	7.1	44.2	270	.527	365.8	192.8
V_2	310	.492	55.9	27.6	290	.508	233.1	168.3
V_3	300	.500	26.8	13.4	300	.500	304.8	152.4
V_4	160	.652	52.4	34.2	440	.406	309.2	125.1
V_5	270	.527	109.2	57.4	330	.476	307.9	146.5
V_6	–	–	0	0	600	.333	141.6	47.2

Table 28. Impact Stresses for the Arch With Cantilever Arms. (in kips)

Member	I	Ratio	L. L.	Stress	I	Ratio	L. L.	Stress
U_2'	–	–	–	–	–	–	–	–
U_1'	250	.546	227.1	123.9	–	–	–	–
U_1	660	.313	475.1	148.8	540	.357	155.8	55.7

U_2	640	.319	550.1	175.5	560	.349	314.4	119.0
U_3	616	.327	595.1	194.5	584	.337	552.4	180.0
U_4	580	.341	549.6	187.1	620	.326	742.8	242.4
U_5	510	.371	348.9	129.4	690	.303	810.0	245.3
U_6	600	.333	204.7	68.2	600	.333	759.7	260.4
L_1'	–	–	–	–	250	.566	256.1	144.9
L_0'	–	–	–	–	300	.500	468.4	234.2
L_0	300	.500	260.9	130.4	900	.250	1410.6	353.1
L_1	376	.444	305.9	135.9	824	.267	1314.5	350.8
L_2	438	.407	381.6	155.1	762	.282	1225.9	345.6
L_3	490	.380	487.6	184.7	710	.297	968.7	287.9
L_4	540	.357	542.9	193.4	660	.313	985.2	289.9
L_5	590	.337	457.3	153.7	610	.330	565.9	186.7

Member	Positive				Negative			
	I	Ratio	L. L.	Stress	I	Ratio	L. L.	Stress
D_2'	250	.545	395	215.4	–	–	–	–
D_1'	300	.500	368.9	184.4	–	–	–	–
D_1	540	.357	346.2	123.4	660	.313	183.4	57.3
D_2	670	.345	329.9	113.9	630	.323	138.2	44.7
D_3	820	.268	318.5	85.3	380	.442	82.1	36.3
D_4	860	.269	357.9	92.7	340	.468	65.4	30.7

D_5	690	.303	634.7	162.1	610	.371	230.5	85.3
D_6	630	.323	679.2	219.5	570	.345	513.5	177.5
V_2'	–	–	–	–	250	.545	445.8	243.4
V_1'	–	–	–	–	300	.500	465.8	232.9
V_0	600	.333	134.4	44.8	600	.333	762.2	254.0
V_1	630	.323	112.9	36.5	570	.345	412.4	141.9
V_2	330	.477	560	267	870	.256	374.3	95.8
V_3	290	.508	26.8	13.6	910	.248	338.7	84.1
V_4	460	.395	81.0	32.0	740	.385	393.6	151.4
V_5	510	.371	179.4	68.5	690	.303	392.2	118.9
V_6	–	–	–	–	1200	.250	141.6	35.5

CHAPTER XIII. WIND STRESSES

The specification requires the span to be designed for a lateral force on the loaded chord of 200 lbs. per feet plus 10% of the specified train load on one track and 200 lbs. per linear foot on the unloaded chord. The panel load for wind effect only is therefore 5 kips for both chords and for lateral forces is. $1 \times 5000 \times 25 = 12.5$ kips for the loaded or upper chord.

The stresses arisen from these lateral forces are come from two sources, the first and most important is due to the direct action of the loads while the other is due to the overturning effect on account of the curvilinear shape of the lower chord. The first mentioned stresses are resisted by the lateral systems and chord. Members of the arch truss while the latter are resisted by the arch truss and sway bracing. There will be two lateral systems, one in top and the other in bottom chords. The top lateral will take the panel load of $12.5 + 5 = 17.5$ kips while the bottom lateral will take the load of 5 kips. Both of these laterals are to be of double intersection warren type and the diagonals are capable of taking both tension and compression.

The stresses in lateral systems are calculated by means of stress coefficients. Since the lower chord is on a curved plane either of two methods must be used: (1) the lateral is developed into a horizontal plane and treated as an ordinary truss: (2) the lateral is projected on a horizontal plane the stresses thus obtained are afterwards corrected by the ratio of actual and projected lengths of the member. The second method will be adopted here because the said projection is exactly the same as the top lateral and the provisional stresses can be obtained from those for the top chords by correcting for the ratio of the panel loads.

The lateral loads will be divided equally among the two

trusses, this being sufficiently close for practical purposes.

In calculating the stresses in laterals the method of coefficients will be used, this being described in full in "Stresses in Simple Trusses." by Merriman and Jacoby, Article 27.

For the other kind of stresses due to overturning moments of wind loads. There are not much data available. The method used in this thesis is that of Merriman and Jacoby, [1] and is based on the following assumptions: (1) the lateral forces on the train are applied at 7' above base of rail, (2) the wind load on upper chord is transmitted to lower chord by sway bracing, (3) the wind load on lower chord is transmitted to abutments by the arch truss.

It has been found in many instances, including two-hinged spandrel braced arches, that the effect of wind loads is negligible if the unit stresses of the material be allowed to increase a certain per cent. For this reason any attempt to calculate the wind stresses to any degree of unnecessary refinement is useless and a total waste of time. Bearing this in mind the wind load will not be considered as moving but distributed uniformly for the arch without cantilever arms and on the arch and on one approach

---

[1]   Higher Structures. Merriman and Jacoby. P. 207.

span for the arch with cantilever arms.

Article 45. Wind Stresses for the Arch without Cantilever Arms.

By the method of stress coefficients the stresses in chord members of the truss are found as follows:

	U_1	U_2	U_3	U_4	U_5	U_6
Stress Coefficient	2.75	7.75	11.75	14.75	16.75	17.75
Stress for S Wind	48.2	135.6	205.5	258.1	293.0	310.4
	L_0	L_1	L_2	L_3	L_4	L_5
Stress for S Wind	16.5	44.1	63.8	77.0	85.2	89.0

The equivalent vertical loads are given in Table 29, in which r is the lever arm of the wind loads, M the overturning moment and E. L. the equivalent vertical load obtained by dividing M by $25'$ the spacing between trusses. In Table (b) "load from T. C. " means the loads transmitted from the train and the top chord. "load from B. C. " that transmitted from the bottom chord panel point that is one panel above the one in consideration, and lastly "W. L. " means the wind load on the panel point itself. These loads set downwards for south winds.

The stresses in arch members are next found for the equivalent loads thus obtained just as for the live load stresses in the previous article. There is one point which, however, requires modification, i. e. , the equivalent loads are not all applied at the top chords, one half of which are found to be acting on the lower chords. As the values of E are different for loads at top and bottom panel points as same distances from the springing the stress coefficients obtained for the live loads cannot be well adopted if theoretical correctness is to be ensured. As the effect of wind is small, all the equivalent loads will be assumed on the top chords but the stresses in verticals are corrected afterwards

Table 29. Equivalent Vertical Loads.

(a) On Top Chords.

Panel Point		VI	V	IV	III	II	I	O
From wind on train	Load	12.5	12.5	12.5	12.5	12.5	12.5	12.5
	r	20.0	21.5	26.0	33.5	44.0	57.5	74.0
	M	250.	269.	325.	418.	550.	718.	925
	E. L.	10.0	10.75	13.0	16.75	22.0	26.75	37.0
From wind on top chord	Load	5.0	5.0	5.0	5.0	5.0	5.0	5.0
	r	12.0	13.5	18.0	25.6	36.0	49.5	86.0
	M	6.0	6.8	9.0	12.8	18.0	24.8	33.0
	E. L.	2.4	2.7	3.6	5.1	7.2	9.9	13.2
	Total	12.4	13.5	16.6	21.8	29.2	38.6	50.2

(b) On Bottom Chords.

Panel Point		VI	V	IV	III	II	I	O
Load From	T. C.	17.5	17.5	17.5	17.5	17.5	17.5	17.5
	B. C.	0.0	11.25	33.75	56.26	78.75	101.26	123.75
	W. L.	5.0	5.0	5.0	5.0	5.0	5.0	5.0
	Total	22.5	33.75	56.25	78.75	101.25	123.75	146.25
	r	1.5	4.5	7.5	10.5	13.5	16.5	0
	M	33.8	152.	422.	827.	1367.	2040.	0
	E. L.	1.38	6.1	16.9	33.1	54.7	81.7	0

for the lower panel loads. Hence the equivalent loads will appear for the present as following:

Panel Point	0	I	II	III	IV	V	VI
Load	50.2	120.3	83.9	54.9	33.5	19.6	13.8

After the panel loads are fixed the next step will be to find the stresses indifferent members produced by these loads. Since the loads are not equal Table 17 cannot be used as for the line loads. To obtain an exact result would call for the use of Table 16 or 19, the coefficients of which are to be multiplied by the different values of loads at different points along the span. This, however is unwarrantable for the preliminary stresses and is not justified with the uncertainty of the stress distribution. A mean value of the loads will therefore be used and treated just like the live load as described in article 44.

Table 30. Wind Stresses for the Arch Without Cantilever Arms. (in kips) (For S. wind)

Member	Positive		Negative		
	Coefficient	Stress	Coefficient	Stress	Total
U_1	.4874	26.1	.8956	48.1	−22.0
U_2	.9421	50.4	1.9740	105.8	−55.4
U_3	1.2808	68.7	3.2490	173.2	−104.5
U_4	1.2221	66.7	4.5265	243.0	−177.3
U_5	.5646	29.7	5.3749	288.0	−258.3
U_6	.0694	3.7	5.6475	302.5	−298.8
L_0	0	0	8.4996	456.0	−456.0
L_1	.1799	9.66	7.7728	417.4	−407.7
L_2	.5660	29.8	7.1302	382.1	−352.3
L_3	1.0863	58.3	6.4371	345.2	−286.9
L_4	1.5422	82.7	5.3915	289.7	−207.0
L_5	1.3470	72.3	3.6242	194.1	−121.8

Member	Positive		Negative		
	Coefficient	Stress	Coefficient	Stress	Total
D_1	1.9866	106.4	1.0701	57.4	49.0
D_2	1.9231	103.2	.8355	44.7	58.5
D_3	1.9156	102.7	.5790	31.1	71.6
D_4	2.1145	113.2	.4622	24.8	88.4
D_5	2.9223	156.8	1.2002	64.3	92.5
D_6	2.5587	190.9	2.4808	132.9	58.0
V_0	.9502	50.9	2.7681	148.5	-97.6
V_1	.6862	36.7	2.5794	138.4	-20.0
V_2	.3946	21.2	2.3498	125.8	-43.5
V_3	.1892	10.1	2.1514	115.3	-72.1
V_4	.3700	19.8	2.1879	117.3	-80.6
V_5	.7720	41.3	2.1725	116.77	-69.3
V_6	0	0	1.0000	53.7	-12.4

The wind stresses for this uniform panel load of 53.7 kips (the mean value of equivalent loads) are given in Table 30.

Article 46. Wind Stresses for the Arch with Cantilever Arms.

In this case the stresses for loads placed inside the arch span are exactly the same as those found in the previous article. For loads placed outside the supports, the laterals will have to carry an additional stress due to wind on the cantilever arm and the suspended span. The total lateral force on the suspended span $= 4(17.5 + 5) = 90$ kips and one half of it being carried to the end of cantilever arm and is equal to 45 kips. The total load at point II' is therefore $45 + \dfrac{17.5}{2} = 53.8$ kips. Load on point I' $= 17.5$ kips.

The stresses in upper chords for south wind are found to be as follows:

U_2'	U_1'	U_1	U_2	U_3	U_4	U_5	U_6
-53.8	-178.9	-11.8	80.8	-156	-213.7	-253.9	-276.5

For bottom lateral the effect of wind on cantilever is very small and may be neglected. The only load that is acting is 5 kips applied at the point I'.

The equivalent vertical loads found by the same method as used in the previous article are tabulated as follows:

Panel point	On Top Chord	On Bottom Chord	Total
II'	87.4	30.4	117.8
I'	38.7	52.2	90.9
0	50.2	0	50.2

I	38. 6	81. 7	120. 3
II	29. 2	54. 7	83. 9
III	21. 8	33. 1	54. 9
IV	16. 6	16. 9	33. 5
V	13. 5	6. 1	19. 6
VI	12. 4	1. 4	13. 8

Mean value = 65 kips.

The wind stresses for the arch with cantilever arms are given is Table 31.

Table 31. Wind Stresses for the Arch With Cantilever Arms. (in kips) (For South wind)

Member	Positive		Negative		
	Coefficient	Stress	Coefficient	Stress	Total
U_2'	0	0	0	0	0
U_1'	.5050	32. 8	0	0	32. 8
U_1	1. 6658	108. 2	.9799	63. 7	44. 5
U_2	2. 1517	139. 6	2. 1539	139. 8	- . 2
U_3	3. 4835	161. 5	3. 5169	228. 5	- 67. 0
U_4	2. 3153	150. 5	4. 8221	313. 4	- 162. 9
U_5	1. 3389	87. 0	5. 5165	359. 1	- 272. 1
U_6	.6345	41. 8	3. 6752	238. 7	- 196. 9

Member	Positive				
L_1'	0	0	.5740	37.3	-37.3
L_0'	0	0	1.3596	88.3	-83.3
L_0	.7569	49.2	9.1004	591.8	-542.6
L_1	.9937	64.6	8.3938	545.2	-480.6
L_2	1.4355	93.2	7.7572	504.3	-411.1
L_3	2.0256	131.9	5.9876	388.8	-257.9
L_4	2.4842	161.6	3.8607	381.0	-219.4
L_5	2.1216	138.0	3.7772	245.0	-107.0

Member	Positive		Negative		
	Coefficient	Stress	Coefficient	Stress	Total
D_2'	.8856	57.6	0	0	57.6
D_1'	1.4004	91.0	0	0	91.0
D_1	2.1736	141.1	1.1625	75.6	65.6
D_2	2.0895	135.8	.8904	57.8	78.0
D_3	2.0521	133.3	.5790	37.6	95.7
D_4	2.2832	148.4	.4622	30.0	118.4
D_5	3.2736	212.5	1.3751	89.2	123.2
D_6	4.0678	264.6	2.9514	191.9	72.7
V_2'	0	0	1.0000	65.0	-34.0
V_1'	0	0	1.7273	112.2	-60.0
V_0	.9502	61.8	4.1855	348.1	-286.3

V_1	.7315	47.5	2.7160	176.3	−128.5
V_2	.3946	25.7	2.4474	159.2	−78.8
V_3	.1892	12.3	2.2500	145.2	−100.8
V_4	.4529	29.5	2.3549	152.7	−106.3
V_5	.9755	63.4	2.413	157.0	−87.5
V_6	0	0	1.0000	65.0	−12.4

CHAPTER XIX. TEMPERATURE STRESSES

In simple trusses, temperature stresses are generally neglected on the plot that one of the ends is free to move although the frictional resistance in end bearings might stress the structure to some extent. For a two hinged arch the span length is fixed, any change resulting from temperature effects is to stress the members and produce a sag at the centre of span. It has been found that these stresses are very important and cannot be safely neglected without affecting the stability of the structure.

The bridge will be designed for a variation of temperature within the range of 75° y. above and below zero. Take coefficient of linear expansion of steel to be .0000065, the total change in length of span will be 1.755 inches. This is the displacement of end hinge if the latter be free to move while the other is fixed or is the values of δ' in the formula for horizontal thrust. We have calculated δ' for a unit horizontal thrust. It is $2\sum T^2 L$ or 5664. 670 in which T is the stress in kips. L the length of members is feet. E and A both assured equal to unity.

The actual displacement in inches for $H = 1$ kip is $\dfrac{56664.67 \times 12}{A \times 29000} = \dfrac{2.344}{A}$ in which A is the mean area in sq. in. For displacement of 1. 755 inches the value of H may be found by proportion, i, e. , $H = 1.755$ divided by $\dfrac{2.344}{A}$ or 0. 74872A kips. To find the value of A which is the weighted mean of all the areas of the members. Professors Merriman and Jacoby suggested a unique method which has been proved to be very nearly correct. "The influence of chord members on H increase very rapidly toward the middle since the lever are of H increases while at the same time the lever arm of the stresses decrees. On account of these considerations the mean value of A will not differ very far from the area of U_6 whose influence on H is the greatest of all the members. "[1] See Diagram 6.

---

[1] Higher Structures. Merriman and Jacoby. P. 278.

Stress is U_6 due to unit $H = 4.5$ kips

Temperature stress in $U_6 = 4.5 \times H = 3.3692A$ kips which is tension for a rise and compression for a fall of temperature.

$$\text{D. L. Stress in } U_6 = 0$$

$$\text{L. L. Stress in } U_6 = -799.7$$

$$\text{Impact Stress in } U_6 = -266.4$$

$$\text{Direct Wind Stress in } U_6 = -276.5$$

$$\text{Overturning Wind Stress in } U_6 = \underline{196.9}$$

$$\text{Total} = 1145.7 \text{ kips}$$

Total negative stress in $U_6 = -1145.7 - 3.3692(A)$ kips.

The cross sectional area may now be determined if the unit stress be known, Take $\dfrac{l}{r}$ equal to 100 which being the limit set up in the specifications, the unit stress equals 9 kips per sq. in.

Hence, $A = \dfrac{1145.7 - 3.3692(A)}{9.000}$

From which $A = 90.396$ sq. in. The horizontal thrust due to temperature changes may now be found by substituting this value of A in the expression $0.74872 \times A$ or $H = 67.68$ kips.

Having obtained H, the temperature stresses may be found from the values of T and are given in Table 32. These stresses will be assumed to apply to both arches with and without cantilever arms.

Table 32. Temperature Stresses. (in kips) (Fall in temperature)

Member	T	Stress	Member	T	Stress
U_1	-.3333	-22.6	D_1	.7394	49.9
U_2	-.8334	-56.4	D_2	.8766	59.3
U_3	-1.5883	-107.4	D_3	1.0783	72.9
U_4	-2.6667	-180.7	D_4	1.3289	89.9
U_5	-3.8889	-263.3	D_5	1.3891	94.0
U_6	-4.5000	-304.4	D_6	.6779	45.9
L_0	1.1982	81.1	V_0	-.6600	-44.7
L_1	1.5153	102.5	V_1	-.7200	-48.7
L_2	1.9885	134.6	V_2	-.7700	-52.1
L_3	2.7022	182.7	V_3	-.7765	-52.5
L_4	3.7255	252.5	V_4	-.5600	-44.7
L_5	4.8977	331.5	V_5	-.2933	-19.8
			V_6	0	0

CHAPTER XX. SUMMATION OF PRELIMINARY STRESSES

Article 47. Preliminary Stresses for the Arch
without Cantilever Arms.

The preliminary stresses for arch without cantilever arms are summarized in Table 33 in which the D. L. stresses are taken from Table 6. L. L. stresses from Table 25, impact stresses from Table 27, tempeature stresses from Table 32 and wind stresses from Table 30. The maximun and minimum stresses are found for two cases in one of which the effect of wind is considered while in the other it is not. Dead load stresses are reduced to two third their values when added to live load stresses of opposite sign, this being done in accordance with specifications. At the bottom of the table are given the stresses due to full live loads with the variation of temperature. Wind stresses, however, are neglected.

Table 33. Summation of Preliminary Stresses for the Arch Without Cantilever Arms. (in kips)

(a) Upper Chords.

Member		U_1	U_2	U_3	U_4	U_5	U_6
D. L.		+14.8	+23.8	+26.1	+20.8	+12.0	0
L. L.		+69.1	+133.2	+181.7	+173.3	+78.5	+9.81
Imp.		+31.5	+62.7	+88.4	+89.5	+46.2	+9.50
Temp. Rise		+22.6	+56.4	+107.4	+180.7	+263.5	+304.4
S. Wind Direct		+48.2	+135.5	+205.5	+238.1	+293.0	+310.4
N. Wind Over		+22.0	+55.4	+104.5	+177.3	+258.3	+298.8
L. L.		−126.9	−279.8	−461.5	−642.5	−761.1	−799.8
Imp.		−70.5	−150.3	−237.0	−311.4	−331.5	−269.1
Temp. Fall		−22.6	−56.4	−107.4	−180.7	−263.3	−304.4
N. Wind Direct		−48.2	−135.6	−205.5	−258.1	−293.0	−310.4
S. Wind Over		−22.0	−53.4	−104.5	−177.3	−258.3	−298.8
Without	Max.	120.0	276.1	403.6	464.3	4004	323.7
Wind	Min.	−210.1	−470.6	−788.5	−1120.8	−1347.6	−1373.3
With	Max.	168.2	411.7	609.1	722.4	653.4	634.1
Wind	Min.	−258.3	−606.2	−994.0	−1378.9	−1640.6	−1683.7
Full L. L.	Temp. Rise	−60.6	−154.0	−294.9	−489.6	−692.6	−745.2

| Stresses | Temp. Fall | − 104. 6 | − 266. 8 | − 509. 7 | − 650. 0 | − 1219. 2 | − 1354. 0 |

(b) Lower Chords.

Member	L_0	L_1	L_3	L_3	L_4	L_5
L. L.	0	+ 25. 5	+ 78. 8	+ 153. 9	218. 4	+ 190. 9
Imp.	0	+ 20. 1	33. 8	+ 94. 1	+ 116. 1	+ 97. 1
Temp. Fall	+ 81. 1	+ 102. 5	+ 134. 6	+ 182. 7	+ 252. 5	+ 331. 5
S. Wind Direct	+ 16. 5	+ 44. 1	63. 8	+ 77. 0	+ 85. 2	+ 89. 0
N. Wind Over	+ 456. 0	+ 407. 7	+ 352. 3	+ 286. 9	+ 207. 0	+ 121. 8
D. L.	− 863. 3	− 801. 4	− 757. 6	− 723. 7	− 710. 2	− 710. 1
L. L.	− 1203. 0	− 1099. 1	− 1011. 2	− 912. 3	− 764. 2	− 513. 4
Imp.	− 401. 5	− 402. 7	− 400. 0	− 385. 2	− 347. 1	− 252. 3
Temp. Rise	− 81. 1	− 102. 5	− 163. 6	− 182. 7	− 252. 5	− 331. 5
N. Wind Direct	− 16. 5	− 44. 1	− 63. 8	− 77. 0	− 85. 2	− 89. 0
S. Wind Over	− 456. 0	− 407. 7	− 352. 3	− 286. 9	− 207. 0	− 121. 8
Without Max.	− 493. 9	− 384. 9	− 237. 8	− 51. 3	115. 0	147. 5
Wind Min.	− 2584. 9	− 2405. 7	− 2303. 4	− 2203. 9	− 2074. 0	− 1807. 3
With Max.	− 37. 9	22. 8	114. 5	235. 6	322. 0	269. 3
Wind Min.	− 3004. 9	− 2813. 4	− 2591. 2	− 2490. 8	− 2281. 0	− 1929. 1
Full Temp. L. L. Rise	− 2548. 9	− 2360. 1	− 2170. 8	− 1955. 9	− 1740. 5	1519. 3

Stresses Temp Fall	-2386.7	-2155.1	-1901.6	-1590.5	-1235.5	-866.3

(c) Diagonals.

Member	D_1	D_2	D_3	D_4	D_5	D_6
D. L.	+33.2	+16.1	-1.8	-5.9	-10.6	-6.9
L. L.	+281.5	+272.7	+271.3	+299.1	+413.7	+503.4
Imp.	+156.1	+143.7	+156.1	+161.0	+213.6	+285.9
Temp. Fall	+49.9	+59.3	+72.9	+89.9	+94.0	+45.9
S. Wind Over	+49.0	+58.5	+71.6	+88.4	+92.5	+58.0
L. L.	-151.7	-118.2	-82.2	-65.4	-170.1	-351.7
Imp.	-68.8	-56.2	36.2	-30.7	-98.1	-185.2
Temp. Rise	-49.9	-59.3	-72.9	-89.9	-84.0	-45.9
N. Wind Over	-49.0	-58.5	-71.6	-88.4	-92.5	-58.0
Without Max.	510.9	491.8	502.1	555.9	731.9	842.1
Wind Min.	-248.3	-223.0	-190.1	-182.1	-355.1	-587.2
With Max.	559.9	550.3	573.7	644.3	824.4	900.1
Wind Min.	-297.3	-281.5	-261.3	-270.5	-447.6	645.2
Full L. L. Temp. Rise	200.4	198.8	237.9	270.0	275.7	213.4
Stresses Temp. Fall	300.2	317.4	383.7	449.8	463.7	-305.2

(d) Verticals.

Member	V_0	V_1	V_2	V_3	V_4	V_5	V_6
L. L.	+134.5	+97.1	+55.9	+26.8	+52.4	+109.2	0
Imp.	+44.7	+44.2	+27.6	+13.4	+34.2	+57.4	0
Temp. Rise	+44.7	+48.7	+52.1	+52.5	+44.7	+19.8	0
N. Wind Over	+97.6	+20.0	+43.5	+72.1	+80.6	+69.3	12.4
D. L.	−72.1	78.8	−60.2	−54.2	−50.2	−50.9	−46.6
L. L.	−391.5	−365.8	−333.1	−304.6	−309.2	−307.9	−141.6
Imp.	−129.3	−192.8	−168.3	−152.4	−125.1	−146.5	47.2
Temp. Fall	−44.7	−48.7	−52.1	−52.5	−44.7	−19.8	0
S. Wind Over	−97.6	−20.0	−43.5	−72.1	−80.6	−69.3	−12.4
Without Max.	175.8	137.5	95.5	56.6	97.9	152.5	0
Wind Min.	−637.6	−686.1	−613.7	−563.9	−529.2	−525.1	−235.4
With Max.	273.4	157.5	139.0	128.7	178.5	221.8	124.0
Wind Min.	−735.2	−706.1	−657.2	−636.0	−609.8	−594.4	−247.8
Full Temp. L. L. Rise	−369.0	−447.4	−426.0	−418.7	−353.2	−318.9	−235.4
Stresses Temp. Fall	−458.4	−544.8	−530.2	−523.7	−442.6	−358.5	−235.4

Article 48. Preliminary Stresses for the Arch with Cantilever Arms.

In Table 34 are given the same summations as described in the previous article. The D. L. Stresses are taken from Table 4,

L. L. Stresses from Table 26, impact stresses from Table 28, temperature stresses from Table 32, and wind stresses from Table 31.

Article 49. Comparison of Stresses in the Arch with and without Cantilever Arms.

To investigate the effect of balancing an arch by the introduction of two cantilever arms without changing the span length between springings, three diagrams are constructed. The first compares maximum stresses, second, minimum stresses and the third, full load stresses.

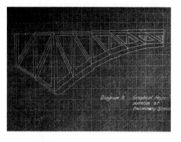

Table 34. Summation of Preliminary Stresses for the Arch With Cantilever Arms. (in kips)

(a) Upper Chords.

Member	U_2'	U_1'	U_1	U_2	U_3	U_4	U_5	U_6
D. L.	0	76.9	131.5	105.6	78.4	47.7	16.8	0
L. L.	0	227.1	475.1	550.1	595.1	549.6	348.9	204.7
Imp.	0	123.9	148.8	175.5	194.5	187.1	129.4	68.2

Temp. Rise	0	0	22.6	56.4	107.4	180.7	363.3	304.4
S. Wind Direct	−53.8	−178.9	−11.8	80.8	156.0	213.7	253.9	276.5
S. Wind Over	0	32.8	44.5	0	0	0	0	0
N. Wind Over	0	0	0	.2	67.6	162.9	272.1	196.9
L. L.	0	0	156.8	341.4	552.4	742.8	810.0	799.7
Imp.	0	0	55.7	119.0	186.0	242.4	245.3	266.4
Temp. Fall	0	0	22.6	56.4	107.4	180.7	263.3	304.4
S. Wind Direct	+53.0	+78.9	+11.8	80.8	156.0	13.7	253.7	276.5
S. Wind Over	0	32.8	44.5	0	0	0	0	0
N. Wind Over	0	0	0	.2	67.6	162.9	272.1	196.9
Without Max.	0	427.9	778.0	887.6	975.4	965.1	758.4	577.3
Wind Min.	0	51.2	−146.4	−446.3	−795.6	−1134.1	−1307.4	−1370.5
With Max.	53.8	606.8	769.8	968.4	1131.4	1178.8	1031.3	853.8
Wind Min.	−53.8	−127.6	−158.2	−527.1	−951.6	−1347.8	−1580.0	−1647.0
Full Temp. L. L. Rise	0	418.4	580.4	486.7	366.8	199.5	7.4	−58.1

| Stresses | Temp. Fall | 0 | 418.4 | 531.2 | 363.3 | 132.1 | −194.9 | −568.6 | 723.6 |

(b) Lower Chords.

Member	L_1'	L_0'	L_0	L_1	L_2	L_3	L_4	L_5
L. L.	0	0	260.9	306.9	381.6	487.6	542.9	457.3
Imp.	0	0	130.4	136.9	165.1	184.7	193.4	153.7
Temp. Rise	0	0	81.1	102.5	134.6	182.7	252.5	331.5
S. Wind Direct	−5.9	−22.7	16.6	44.1	63.8	77.0	85.2	89.0
N. Wind Over	37.3	88.3	542.6	480.6	411.1	257.9	219.4	107.0
D. L.	87.3	188.5	823.0	750.0	690.0	635.0	587.0	548.0
L. L.	256.1	468.4	1410.6	1314.5	1225.9	968.7	925.2	565.9
Imp.	144.9	234.2	353.1	350.8	345.6	287.9	289.9	186.7
Temp. Rise	0	0	81.1	102.5	134.6	182.7	252.5	331.5
N. Wind Direct	+5.9	+22.7	16.5	44.1	63.8	77.0	85.2	89.0
S. Wind Over	37.3	86.3	542.5	480.6	411.1	257.9	219.4	107.0
Without Wind Max.	−58.2	−125.6	−76.0	44.3	212.3	433.0	597.6	577.1
Without Wind Min.	−488.3	−891.1	−2667.8	−2517.8	−2396.1	−2074.3	−2054.6	−1632.1

416

With Max.	-20.9	37.3	466.6	524.9	623.4	690.9	817.0	684.1
Wind Min.	-531.5	-1002.1	-3210.4	-2998.4	-2807.2	-2332.2	-2274.0	-1739.0
Full L.L. Temp. Rise	-476.3	-868.7	-2266.7	-2155.6	-1939.0	-1757.9	-1550.4	-1368.0
Stresses Temp. Fall	-476.3	-868.7	-2089.5	-1931.4	-1644.8	-1358.5	-999.2	-643.8

(c) Diagonals.

Member	D_2'	D_1'	D_1	D_2	D_3	D_4	D_5	D_6
D. L.	134.5	189.9	57.8	45.0	38.9	37.9	35.1	16.5
L. L.	395.0	368.9	346.2	329.9	318.5	357.9	534.7	679.2
Imp.	215.4	184.4	123.4	113.9	85.3	92.7	162.1	219.5
Temp. Fall	0	0	49.9	59.3	72.9	89.9	94.0	45.9
S. Wind Over	57.6	91.0	65.5	78.0	95.7	118.4	123.2	72.7
L. L.	0	0	183.4	138.2	82.1	65.4	230.5	513.5
Imp.	0	0	57.3	44.7	36.3	30.7	85.3	177.5
Temp. Rise	0	0	49.9	59.3	72.9	89.9	94.0	45.9
N. Wind Over	53.6	91.0	65.5	78.0	95.7	118.4	123.2	72.7
Without Max.	744.9	741.2	577.2	548.1	515.6	578.4	825.9	961.1
Wind Min.	89.7	126.1	-252.1	-212.1	-165.4	-160.8	-386.4	-725.9
With Max.	802.5	834.2	643.8	626.1	611.3	697.3	949.1	1033.8
Wind Min.	32.1	27.9	-317.6	-290.2	-261.1	-279.7	-509.6	-798.6
Full L.L. Temp. Rise	734.1	719.2	161.4	151.4	134.4	186.9	201.1	125.8

| Stresses Temp. Fall | 734.1 | 719.2 | 270.6 | 280.8 | 293.8 | 383.3 | 406.4 | 226.2 |

(d) Verticals.

Member	V_2'	V_1'	V_0	V_1	V_2	V_3	V_4	V_5	V_6
L. L.	0	0	134.4	112.9	56.0	26.8	81.0	179.4	0
Imp.	0	0	44.0	36.5	26.7	13.6	32.0	66.5	0
Temp. Rise	0	0	44.7	48.7	52.1	52.5	44.7	19.8	0
N. Wind Over	34.6	60.0	286.3	128.5	78.8	100.8	106.3	67.5	12.4
D. L.	121.3	177.0	299.0	103.8	89.6	79.8	71.9	61.0	46.6
L. L.	445.8	465.8	762.2	412.4	374.3	338.7	393.6	392.2	141.6
Imp.	243.4	232.9	254.0	141.9	95.8	84.1	151.4	118.9	10.4
Temp. Fall	0	0	44.7	48.7	52.1	52.5	44.7	19.8	0
S. Wind Over	34.6	60.0	286.5	128.5	78.8	100.8	106.3	87.5	12.4
Without Max.	−80.9	−117.8	24.9	128.9	75.3	39.7	109.9	225.1	−31.0
Wind Min.	−81.0	−875.7	−1359.9	−706.8	−611.7	−555.1	−661.6	−591.99	−198.6
With Max.	−46.3	−65.6	311.2	259.4	164.1	140.5	216.2	312.6	−18.6
Wind Min.	−845.7	−935.7	−1646.2	−835.3	−690.5	−655.9	−767.9	−679.4	−211.0
Full L.L. Temp. Rise	−609.3	−855.2	−789.9	−332.1	−331.3	−342.2	−311.7	−264.9	−256.6
Stresses Temp. Fall	−809.3	−856.2	−896.5	−438.5	−445.1	−499.0	−409.2	−308.1	−256.6

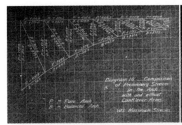

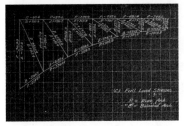

In those diagrams the effect of wind is neglected and a rise of temperature considered in the full load stresses. B stands for balanced arch or the arch with cantilever arms and P for the pure arch or the arch without cantilever arms. Members of cantilever arms not entered into comparison.

It will be seen that for maximum and minimum stresses there is not much difference between the two types while for full load stresses the balanced type has a tremendous advantage over those for the pure arched structure.

CHAPTER XXI. PRELIMINARY DESIGN

Article 50. Verticals.

The verticals will be designed to secure the greatest moment of inertia for the smallest overall dimensions. For this reason square box form with open lacing is adopted as shown in the accompanying figure. [1]

Plates and angles are the principal sections used: fills, covers and flats are used to provide additional

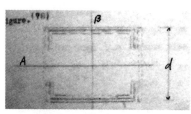

arms. In the figure, dashed lines represent additional sections to be used for heavy members while dotted lines indicated lacing.

To enable all the floor beams to be made of the same length, the distance d perpendicular to the plane of truss will be made constant and as small as possible to reduce the secondary stresses. [2]

The sections are designed without taking into account the effect of wind loads which are provided for by an increased unit stress as 25% above normal.

Member V_0

Greatest stresses: Without wind = -1359.9 kips

With wind = -1646.2 kips

① Steel Railway Bridge. Dilworth. P. 12.
② Modern Structures. Part III. P. 105.

$$2 \text{ webs } 25'' \times \frac{7}{8}'' \qquad = 43.7$$

$$4 \lfloor S \ 6'' \times 4'' \times \frac{7}{8}'' \qquad = 32.0$$

$$2 \text{ fills } 13'' \times \frac{7}{8}'' \qquad = 22.7$$

$$2 \text{ side pl. } 23\frac{1}{2}'' \times \frac{1}{8}'' \quad = 23.25$$

$$6 \text{ flats } 5 \times \frac{1}{2} \qquad = \underline{20.0}$$

$$\text{Area supplied} \qquad = 141.65 \text{ sq. in.}$$

Least "r" about axis $A = 8.8''$

$$\text{Length} = 792''$$

$$\text{Unit Stress} = 9.7 \text{ kips per sq. in.}$$

Area required: —

Without wind stresses $= 140$ sq. in.

With wind Stresses $= 135.8$ sq. in.

Member V_1

Greatest Stresses: — Without wind $= -706.8$ kips

with wind $= -635.3$ kips

$$2 \text{ webs } 25'' \times \frac{9}{16}'' \qquad = 28.1$$

$$4 \lfloor S \ 6'' \times 4'' \times \frac{3}{4}'' \qquad = 27.8$$

$$4 \text{ flats } 5'' \times \frac{1}{2}'' \qquad\qquad = \underline{10.0}$$

Area Supplied $= 65.9$ sq. in.

Least "r" about axis A $= 9.02$ in.

Length $= 593$ in.

Unit stress $= 11.39$ kips per sq. in.

Area required: −

Without wind stresses $= 62.1$ sq. in.

With wind stresses $= 58.5$ sq. in.

Member V_2

Greatest stresses: Without wind $= -611.7$ kips

With wind $= -690.5$ kips

$$2 \text{ webs } 25'' \times \frac{1}{3} \quad = 25.0$$

$$4 \lfloor S \ 6'' \times 4'' \times \frac{3}{4}'' \quad = \underline{27.8}$$

Area supplied $= 52.8$

Least "r" about axis $S = 9.05$ in.

Length $= 432$ in.

Unit stress $= 12.66$ kips per sq. in.

Area required: −

Without wind stresses $= 48.2$ sq. in.

With wind stresses $= 13.6$ sq. in.

Member V_3

Greatest stresses: −

Without wind $= -555.1$ kips

With wind $= -655.9$ kips

2 webs $25'' \times \dfrac{1}{2}''$ $= 25.0$

4 $\underline{|S}$ $6'' \times 4'' \times \dfrac{1}{2}''$ $= \underline{19.0}$

Area supplied $= 44.0$ sq. in.

Least "r" about axis $B = 8.57$ in.

Length $= 306$ in.

Unit stress $= 13.5$ kips per sq. in.

Area required: −

Without wind stresses $= 42.7$ sq. in.

With wind stresses $= 38.8$ sq. in.

Member V_4

Greatest stresses: −

Without wind stresses $= -661.6$ kips

With wind stresses $= -767.9$ kips

2 webs $25'' \times \dfrac{1}{2}'' = 25.0$

4 $\underline{|S}$ $6'' \times 4'' \times \dfrac{5}{8}'' = \underline{23.4}$

Area supplied $= 48.4$ sq. in.

Least "r" about axis $B = 9.02$ in.

Length $= 216$ in.

Unit stress $= 14$ kips per sq. in.

Area required: –

Without wind stresses $= 46.2$ sq. in.

With wind stresses $\quad = 43.8$ sq. in.

Member V_5

Greatest stresses: –

Without wind stresses $= -591.9$ kips

With wind stresses $\quad = -679.4$ kips

2 webs $25'' \times \dfrac{1}{2}''$ $\quad = 25.0$

4 $\lfloor S \; 6 \times 4'' \times \dfrac{1}{2}$ $\quad = \underline{19.0}$

Area supplied $\quad = 44.0$ sq. in.

Least "r" about axis $B = 8.57$ in.

Length $= 162$ in.

Unit stress $= 14.7$ kips per sq. in.

but specification limits to 14 kips per sq. in.

Area required: –

Without wind stresses $= 40.3$ sq. in.

With wind stresses   = 38. 9 sq. in.

Member V_6

Greatest stresses:

      Without wind stresses   = $-198. 6$ kips

      With wind stresses   = $-211. 0$ kips

      2 webs $20'' \times \dfrac{3}{8}''$ = 15. 0

      4 ⌊S $6'' \times 4'' \times \dfrac{3}{8}''$ = 14. 5

      Area supplied   = 29. 5 sq. in.

      Least "r" about axis B = 6. 95 in.

      Length   = 144 in.

       Unit stress = 14. 55 kips per sq. in.

Use 14 kips per sq. in.

Area required: –

      Without wind stresses = 13. 6 sq. in.

      With wind stresses   = 12. 1 sq. in.

Member V_1'

Greatest stresses: –

      Without wind   = $-875. 7$ kips

      With wind   = $-935. 7$ kips

      2 webs $25'' \times \dfrac{3}{4}''$ = 37. 5

$$4 \lfloor S \quad 6'' \times 4'' \times \frac{3}{4}'' \qquad = 27.8$$

$$4 \text{ flats } 5'' \times \frac{5}{8}'' \qquad = \underline{12.5}$$

Area supplied $\qquad = 77.8$ sq. in.

Least "r" about axis $A = 9.13$ in.

Length $\qquad = 593$ in.

Unit stress $= -11.95$ kips per sq. in.

Area required: –

Without wind stresses $= 76.3$ sq. in.

With wind stresses $\quad = 65.5$ sq. in.

Member V_2'

Greatest stresses: –

Without wind $= -810$ kips

With wind $\quad = -845.1$ kips

$$2 \text{ webs } 25'' \times \frac{3}{4}'' \quad = 37.5$$

$$4 \lfloor S \quad 6'' \times 4'' \times \frac{3}{4}'' = \underline{27.6}$$

Area supplied $\qquad 65.3$ sq. in.

Least "r" about axis $B = 5.8$ in.

Length $= 432$ in.

Unit stress $= 12.57$ kips per sq. in.

Area required: —

> Without wind stresses = 64.6 sq. in.
>
> With wind stresses   = 54.0 sq. in.

Article 51. Diagonals.

The same section as for verticals will be used for the diagonals. Both tension and compression are investigated.

Member D_2'

Maximun stresses: —

> Without wind = +744.9 kips
>
> With wind   = +802.3 kips

Minimun stresses: —

> Without wind   = +69.7 kips
>
> With wind   = +32.1 kips
>
> 2 webs $20'' \times \dfrac{3}{4}''$ = 26.37

Deduct 4 1" holes

$$4 \underline{\lfloor S} \ 6'' \times 4'' \times \dfrac{7}{8}'' = 25.00$$

> Deduct 8 1" holes ———
>
> Net area supplied = 51.37 sq. in.

Net area required: —

> Without wind stresses = 46.8 sq. in.
>
> With wind stresses   = 40.0 sq. in.

Least "r" about axis $A = 6.95$ in.

Length $= 626$ in.

$$\frac{1}{r} = 75.7$$

Member D_1'

Maximum stresses: —

Without wind $= +743.2$ kips

With wind $\quad = +834.2$ kips

Minimum stresses: —

Without wind $= 89.7$ kips

With wind $\quad = 32.1$

Same section as D_2' used.

Net area required: —

Without wind stresses $\quad = 46.5$ sq. in.

With wind stresses $\quad = 41.7$ sq. in.

Least "r" about axis $A = 6.95$ in.

Length $= 665$ in.

$$\frac{1}{r} \quad = 95.7$$

Member D_1

Maximum stresses: —

Without wind $= +577.3$ kips

With wind $\quad = +543.8$ kips

Minimum stresses: −

$$\text{Without wind} \quad = -252.1 \text{ kips}$$

$$\text{With wind} \quad = -317.6 \text{ kips}$$

$$2 \text{ webs } 20'' \times \frac{1}{2}'' = 18.0$$

Deduct 4 − 1″ holes

$$4 \lfloor S \ 6'' \times 4'' \times \frac{3}{4}'' = 20.13$$

Deduct 8 − 1″ holes _____

$$\text{Net area supplied} = 38.13 \text{ sq. in.}$$

Net area required: −

$$\text{Without wind stresses} \ = 36.0 \text{ sq. in.}$$

$$\text{With wind stresses} \quad = 32.2 \text{ sq. in.}$$

$$\text{Least "}r\text{" about axis } A = 7.11 \text{ in.}$$

$$\text{Length} = 665 \text{ in.}$$

$$\text{Unit stress} = 9.45 \text{ kips per sq. in.}$$

Area required for compression: −

$$\text{Without wind stresses} = 26.7 \text{ sq. in.}$$

$$\text{With wind stresses} \quad = 35.8 \text{ sq. in.}$$

Member D_2

Maximum stresses: −

$$\text{Without wind} = +548.1 \text{ kips}$$

$$\text{With wind} \quad = +626.1 \text{ kips}$$

Minimum stresses: −

$$\text{Without wind} \quad = -212.1 \text{ kips}$$

$$\text{With wind} \quad = -290.2 \text{ kips}$$

$$2 \text{ webs } 20'' \times \frac{3}{8}'' = 15.50$$

Deduct 4 − 4″ holes

$$4 \lfloor \underline{S} \ 6'' \times 4'' \times \frac{3}{4}'' = 20.13$$

Deduct 6 − 1″ holes _____

Net area supplied = 35.63 sq. in.

Net area required: −

$$\text{Without wind stresses} \quad = 34.3 \text{ sq. in.}$$

$$\text{With wind stresses} \quad = 31.2 \text{ sq. in.}$$

$$\text{Least "} r \text{" about axis } A = 7.19 \text{ in.}$$

$$\text{Length} = 527 \text{ in.}$$

$$\text{Unit stress} = 10.88 \text{ kips per sq. in.}$$

Area required for compression: −

$$\text{Without wind stresses} = 19.5 \text{ sq. in.}$$

$$\text{With wind stresses} \quad = 21.3 \text{ sq. in.}$$

<u>Member D_3</u>

Maximum stresses: −

$$\text{Without wind} = +515.6 \text{ kips}$$

$$\text{With wind} \quad = +611.3 \text{ kips}$$

Minimum stresses: −

        Without wind = − 165.4 kips

        With wind   = − 251.1 kips

Same section as D_2 used.

Net area required: −

        Without wind stresses = 35.6 sq. in.

        With wind stresses   = 20.5 sq. in.

        least "r" about axis A = 7.19 in.

                Length = 428 in.

        Unit stress = 12.52 kips per sq. in

Area required for compression: −

        Without wind = 13.2 sq. in.

        With wind   = 15.8 sq. in.

Member D_4

Maximum stresses: −

        Without wind = + 578.4 kips

        With wind   = + 697.3 kips

Minimum stresses: −

        Without wind   = − 160.8 kips

        With wind   = − 279.7 kips

        2 webs $20'' \times \dfrac{1}{2}'' = 18.0$

Deduct 4 – 1″ holes

$$4 \lfloor S \ 6'' \times 4'' \times \frac{3}{4}'' = 20.13$$

Deduct 8 – 1″ holes
_____

Net area supplied   38. 13 sq. in.

Net area required: –

        Without wind stresses = 36. 1 sq. in.

        With wind stresses   = 34. 8 sq. in.

        Least "r" about axis A = 7. 11 in.

            Length = 370 in.

Unit stress = 12. 43 kips per sq. in.

Area required for compression: –

        Without wind stresses = 12. 9 sq. in.

        With wind stresses   = 18. 0 sq. in.

Member D_5
_____

Maximum stresses: –

        Without wind = + 825. 9 kips

        With wind   = + 949. 1 kips

Minimum stresses: –

        Without wind = – 386. 4 kips

        With wind   = – 509. 6 kips

$$2 \text{ webs } 20'' \times \frac{13''}{16} = 28.86$$

Deduct 4 – 1″ holes

$$4 \underline{|S}\ 6'' \times 4'' \times \frac{7}{8}'' = 25.0$$

Deduct 8 – 1″ holes

Net area supplied = 33. 86. sq. in.

Net area required：–

Without wind stresses = 51. 7 sq. in.

With wind stresses   = 47. 4 sq. in.

Least "r" about axis A = 6. 87 in.

Length = 342 in.

Unit stress = 12. 52 kips per sq. in.

Area required for compression：–

Without wind stresses = 30. 9 sq. in.

With wind stresses   = 32. 5 sq. in.

Member D_6

Maximum stresses：–

Without wind = + 961. 1 kips

With wind   = + 1033. 8 kips

Minimum stresses：–

Without wind = – 725. 9 kips

With wind   = – 798. 6 kips

$$2\ \text{webs}\ 20'' \times \frac{7}{8}'' = 38.05\ \text{sq. in.}$$

Deduct 4 1″ holes

$$4 \lfloor S \ 6'' \times 4'' \times \frac{7}{8}'' = 25.00$$

Deduct 8 1″ holes

Net area supplied   =   63.05 sq. in.

Net area required: −

Without wind stresses = 60.0 sq. in.

With wind stresses   = 51.9 sq. in.

Least "r" about axis A = 6.78 in.

Length = 332 in.

Unit stress = 12.57 kips per sq. in.

Area required for compression: −

Without wind stresses = 57.8 sq. in.

With wind stresses   = 50.8 sq. in.

Article 52. Lower Chords.

The lower chords will be designed for compression only. As the verticals are to come into chords the distance c is made constant and is equal to the width of the verticals plus 2″ for gusset plates. Heavy lacing will be used at both the upper and lower faces of the member.

Member L_0

Greatest stresses: −

Without wind = -2667.8 kips

With wind　 = -3210.4 kips

2 webs $40'' \times \dfrac{3}{4}''$　　= 60

2 webs $40'' \times \dfrac{3}{4}''$　　= 60

4 $\underline{\lfloor S}$ $6'' \times 4'' \times \dfrac{3}{4}''$　= 27.8

2 fills $32'' \times \dfrac{3}{4}''$　　= $\underline{48.0}$

Area supplied　　　= 195.8 sq. in.

Least "r" about axis $A = 10.3$ in.

Length = 360 in.

Unit stress = 13.65 kips per sq. in.

Area required: −

Without wind stresses = 195.5 sq. in.

With wind stresses　 = 188.0 sq. in.

Member L_1

Greatest stresses: −

Without wind　 = -2517.8 kips

With wind　　 = -2998.4 kips

2 webs $40'' \times \dfrac{3}{4}''$ = 60

$$2 \text{ webs } 40'' \times \frac{5''}{8} = 50$$

$$4 \ 6'' \times 4'' \times \frac{3''}{4} = 27.8$$

$$2 \text{ fills } 32'' \times \frac{3''}{4} = 48.0$$

Area supplied $\quad = \quad\quad\quad 185.8$ sq. in.

Length $= 341$ in.

Unit stress $= 13.7$ kips per sq. in.

Area required: –

Without wind stresses $= 183.5$ sq. in.

With wind stresses $\quad = 175.0$ sq. in.

Member L_2

Greatest stresses: –

Without wind $\quad\quad = -2396.1$ kips

With wind $\quad\quad\quad = -2607.2$ kips

$$4 \text{ webs } 40'' \times \frac{5''}{8} = 100.00$$

$$4 \ \underline{|S} \ 6'' \times 4'' \times \frac{3''}{4} = 27.8$$

$$2 \text{ fills } 32'' \times \frac{3''}{4} = \underline{43.0}$$

Area supplied $\quad = 175.8$ sq. in.

Least "r" about axis $A = 10.25$ in.

$$\text{Length} = 325 \text{ in.}$$

$$\text{Unit stress} = 13.7 \text{ kips per sq. in.}$$

Area required: –

$$\text{Without wind stresses} = 175.0 \text{ sq. in.}$$

$$\text{With wind stresses} \quad = 164.0 \text{ sq. in.}$$

Member L_3

Greatest stresses: –

$$\text{Without wind} \qquad = -2074.3 \text{ kips}$$

$$\text{With wind} \qquad = -2332.2 \text{ kips}$$

$$2 \text{ webs } 40'' \times \frac{1}{2}{}'' \quad = 40$$

$$2 \text{ webs } 40'' \times \frac{5}{8}{}'' \quad = 50$$

$$4 \underline{\lfloor S} \ 6'' \times 4'' \times \frac{5}{8}{}'' = 23.4$$

$$2 \text{ fills } 32'' \times \frac{5}{8}{}'' \quad = 40$$

$$\text{Area supplied} \qquad = 153.4 \text{ sq. in.}$$

$$\text{Least } "r" \text{ about axis A} = 10.3 \text{ in.}$$

$$\text{Length} = 313 \text{ in.}$$

$$\text{Unit stress} = 13.87 \text{ kips per sq. in.}$$

Area required: –

$$\text{Without wind stresses} = 150.0 \text{ sq. in.}$$

With wind stresses   = 134. 0 sq. in.

Member L_4

Greatest stresses : –

Without wind = – 2054. 6 kips

With wind   = – 2274. 0 kips

Same section as L_3 used.

Least " r " about axis A = 10. 3 in.

Length = 305 in.

Unit stress = 13. 97 kips per sq. in.

Area required : –

Without wind stresses = 147. 0 sq. in.

With wind stresses   = 130. 0 sq. in.

Member L_5

Greatest stresses : –

Without wind   = – 1632. 1 kips

With wind   = – 1739. 0 kips

2 webs $40'' \times \dfrac{1}{16}''$ = 55

4 $\underline{|S}$ $6'' \times 4'' \times \dfrac{5}{8}''$ = 23. 4

2 fills $32'' \times \dfrac{5}{8}''$ = 40

Area supplied $\qquad$ = 118. 4 sq. in.

Least "r" about axis A = 0. 9 in.

Length = 300. in.

Unit stress = 13. 55 kips per sq. in.

Area required: −

Without wind stresses = 118. 0 sq. in.

With Stresses $\qquad$ = 100. 5 sq. in.

Members L_0'

Greatest stress: −

Without wind $\qquad$ = −891. 1 kips

With wind $\qquad$ = −1002. 1 kips

2 webs $40'' \times \dfrac{5}{8}'' = 50$

$4 \ 6'' \times 4'' \times \dfrac{5}{8}'' = 23. 4$

Area supplied = 73. 4

Least "r" about axis A = 10. 1 in.

Length = 360 in.

Unit stress = 13. 5 kips per sq. in.

Area required: −

Without wind stresses = 66. 9 sq. in.

With wind stresses = 60. 4 sq. in.

Member L_1'

Greatest stresses: −

Without wind	$= -488.3$ sq. in.
With wind	$= -531.5$ sq. in.
2 webs $40'' \times \dfrac{1}{2}''$	$= 30.0$
4 $5'' \times 4 \times \dfrac{1}{2}''$	$= 19.0$
Area supplied	$= 49.0$ sq. in.

Least "r" about axis $A = 9.73$ in.

Length $= 341$ in.

Unit stress $= 13.57$ kips per sq. in.

Area required: −

Without wind stresses	$= 36.0$ sq. in.
With wind stresses	$= 31.3$ sq. in.

Article 53. Upper Chords.

The sections of upper chords are made as shown in which the distance c is kept constant and is equal to the widths of verticals plus $2''$ for thickness of gusset plates.

The neutral axis is kept as near to center as possible. In order to make the radius of gyration in the transverse direction not less than that in the planes of the truss, the depth of web will be made about $\dfrac{4}{3}$ the width inside of the webs or 25″.

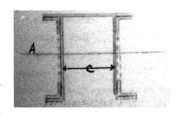

The thickness of cover plates and webs will be made as uniform as possible to enable the rivet gauge in the horizontal flanges and cover plates to be multiple panels.

Member U_6

Greatest stresses: −

Without wind	= −1370.5 kips
With wind	= −1647.0 kips
2 webs $30'' \times \dfrac{1}{2}''$	= 30
4 $\lfloor S \ 6'' \times 4'' \times \dfrac{1}{2}$	= 19
1 cover $35'' \times \dfrac{1}{2}'' = 17.7$	
2 fills $22 \times \dfrac{1}{2}''$	= 22.
2 flats $6'' \times \dfrac{3}{4}''$	= 10.5
Area supplied	= 99.2 sq. in.

Least "r" about axis A which is 1.13 in. Above center line of section = 11.5 in.

Length = 300 in.

Unit stress = 14.17 kips per sq. in.

Use 14. kips per sq. in.

Area required: −

Without wind stresses = 97.8 sq. in.

With wind stresses = 94.0 sq. in.

Members U_5 and U_4 will be made the same as U_6. As their stresses are smaller and the lengths are the same there is no need of further investigation.

Member U_3

Maximum stresses: −

Without wind = +975.4 kips

With wind = +1131.4 kips

Minimum stresses:

Without wind = −795.5 kips

With wind = −951.5 kips

2 webs $30'' \times \dfrac{1}{2} = 30$

(Deduct 12 − 1″ holes = 24)

1 cover $35'' \times \dfrac{1}{2}$ = 17.7

(Deduct 2 − 1″ holes = 16.7)

$$4 \underline{\underline{S}}\ 6'' \times 4'' \times \frac{1}{2} \qquad = 19.0$$

(Deduct 8 – 1″ holes = 15)

$$2 \text{ flats } 6\frac{1}{2}'' \times \frac{3}{4}'' \qquad = 10.5$$

(Deduct 2 – 1″ holes = 9.5)

Area supplied	= 77.2 sq. in.
Net area	= 65.2 sq. in.

Net area required: –

Without wind stresses = 61.0 sq. in.

With wind stresses   = 56.4 sq. in.

Least "r" about axis A which in 1.45 in. above center line of section = 12.9 in.

Length = 300 in.

Unit stress = 14.3 kips per sq. in.

Use 14.0 kips per sq. in.

Area required for compression: –

Without wind stresses = 66.8 sq. in.

With wind stresses   = 54.3 sq. in.

Members U_2, U_1, U_1' and U_2' will be made the same sections as U_3.

From the above design of the members of the arch it may be

safely said that the effect of wind is general negligible if the unit stresses are allowed to increase a certain percent. In this case 25% was added to the normal stresses.

Another thing that may require notice is the uniformity of the stresses and consequently the sections of the lower chords. This is plainly seen in Diagram 9.

CHAPTER XII. STRESSES AND DESIGN OF LATERAL SYSTEM

Article 54. Stresses.

In designing lateral systems the wind loads of 17.5 kips applied at upper panel points of truss and 5 kips at lower panel points are considered as moving. The stresses are found by means of coefficients as described in Chapter XVIII. For diagonals within the arch span the maximum stresses occur when the wind is moving inside the supports and are as follow (in kips) :

Panel	0 – I	I – II	II – III	III – IV	IV – V	V – VI
Top Lateral	68.1	56.7	46.3	37.1	28.8	21.6
Bottom Lateral	21.4	17.3	13.7	10.8	8.2	6.2

For diagonals of the cantilever arm the stresses are maximum

when the wind is blowing on the suspended span and on the arm.
The stresses are as follows:

Panel	0 – I′	I – II′
Top Lateral	38. 1	25. 7

Sway bracing is provided between every pair of opposite verticals of the two standing trusses. It is made of a rectangular form with two diagonals crossing each other; the upper and lower beams are the struts of the top and bottom laterals while the side posts are the verticals of the truss. On account of the great depth of the truss near the ends, the bracing for the last two pairs of verticals will be double paneled, i. e. , there will be a cross strut at the middle of the frame and there will be two pairs of diagonals, one in the upper and the other in the lower panels, In finding the stresses in the braces, the following external forces will be considered.

(1) Distortion due to unequal wind panel loads applied at the top and bottom struts of the bracing.

According to Merriman and Jacoby.

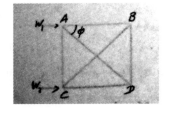

stresses in $AB = \frac{1}{2}(W_1 - W_2)$, stress in $C = \frac{1}{2}(W_1 - W_2)$

stress in $AD = \frac{1}{2}(W_1 - W_2) \sec \phi$, stress in $AC = \frac{1}{2}(W_1 - W_2) \tan \phi$.

(2) Distortion due to eccentricity of applied train loads when only one track is loaded.

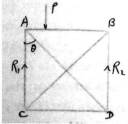

Stress in $AB = \frac{1}{2}(R_1 - R_2) \tan \theta$,

Stress in $AD = \frac{1}{2}(R_1 - R_2) \sec \theta$,

Stress in $BC = \frac{1}{2}(R_1 - R_2) \sec \theta$.

(3) Distortion due to overturning of wind on train.

Replace w by at B and R_1 and R_2 as shown.

$$R_1 = -R_2 = \frac{wh}{b}.$$

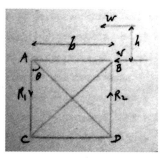

$$\text{Shear} = \frac{1}{2}(R_1 - R_2),$$

Stress in $AD = \frac{1}{2}(R_1 + R_2) \sec \theta$.

Stress due to w same as for the previous cases.

Since the diagonals are the only members that belong properly to the away bracing. only the stresses in them will be calculated. It is assumed and is a fact that the struts of the lateral system and the verticals of the truss are amply provided in

446

strength to resist the stresses due to distortion of the sway bracing.

As a general rule, the members of the sway bracing are largely determined by the ratio of $\dfrac{l}{r}$ to keep it below a certain specified limit rather by the magnitude of stresses which they care to carry. For this reason, any attempt to go to mathematical correctness of the stresses is not warrantable and refined accuracy is not necessary.

(1) Upper wind panel load = 5 kips.

Lower wind panel load = 3 kips.

Hence distortion of first cans is nil.

(2) Maximum floor beam reaction

= 2 × 94.54 − 189.1 kips applied at 6′ from plane of truss.

$$R_1 = 189.1 \times \frac{19}{25} = 144 \text{ kips},$$

$$R_2 = 189.1 \times \frac{6}{25} = 45.3 \text{ kips}.$$

(3) Lateral force on train = 12.5 kips per panel applied at 8′ above top strut.

$$R_1 = -R_2 = 12.5 \times \frac{8}{25}$$

$$= 4 \text{ kips},$$

$$w = 12.5 \text{ kips}.$$

Combining all the cases, for train on left track,

$$R_1 = 144 + 4 = 148.0 \text{ kips},$$

$$R_2 = 45.3 - 4 = 41.3 \text{ kips}.$$

$$\text{Vertical shear} = \frac{1}{2}(R_1 - R_2)$$

$$= \frac{1}{2}(148 - 41.3)$$

$$= 53.4 \text{ kips}.$$

$$\text{Horizontal shear} = w = 12.5 \text{ kips}.$$

For train on right track.

$$R_1 = 45.3 + 4 = 49.3 \text{ kips},$$

$$R_2 = 144 - 4 = 140 \text{ kips}.$$

$$\text{Vertical shear} = \frac{1}{2}(140 - 49.3) = 49.8 \text{ kips}.$$

$$\text{Horizontal shear} = 12.5 \text{ kips}.$$

When the direction of wind is reversed, the shear changes sign accordingly. The diagonals will be made stiff to carry both tension and compression.

Vertical shear carried by each diagonal

$$= \frac{1}{2} \times 53.4 = 26.7 \text{ kips}.$$

Horizontal shear carried by each diagonal

$$= \frac{1}{2} \times 12.5 = 6.3 \text{ kips.}$$

Stresses in Diagonals

Panel Point	Vertical Shear		Horizontal Shear		Total
	sec θ	Stress	sec θ	Stress	
VI	2.31	61.7	1.11	7.0	68.7
V	2.10	56.1	1.14	7.2	63.3
IV	1.71	45.7	1.23	7.7	53.4
III	1.40	37.4	1.43	9.0	46.4
II	1.22	32.6	1.75	11.0	43.6

For the last two bracings, assume vertical shear carried across the panel $= \frac{1}{2} \times 26.7 = 13.4$ kips.

For point I.

$$\sec \theta = 1.425,$$

$$\sec \theta = 1.414.$$

Stress due to vertical shear

$$= 13.4 \times 1.425 = 1.91 \text{ kips.}$$

Stress due to horizontal shear

$$= 6.3 \times 1.414 = 8.8 \text{ kips.}$$

Total stress in diagonals

$$= 27.9 \text{ kips.}$$

For point 0,

$$\sec \theta = 1.250,$$

$$\sec \theta = 1.660.$$

Stress due to vertical shear

$$= 1.25 \times 13.4 = 16.9 \text{ kips.}$$

Stress due to horizontal shear

$$= 1.66 \times 6.3 = 10.5.$$

Total stress in diagonals

$$= 87.4 \text{ kips.}$$

Article 55. Design.

(a) Top Laterals.

The section will be composed of four angles, forming an I – section. Since the stresses are small, and the laterals are riveted to the stringers $4 \lfloor S \ 3 \frac{1}{2}'' \times 3 \frac{1}{2}'' \times \frac{3}{8}''$ will be sufficient for all members. The unsupported length is less than 5', and greatest, stress is 68.7 kips requiring only 4 or 5 sq. in. The lateral will be arranged to form a thrust frons, as shown. [1]

The upper two angles of the section are considered effective

---

[1] Bridge Engineering. Waddell. P.400.

in resisting the traction load. The angles will have lacing as webs. the distance out to out of the angles being the same as the distance between the bottom of floor beams and the

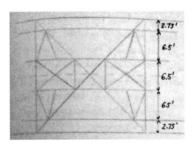

bottom of stringers. The bottom two angles are connected to the floor beams by connecting plates while the upper two angles are connected by connecting angles, riveted to the webs of the floor beams,[1] and the vertical posts.

A connecting plate is required at the bottom of stringers to connect the diagonal struts of the thrust frame. The strut frame will be made of

$$1 \lfloor S \ 3\frac{1}{2}'' \times 3\frac{1}{2}'' \times \frac{3}{8}'' \text{ sections.}$$

(b) Bottom Laterals.

The diagonals of bottom lateral will be made of $4 \lfloor S \ 6'' \times 4'' \times \frac{3}{8}''$ laced into the form of an I-section whose depth is made equal to that of the lower chord of the truss so that both faces of

① Bridge Engineering. Waddell. P. 399.

the latter may be gripped of actively.

Longest length of diagonals

$$= 39.06' = 468''.$$

Unsupported length in transverse direction

$$= \frac{1}{2} \times 468 = 234''.$$

Least radius of gyration of the I section about an axis passing through the web $= 1.95$ in.

$$\frac{l}{r} = 120.$$

Since the stresses are very low the design is principally governed by the requirement of stiffness. The section which is safe for the longest diagonal is therefore more than sufficient for all the other diagonals.

The struts also form the cross beam of the same bracing. They are made of 4 $\lfloor S$ $3 \frac{1}{2}'' \times 3 \frac{1}{2}'' \times \frac{3}{8}''$ laced to form a box section of 40'' deep and 25'' wide.

(c) Sway Bracing.

Since the sway bracing is provided primarily to stiffen the structure rather than to carry the stresses which are generally small, the design is entirely governed by the ratio of stiffness,

i. e. , $\dfrac{l}{r}$.

The top strut of the bracing is the floor beam while the bottom strut in the cross beam of the bottom lateral. $4 \lfloor S \ 6'' \times 4''$ $\times \dfrac{3}{8}''$ laced into an I-section will be used for the diagonals, the depth of which is made the same as the width of the verticals of the truss.

CHAPTER XIII. STRESSES AND DESIGN OF SUSPENDED SPAN

Article 56. Stresses.

The suspended span of $100'$ is trussed like a portion of the arch so as to suggest a continuity of the truss with the cantilever arms. Its dimensions are exactly the same as the middle four panels of either half of the arch, the elements of members are therefore given in Table 1. A new set of notations is used as shown in the accompanying diagram.

The stress coefficients for the suspended span are found from the stresses for unit reactions in Table 35, and are given in Table 36. The dead load is assumed at 3000 lbs. per ft. of span or 37. 5 kips per panel of truss. The equivalent uniform live load for equal moments at quarter points as wheel loads is found to be

3380 lbs. per ft. for one rail or 16. 9 kips per panel of truss.
Stresses for live loads are given in Table 37, in which the positive and negative stress coefficients are summed up to give

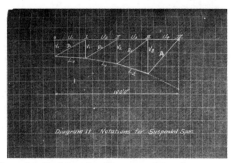

Diagram 11. Notations for Suspended Span.

the maximun effect and multiplied by the live panel load.

Table 35. Stresses in Suspended Span for Unit Reactions.
(in kips)

(a) For Unit Left Reaction.

Member	x	t	Stress	Member	x	t	Stress
U_1	0	12	0	D_1	−200	97. 35	−2. 05
U_2	25	13. 5	−1. 35	D_2	−50	47. 51	−1. 05
U_3	60	18	−2. 78	D_3	−10	49. 67	−. 202
L_0	25	13. 48	1. 852	V_0	−	−	0
L_1	50	17. 73	2. 82	V_1	−200	225.	. 89
L_3	73	34. 42	3. 06	V_2	−50	100.	. 50
				V_3	−10	85.	. 118

(b) For Unit Right Reaction.

Member	x'	t	Stress	Member	x'	t	Stress
U_1	100	12	-8.33	D_1	300	227.36	3.08
U_2	75.	13.5	-5.56	D_2	150	47.51	3.16
U_3	90	18.	-2.78	D_3	110	49.87	2.21
U_4	–	–	-9.87	D_4	–	–	1.41
L_0	75	13.48	5.56	V_0	–	–	0
L_1	50	17.72	2.83	V_1	300	225	-1.33
L_2	25	24.42	1.02	V_2	150	100	-1.50
				V_3	110	85	-1.29

Table 36. Stress Coefficients for Suspended Span.

Member	Load at		
	I	II	III
U_1	0	0	0
U_2	-1.39	$-.93$	-0.46
U_3	-0.70	-1.39	-0.70
U_4	-0.25	-0.40	-0.74
L_0	1.39	0.93	0.46
L_1	0.71	1.41	0.71
L_2	0.26	0.51	0.77
D_1	-1.54	-1.03	$-.61$
D_2	.79	$-.53$	$-.26$
D_3	$-.55$	1.10	$-.05$

D_4	$-.36$	$.71$	1.05
V_0	0	0	0
V_1	$-.33$	$.45$	$.22$
V_2	$-.38$	$-.75$	$.12$
V_3	$-.35$	$-.65$	$-.97$

Table 37. Live and Dead Load Stresses in Suspended Span.
(in kips)

Member	Positive L. L.		Negative L. L.		Full L. L.	D. L. Stress
	Coefficient	Stress	Coefficient	Stress		
U_1			0	0	0	0
U_2			2.78	470	-470	-104.2
U_3			2.78	470	-470	-104.2
U_4			1.477	250	-250	-55.5
L_0	2.78	470			470	104.2
L_1	2.82	477			477	105.9
L_2	1.53	250			259	57.3
D_1			3.06	521	-521	-105.6
D_2	$.79$	63	$.79$	13	0	0
D_3	1.65	279	$.06$	8.4	270.6	61.1
D_4	2.11	357			357	79.2
V_0			1.00	169	-169	-37.5
V_1	$.67$	11	$.33$	35.7	-44.7	-9.8

| V_2 | . 12 | 20 | 1. 13 | 191 | − 171 | − 37. 9 |
| V_3 | | | 1. 94 | 328 | − 328 | 72. 8 |

Table 38. Impact Stresses in Suspended Span. (in kips)

Member	Positive				Negative			
	I	Ratio	L. L.	Stress	I	Ratio	L. L.	Stress
U_1					200	. 6	0	0
U_2					200	. 6	470	282
U_3					200	. 6	470	282
U_4					200	. 6	250	150
L_0	200	. 6	470	282				
L_1	200	. 6	477	286				
L_2	200	. 6	259	155				
D_1					175	0. 63	521	328
D_2	75.	. 80	13	10. 4	125	. 705	13	9. 1
D_3	125.	. 705	279	197	76	. 80	8. 4	6. 7
D_4	175.	. 63	357	225				
V_0					200	. 60	169	101
V_1	75	. 80	11	8. 8	125	. 705	55. 7	3. 92
V_2	125	. 785	20	16	75	. 80	171	137
V_3					200	. 60	328	197

The dead load stresses given in the same table are obtained from the full live load stresses by multiplying by the ratio of respective panel loads. Impact stresses are given in Table 38. The wind panel load for the top chord is 17. 8 kips and for the lower chord is 5 kips. For top lateral system.

Panel	0 – I	I – II		
Stresses in Chords	13. 1	30. 6		
Stresses in Diagonals	18. 5	9. 3		
For Bottom Lateral system/				
Panel	0 – I	I – II	II – III	III – IV
Stresses in Chords	3. 8	8. 88	9. 13	4. 08
Stresses in Diagonals	5. 33	2. 68	2. 72	5. 54

The summation of stresses for the suspended span is given in Table 39.

Table 39. Summation of Stresses in Suspended Span. (in kips)

	Positive				Negative					
	D. L.	L. L.	Imp.	Wind	D. L.	L. L.	Imp.	Wind	Max.	Min.
U_1				13	0	0	0	13	· 13	– 13
U_2				31	104	470	282	31	31	– 887
U_3				31	104	470	282	31	31	– 887
U_4				13	35. 5	260	150	13	13	– 469
D_1	104	470	282	4				4	860	– 4

D_2	106	477	286	9				9	878	-9
D_3	57.3	25.9	135	9				9	480	-9
L_1					106.	54	328		-	-955
L_2	0	13	10		0	13	9.1		23	-21
L_3	61	279	197			8.4	6.7		537	25
L_4	79	357	325						661	0
V_0					38	169	101		-	-308
V_1		11	9		10	56	39		13	-105
V_2		20	15		38	19.1	137		10	-194
V_3					73	328	197		-	-598

Article 57. Design.

(a) Verticals.

The verticals are made of the same form as those for the arch but instead of a rectangular section they are made of $20''$ squares.

Member V_3

$$\text{Stress} = -698 \text{ kips}$$

$$2 \text{ webs } 20'' \times \frac{3}{4}'' \qquad 30$$

$$4 \lfloor \underline{S} \ 6'' \times 4'' \times \frac{1}{2}'' \qquad \underline{19}$$

Area supplied $= 49$ sq. in.

Least "r" about axis $A = 8.02$ in.

Length = 306 in.

Unit stress = 13. 33 kips per sq. in.

Area required = 45 sq. in.

Member V_4

Stress = -194 kips

2 webs $20'' \times \dfrac{3}{8}''$ 15

$4 \lfloor S \; 6'' \times 4'' \times \dfrac{3}{8}''$ 14. 5

Area supplied = 29. 5 sq. in.

Least "r" about axis $S = 6.95$ in.

Length = 215 in.

Unit stress = 13. 8 kips per sq. in.

Area required = 14. sq. in.

Members V_1 and V_0 are made the same as V_2 which has the smallest thickness allowed in the coefficients.

(b) Diagonals.

These are made of the same sections as the diagonals of the truss but a rectangular form of $16''$ width and $20''$ deep in used.

Member D_1

Stress = -955 kips

2 webs $16'' \times \dfrac{7}{16}''$ 28. 0

$$4 \underline{\lfloor S} \ 6'' \times 4'' \times \frac{3}{4}'' \qquad 27.8$$

$$4 \ \text{flats} \ 7'' \times \frac{3}{4}'' \qquad \underline{21.0}$$

Area supplied = 76.8 sq. in.

Least "r" = 6.65 in.

Length = 332 in.

Unit stress = 12.5 kips per sq. in.

Area required = 76.2 sq. in.

Member D_2

Stress = 21 kips

$$2 \ \text{webs} \ 12'' \times \frac{1}{7}'' \qquad 9.0$$

$$4 \underline{\lfloor S} \ 6'' \times 4'' \times \frac{3}{8}'' \qquad \underline{14.4}$$

Area supplied $\qquad$ 23.4

Least "r" = 4.06 in.

Length = 341 in.

$$\frac{l}{r} = 84$$

Member D_3

Stress = 637 kips

$$2 \ \text{webs} \ 16'' \times \frac{3}{4}'' \qquad 20.$$

(Deduct 5 1″ holes)

$$4 \ \llcorner S \ \ 6'' \times 4'' \times \frac{7}{8}'' \qquad \underline{25.}$$

(Deduct 8 1″ holes)

Net area supplied   = 45. 4

Net area required = 33. 5 sq. in.

Member D_4

Stress = 631 kips

Same section as D_3 used. Net area required

= 41. 3 sq. in.

Stress = 21 kips

$$2 \text{ webs } 12'' \times \frac{1}{7}'' \qquad\qquad 8. 0$$

$$4 \ \llcorner S \ \ 6'' \times 4'' \times \frac{3}{8}'' \qquad \underline{14. 4}$$

Area supplied

Least "r" = 4. 06 in.

Length = 341 in.

Stress = 637 kips

$$2 \text{ webs } 16'' \times \frac{3}{4}'' \quad 20.$$

(Deduct 8 1″ holes)

$$4 \ \llcorner S \ \ 6'' \times 4'' \times \frac{7}{8}'' \quad \underline{25.}$$

(Deduct 8 1″ holes)

Area Supplied = 46.4

Net area required = 33.5 sq. in.

Stress = 861 kips

Same section as D_3 used. Net area required

= 41.8 sq. in.

(c) Lower Chords.

Same section will be used here as for the lower chords of the arch. The depth of sections however is made 30″ instead of 40″.

Member D_0

Stress = 860 kips

2 webs $30″ \times \dfrac{3}{4}$ 37.5

(Deduct 10 1″ holes)

4 ⌊S $6″ \times 4″ \times \dfrac{5}{8}$″   20.9

(Deduct 4 1″ holes)   _____

Net area supplied   = 58.4 sq. in.

Net area required   = 53.7 sq. in.

Member L_1

Stress = 878 kips

Same section as L_0 used. Net area required

$$= 55.0 \text{ sq. in.}$$

Member L_2

$$\text{Stress} = 480 \text{ kips}$$

2 webs $30'' \times \dfrac{3}{8}''$ 18.8

(Deduct 10 1'' holes)

4 $\lfloor S$ $6'' \times 4'' \times \dfrac{3}{8}''$ 12.9

(Deduct 4 1'' holes)

Net area supplied	$= 31.1 \text{ sq. in.}$
Net area required	$= 30 \text{ sq. in.}$

Member L_3

Same section as L_2.

(d) Upper Chords.

All the upper chords will be made of the same section as U_3 of the area truss. It is found that it is more than sufficient for all the chords of the suspended span.

(e) Lateral System.

The lateral systems of suspended span are made of the same sections as those of the arch truss.

CHPTER XⅣ. ESTIMATE OF WEIGHT

Article 58. Arch Span.

(a) Verticals.

(V_2') 2 webs $25 \times \dfrac{3}{4}$ $= 127.6$

$4 \lfloor S \ 6 \times 4 \times \dfrac{3}{4}$ $= \underline{94.5}$

wt. per ft. $= 222.1$ lbs.

Total $= 8000$ lbs.

(V_1') 2 webs $25 \times \dfrac{3}{4}$ $= 127.6$

$4 \lfloor S \ 6 \times 4 \times \dfrac{3}{4}$ $= 94.5$

4 flats $5 \times \dfrac{5}{8}$ $= \underline{42.5}$

wt. per ft. $= 264.7$ lbs.

Total $= 13100$ lbs.

(V_0) 2 webs $25 \times \dfrac{7}{8}$ $= 148.0$

$4 \lfloor S \ 6 \times 4 \times \dfrac{7}{8}$ $= 108.8$

2 fills $13 \times \dfrac{7}{8}$ $= 77.3$

2 side pl. $23\frac{1}{2} \times \frac{1}{2}$ $= 79.0$

8 flats $5 \times \frac{1}{2}$ $= \underline{68.0}$

wt. per ft. $= 481.9$

Total $= 31700$ lbs.

(V_1) 2 webs $25 \times \frac{9}{16}$ $= 92.4$

4 $\lfloor S$ $6 \times 4 \times \frac{3}{4}$ $= 94.6$

4 flats $5 \times \frac{1}{2}$ $= \underline{34.0}$

wt. per ft. $= 220.9$ lbs.

Total $= 10900$ lbs.

(V_2) 2 webs $25 \times \frac{1}{2}$ $= 85.0$

4 $\lfloor S$ $6 \times 4 \times \frac{3}{4}$ $= \underline{94.5}$

wt. per ft. $= 179.5$ lbs.

Total $= 6470$ lbs.

(V_3) 2 webs $25 \times \frac{1}{2}$ $= 85.0$

4 $\lfloor S$ $6 \times 4 \times \frac{1}{2}$ $= \underline{64.8}$

wt. per ft. $= 149.8$ lbs.

	Total	$= 3820$ lbs.

(V_4) 2 webs $25 \times \dfrac{1}{2}$ $= 85.0$

4 $\underline{\text{S}}$ $6 \times 4 \times \dfrac{5}{8}$ $\underline{80.0}$

wt. per ft. $= 165.0$ lbs.

Total $= 2970$ lbs.

(V_5) Same as V_3

wt. per ft. $= 149.8$ lbs.

Total $= 2020$ lbs.

(V_6) 2 webs $20 \times \dfrac{3}{8}$ $= 51.0$

4 $\underline{\text{S}}$ $6 \times 4 \times \dfrac{1}{8}$ $= \underline{49.3}$

wt. per ft. $= 100.3$ lbs.

Total $= 1203$ lbs.

(b) Diagonals.

(D_2') 4 $\underline{\text{S}}$ $6 \times 4 \times \dfrac{5}{8}$ $= 108.8$

2 webs $20 \times \dfrac{3}{4}$ $= \underline{122.5}$

wt. per ft. $= 231.3$ lbs.

Total $= 10300$ lbs.

(D_1') Same section as (D_2')

	wt. per ft.	= 231. 3 lbs.
	Total	= 12800 lbs.

(D_1) 4 $\lfloor S$ 6 × 4 × $\dfrac{3}{4}$ = 94. 3

2 webs 20 × $\dfrac{1}{2}$ = 74. 0

wt. per ft.   = 168. 3 lbs.

Total   = 9320 lbs.

(D_2) 4 $\lfloor S$ 6 × 4 × $\dfrac{3}{4}$ = 94. 3

2 webs 20 × $\dfrac{3}{8}$ = 65. 5

wt. per ft.   = 159. 8 lbs.

Total   = 7010 lbs.

(D_3) Same section as D_2

wt. Per ft.   = 159. 8 lbs.

Total   = 5700 lbs.

(D_4) 4 $\lfloor S$ 6 × 4 × $\dfrac{3}{4}$ = 94. 3

2 webs 20 × $\dfrac{1}{2}$ = 74. 0

wt. per ft.   = 168. 3 lbs.

Total   = 5180 lbs.

(D_5) $4 \lfloor \underline{S} \ 6 \times 4 \times \dfrac{7}{8}$ $= 108.8$

          2 webs $20 \times \dfrac{13}{16}$ $= 131.0$

          wt. per ft. $= 239.8$ lbs.

                     Total $= 6820$ lbs.

(D_6) $4 \lfloor \underline{S} \ 6 \times 4 \times \dfrac{7}{8}$ $= \underline{108.8}$

          2 webs $20 \times \dfrac{7}{8}$ $= 164.3$

          wt. per ft. $= 273.1$ lbs.

                     Total $= 7560$ lbs.

(c) Lower Chords.

(L_1') 2 webs $40 \times \dfrac{1}{2}$ $= 136.0$

          $4 \lfloor \underline{S} \ 6 \times 4 \times \dfrac{1}{2}$ $= \underline{64.8}$

          wt. per ft. $= 220.8$ lbs.

                     Total $= 5680$ lbs.

(L_0') 2 webs $40 \times \dfrac{5}{8}$ $= 170$

          $4 \lfloor \underline{S} \ 6 \times 4 \times \dfrac{5}{8}$ $\underline{80}$

              wt. per ft. $= 250$ lbs.

| | Total | = 7500 lbs. |

$$(L_0) \quad 2 \text{ webs } 40 \times \frac{3}{4} \qquad = 204.0$$

$$4 \lfloor S \ 6 \times 4 \times \frac{3}{4} \qquad = 94.5$$

$$2 \text{ fills } 32 \times \frac{3}{4} \qquad = \underline{163.2}$$

| | wt. per ft. | = 665.7 lbs. |
| | Total | = 20000 lbs. |

$$(L_1) \quad 2 \text{ webs } 40 \times \frac{3}{4} \qquad = 204.0$$

$$2 \text{ webs } 40 \times \frac{5}{8} \qquad = 170.0$$

$$4 \lfloor S \ 6 \times 4 \times \frac{3}{4} \qquad = 94.5$$

$$2 \text{ fills } 32 \times \frac{3}{4} \qquad = \underline{163.2}$$

| | wt. per ft. | = 631.7 lbs. |
| | Total | = 17900 lbs. |

$$(L_2) \quad 4 \text{ webs } 40 \times \frac{5}{8} \qquad = 340.0$$

$$4 \lfloor S \ 6 \times 4 \times \frac{3}{4} \qquad = 94.5$$

$$2 \text{ fills } 32 \times \frac{3}{4} \qquad = \underline{163.2}$$

| | wt. per ft. | = 597.7 lbs. |
| | Total | = 16150 lbs. |

(L_3)　2 webs $40 \times \dfrac{1}{2}$ 　　　= 136

　　　2 webs $40 \times \dfrac{5}{8}$ 　　　= 170

　　　4 $\lfloor S$ $6 \times 4 \times \dfrac{5}{8}$ 　　= 80

　　　2 fills $32 \times \dfrac{5}{8}$ 　　　= $\underline{136}$

| | wt. per ft. | = 522 lbs. |
| | Total | = 13630 lbs. |

(L_4)　Same section as L_3.

| | wt. per ft. | = 522 lbs. |
| | Total | = 13250 lbs. |

(L_5)　2 webs $40 \times \dfrac{11}{16}$ 　　= 187.0

　　　4 $\lfloor S$ $6 \times 4 \times \dfrac{5}{8}$ 　　= 80.0

　　　2 fills $32 \times \dfrac{5}{8}$ 　　　= $\underline{136.0}$

| | wt. per ft. | = 403.0 lbs. |
| | Total | = 10100 lbs. |

(d) Upper Chords.

(U_2') (U_1') (U_1) (U_2) (U_3) all of same section.

$$2 \text{ webs } 30 \times \frac{1}{2} \qquad = 102$$

$$1 \text{ cover } 35\frac{1}{2} \times \frac{1}{2} \qquad = 60.4$$

$$4 \lfloor S \ 6 \times 4 \times \frac{1}{2} \qquad = 64.8$$

$$2 \text{ flats } 6\frac{1}{2} \times \frac{3}{4} \qquad = \underline{33.2}$$

wt. per ft. $\qquad = 260.4$ lbs.

Total $\quad = 25 \times 260.4 = 6510$ lbs.

(U_4) (U_5) (U_6) same sections.

$$2 \text{ webs } 30 \times \frac{1}{2} \qquad = 102.0$$

$$4 \lfloor S \ 6 \times 4 \times \frac{1}{2} \qquad = 64.8$$

$$1 \text{ cover } 35\frac{1}{2} \times \frac{1}{2} \qquad = 60.4$$

$$2 \text{ fills } 22 \times \frac{1}{2} \qquad = 74.8$$

$$2 \text{ flats } 6\frac{1}{2} \times \frac{3}{4} \qquad = -\underline{33.2}$$

wt. per ft. $\qquad = 335.2$

Total $\qquad = 8370$ lbs.

(e) Summation.

Total Weight of Verticals	$= 80180$
Total Weigh of Diagonals	$= 64690$
Total Weight of Lower Chords	$= 10420$
Total Weight of Upper Chords	$= \underline{57650}$
Total wt. of $\frac{1}{2}$ truss	$= 306740$ lbs.

(include vertical at center)

Total weight of one truss

$$= 2 \times 306740 - \frac{1}{2} \text{ weight of } V_6$$

$$= 612,885 \text{ lbs.}$$

Add 32% for [1] Detail.	$= \underline{196000}$ lbs.
Grand Total	$= 810000$ lbs.
	$= 2030$ lbs. per ft.
wt. assumed	$= 2500$ lbs. per ft.

Article 59. Suspended Span.

(a) Verticals.

(V_3)	2 webs $20 \times \frac{3}{4}$	$= 102.0$
	$4 \lfloor S \ 6 \times 4 \times \frac{1}{2}$	$= \underline{64.8}$

---

① Bridge Engineering. Waddell. P. 1239.

| | | wt. per ft. | $= 166.8$ |
| | | Total | $= 4250$ lbs. |

(V_2) 2 webs $20 \times \dfrac{3}{8}$ $= 51.0$

           $4 \lfloor S \;\; 6 \times 4 \times \dfrac{3}{8}$ $= \underline{49.2}$

                    wt. per ft. $= 100.2$

                    Total $= 1805$ lbs.

(V_1) 2 webs $20 \times \dfrac{3}{8}$ $= 51$

           $4 \lfloor S \;\; 6 \times 4 \times \dfrac{3}{8}$ $= \underline{49.2}$

                    wt. per ft. $= 100.2$

                    Total $= 1352$ lbs.

(V_0) Same section as (V_1)

                    Total $= 1202$ lbs.

(b) Diagonals.

(D_1) 2 webs $16 \times \dfrac{7}{16}$ $= 47.6$

           $4 \lfloor S \;\; 6 \times 4 \times \dfrac{3}{4}$ $= \underline{94.5}$

                    wt. per ft. $= 213.5$ lbs.

                    Total $= 5920$ lbs.

(D_2) 4 $\lfloor S$ $6 \times 4 \times \dfrac{3}{8}$ $= 49.2$

         2 webs $12 \times \dfrac{3}{8}$ $= \underline{30.6}$

               wt. per ft. $= 79.8$

                 Total $= 2841 \times 79.8$

                            $= 2270$ lbs.

(D_3) 4 $\lfloor S$ $6 \times 4 \times \dfrac{3}{8}$ $= 108.8$

         2 webs $16 \times \dfrac{3}{4}$ $= \underline{81.6}$

               wt. per ft. $= 190.4$

                 Total $= 4930$ lbs.

(D_4) 4 $\lfloor S$ $6 \times 4 \times \dfrac{5}{8}$ $= 80.8$

         2 webs $16 \times \dfrac{3}{4}$ $= \underline{81.6}$

               wt. per ft. $= 161.6$

                 Total $= 6500$ lbs.

(c) Lower Chords.

(L_0) 2 webs $30 \times \dfrac{3}{4}$ $= 153.0$

         4 $\lfloor S$ $6 \times 4 \times \dfrac{5}{8}$ $= \underline{80.0}$

$$\text{wt. per ft.} = 233.0$$

$$\text{Total} = 5820 \text{ lbs.}$$

(L_1) Same section as (L_0)

$$\text{Total} = 5920 \text{ lbs.}$$

(L_2) 2 webs $30 \times \dfrac{3}{8}$ $= 76.5$

         $4 \lfloor S \; 6 \times 4 \times \dfrac{3}{8}$ $= \underline{49.2}$

$$\text{wt. per ft.} = 125.7$$

$$\text{Total} = 3270 \text{ lbs.}$$

(L_3) Same section as (L_2)

$$\text{Total} = 3400 \text{ lbs.}$$

(d) Upper Chords.

All sections are the same.

         2 webs $30 \times \dfrac{1}{2}$ $= 102.0$

         1 cover $35 \dfrac{1}{2} \times \dfrac{1}{2}$ $= 60.4$

         $4 \lfloor S \; 6 \times 4 \times \dfrac{1}{2}$ $= 64.8$

         2 flats $6 \dfrac{1}{2} \times \dfrac{3}{4}$ $= \underline{33.2}$

$$260.4 \text{ lbs.}$$

$$\text{Total} = 26040 \text{ lbs.}$$

(e) Summation.

Total Weight of Verticals	$= 8609$
Total Weight of Diagonals	$= 19620$
Total Weight of Lower Chords	$= 18460$
Total Weight of Upper Chords	$= \underline{26040}$
Total wt. of Suspended Span	$= 72679$ lbs.

Add 50% for laterals

And details	$= \underline{36340}$
Total	$= 109000$ lbs.
Floor system	$= \underline{82000}$ lbs.
Grand Total	$= 191200$ lbs.
End Reaction	$= 95$ kips
End Reaction assumed	$= 75$ kips

Article 60. Lateral System.

(a) Top Lateral.

Since the sections are made the same for the whole system the weight concentrated at any panel point may be found by summing the lengths of members carried to that point multiplied by the weight per unit length of the section.

$$4 \lfloor S \ 3\frac{1}{2} \times 3\frac{1}{2} \times \frac{3}{8} \qquad = 34 \text{ lbs. per ft.}$$

Length of diagonal $\qquad = 1.41425$

$$= 35.3 \text{ ft.}$$

Weight of lateral system carried to any upper chord panel point

$$= 35.3 \times 34 = 1200 \text{ lbs.}$$

plus 10% for thrust frame $\qquad = 1320 \text{ lbs.}$

(b) Lower Lateral System.

Weight of diagonals $\left(4 \ \underline{|S} \ 6 \times 4 \times \dfrac{3}{8} \right) = 49.2$ lbs. per ft.

Panel Concentration $= 49.2$ length of member carried to the panel point.

Weight of struts $\left(4 \ \underline{|S} \ 3 \ \dfrac{1}{2} \times 3 \ \dfrac{1}{2} \times \dfrac{3}{8} \right) = 34$ lbs. per ft.

Panel concentration $= 34 \times \dfrac{25}{2} = 425$ lbs.

Panel Point	II′	I′	0	I	II	IIII	IV	V	VI
Length Carried	18.9	38.43	39.06	38.43	37.4	36.5	35.9	35.5	35.4
wt. of Diagonal	928	1890	1918	1890	1840	1795	1765	1747	1740
Total wt.	1353	2315	2343	2315	2265	2220	2190	2172	2165

(c) Sway Bracing.

wt. of diagonals $\left(4 \ \underline{|S} \ 6 \times 4 \times \dfrac{3}{8} \right) = 49.2$ lbs. /ft.

Panel Point	II′	I′	0	I	II	III	IV	V	VI
Length Carried	21.90	35.4	41.3	35.4	21.9	17.8	15.4	16.2	13.9
Total Weight		1080	1953	2243	1953	1080	875	758	698	684

For bracings at points I′, 0 and I′ one half the weight of the central strut (= 213 lbs.) is added to the weight of diagonals.

Article 61. Revised Dead Panel Loads.

The revised dead panel loads are given in Table 39a and Diagram 12. In Table 40 are given the assumed and revised weights and their comparison. It is evident that the assumed weight is much in excess than that revised for the upper chord panel points, but is about the same for the lower chord panel points. Although the revised weight is not taken from detail drawings it is not far from actual and it certainly proves the statement that "two hinged spandrel braced arches are lighter than simple trusses".

The variation of panel loads at different panel points is very remarkable. It shows that the assumption of uniform panel loads is entirely not justifiable and may give errors that are too large to be allowed. Although the law of variation is uncertain the fact that the weight is approximately proportional to the concentration

of lengths of members is apparent. This is largely due to the use
of uniform sections for the different members.

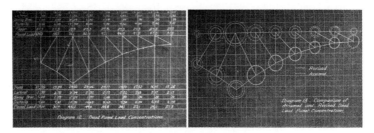

Table 39a. Revised Dead Panel Loads.

Point	Truss (lbs)	Lateral (lbs)	Sway Bracing (lbs)	Total (kips)	Plus detail 32% (kips)	Plus detail and floor (kips)
II′	7250	660	1080	8.99	11.8	120.3
I′	18210	1320	2243	36.98	48.8	76.1
I	15470	1320	1953	18.74	24.7	52.0
II	12590	1320	1080	15.00	19.8	47.1
III	11940	1320	875	14.13	18.7	46.0
V	13160	1320	698	15.17	20.0	47.3
II′	11990	1353	1080	14.42	19.0	19.0
I′	19540	2315	1953	23.80	31.4	31.4
0	29600	2343	2243	34.18	45.2	45.2
I	29060	2315	1953	33.33	44.0	44.0

II	23770	2265	1080	27. 11	35. 8	35. 8
III	19700	2220	875	22. 80	30. 1	30. 1
IV	17520	2190	768	20. 47	27. 1	27. 1
V	16100	2172	698	18. 97	25. 0	25. 0
VI	18260	2165	684	21. 10	27. 8	27. 8

Table 40. Comparison of Weights.

Point	Assumed		Revised		Difference	Revised
	Truss	Lateral	Truss	Lateral		
II′	19. 0		11. 8		7. 2	60.
I′	39. 4		28. 4		11. 0	38. 7
0	50. 1		48. 8		11. 3	23. 2
I	39. 4		24. 7		14. 7	59. 5
II	34. 2		19. 8		14. 4	72. 0
III	30. 4		18. 7		11. 7	64. 0
IV	27. 9		20. 2		7. 7	37. 1
V	26. 6		20. 0		6. 6	33. 0
VI	19. 3		14. 5		4. 8	33. 0
II′	30. 9		19. 0		11. 9	62. 7
I′	44. 6		31. 4		13. 2	42. 0
0	35. 5		45. 2		− 9. 7	21. 5
I	44. 6		44. 0		0. 6	1. 36

II	37.6	35.8	1.8	5.02
III	32.5	30.1	2.4	7.97
IV	28.8	27.1	1.7	6.27
V	26.9	25.0	1.9	7.6
VI	33.3	27.8	5.5	19.7

·

PART Ⅳ. REVISED AND FINAL STRESSES

CHAPTER XXV. REVISED LIVE LOAD STRESSES

After the preliminary design is made a closer value of horizontal thrust may be obtained from the sectional areas of members. In Table 41 are given the values of $\dfrac{S'TL}{A}$ and $\dfrac{T^2L}{A}$, and in Table 42, their summations and the values of horizontal trust. The graphical values of H obtained from axial deformations of Table 43 and displacement diagrams (shown in full lines) of Plats III are found as follows:

Wind at	II′	I′	0,	I,	II,	III,	IV,	V,	VI
H	-0.430	-220	0.005	0.243	0.468	0.681	0.864	0.997	1.050

The distribution of $\dfrac{S'TL}{A}$ among the members of the truss is shown in Diagram 14, in which that of $S'TL$ is also shown in dotted lines. It will be seen that the effect of upper chords and verticals is very small on the final values of H but that of lower chords and diagonals is very conspicuous.

In diagrams 15 and 16 are shown the influence lines of when the effect of web members is neglected, the former gives the preliminary value while the latter gives the same revised.

Table 41. Values of $\dfrac{S'TL}{A}$ and $\dfrac{T^2L}{A}$.

(a) Upper Chords.

	U_1	U_2	U_3	U_4	U_5	U_6	
A	77.2	77.2	77.2	99.2	99.2	99.2	
$\dfrac{T^2L}{A}$	.03598	.22491	.81692	1.79216	3.81145	5.10325	11.78467
II′	.10902	.37484	1.00854	1.86680	3.63000	4.72522	11.71442
I′	.05451	.18742	.50427	.93340	1.81500	2.36261	5.85721
0	0	0	0	0	0	0	0
I	.05451	.18742	.50427	.93340	1.81500	2.36261	5.85721
II	.05451	.37484	1.00854	1.86680	3.63000	4.72522	11.65991
III	.05451	.37484	1.51279	2.80020	5.44500	7.08783	17.27517
IV	.05451	.37484	1.51279	3.73372	7.26000	9.45044	22.38630
V	.05451	.37484	1.51279	3.73372	9.07480	11.81305	26.56371

| VI | .05451 | .37484 | 1.51279 | 3.73372 | 9.07480 | 14.17580 | 28.92646 |

(b) Lower Chords.

	L_0	L_1	L_2	L_3	L_4	L_5	
A	195.8	185.8	175.8	153.4	153.4	118.4	
$\dfrac{T^2L}{A}$	0.21963	0.35115	0.60991	1.24243	2.29831	5.07400	9.79543
II′	0.16638	0.26602	0.48204	0.94126	1.74112	3.84400	7.42084
I′	0.08319	0.13301	0.23102	0.47063	.87056	1.92200	3.71041
0	0	0	0	0	0	0	0
I		.13301	.23102	.47063	.87056	1.92200	3.62722
II		.13301	.46206	.94126	1.741112	3.84400	7.12145
III		.13301	.46206	1.41187	2.61168	5.76600	10.37462
IV		.13301	.46206	1.41187	3.48233	7.68800	13.17727
V		.13301	.46206	1.41187	3.48233	9.61000	15.09927
VI		.13301	.46206	1.41187	3.48233	9.61000	15.09927

(c) Diagonals.

	D_1	D_2	D_3	D_4	D_5	D_6	
A	38.13	35.63	35.63	38.13	53.86	63.05	
$\dfrac{T^2L}{A}$	.79513	.94522	1.16541	1.42677	1.01791	.20211	5.55255
II′	.60236	.71600	.88298	1.08092	.77138	.15312	4.20676
I′	.30118	.33800	.44149	.54046	.38669	.07656	2.10338

0	0	0	0	0	0	0	0
I	1.20475	.35800	.44149	.54046	.38569	.07656	3.00695
II	1.20475	1.67104	.88298	1.08092	.77138	.15312	5.76419
III	1.20475	1.67104	2.39639	1.62138	1.15707	.22968	8.28031
IV	1.20475	1.67104	2.39639	3.45885	1.54276	.30624	10.58003
V	1.20475	1.67104	2.39639	3.45885	3.08460	.38280	12.19843
VI	1.20475	1.67104	2.39639	3.45885	3.08460	1.07180	12.88743

(d) Verticals.

	V_0	V_1	V_2	V_3	V_4	V_5	V_6	
A	141.65	65.9	32.8	44.0	48.4	44.0	29.5	
$\dfrac{T^2 L}{A}$	.20295	.38940	.40425	.34941	.16199	.02640	0	1.53440
II'	0	.29502	.30628	.26468	.12272	.20000	0	1.18870
I'	.15376	.14751	.15314	.13234	.06136	.1000	0	.74810
0	.30751	0	0	0	0	0	0	.30751
I	.30751	.68843	.15314	.13234	.06136	.10000	0	1.44278
II	.30751	.68843	.83124	.26468	.12272	.20000	0	2.41458
III	.30751	.68843	.83124	.84706	.18408	.30000	0	3.15832
IV	.30751	.68843	.83124	.84706	.49090	.40000	0	3.56514
V	.30751	.68843	.83124	.84706	.49090	.14000	0	3.30514
VI	.30751	.68843	.83124	.84706	.49090	.14000	0	3.30514

Table 42. Values of Revised Horizontal Thrust.

S'T L for load at

	Upper Chords	Lower Chords	Diagonals	Verticals	Total	H
$\dfrac{T^2 L}{A}$	11.78467	9.79543	5.55255	1.53440	28.66705	
II′	11.71442	7.42082	4.20676	1.18870	24.53070	−.42786
I′	5.85721	3.71041	2.10338	.74810	12.41910	−.21661
0	0	0	0	.30751	.30751	.00536
I	5.85721	3.62722	3.00695	1.44278	13.93416	.24304
II	11.65991	7.12145	5.76419	2.41458	26.96013	.47023
III	17.27517	10.38462	8.28031	3.15832	39.09842	.68195
IV	22.38630	13.17727	10.58003	3.56514	49.70874	.86700
V	25.56371	15.09927	12.19843	3.30514	57.16655	.99708
VI	28.92646	15.09927	12.88743	3.30514	60.21830	1.05032

Table 43. Revised Axial Deformations. (in inches)

E 1.

Member	A	TL	$\dfrac{TL}{A}$
U_1	77.2	100	1.295
U_2	77.2	250	3.245
U_3	77.2	476	6.170
U_4	99.2	800	8.070
U_5	99.2	1167	11.750
U_6	99.2	1350	13.62
L_0	195.8	431	2.200

L_1	185.8	517	2.78
L_2	175.8	647	3.680
L_3	153.4	846	5.520
L_4	153.4	1135	7.390
L_5	118.4	1472	12.430
D_1	38.13	493	12.93
D_2	35.63	461	12.91
D_3	35.63	462	12.94
D_4	38.13	491	12.88
D_5	53.86	474	8.80
D_6	63.05	226	3.59
V_0	141.65	523	3.70
V_1	65.9	427	6.48
V_2	52.8	332	6.28
V_3	44.0	238	5.41
V_4	48.4	143	3.96
V_5	44.0	48	1.09
V_6	29.5	0	0

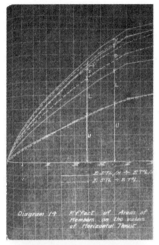

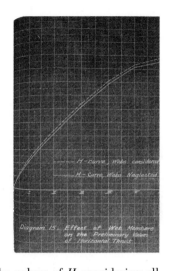

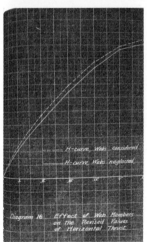

The values of H considering all the members of the truss are also shown in both cases, in dotted lines. It is found that for the preliminary values the effect of web members is very small and may be safely neglected but for the revised values it cannot be done without cation.

Having obtained the values of H the next thing will be to compute the stress coefficients. In Table 44 the stresses due to revised horizontal thrusts are given and in Table 43 the required stress coefficients. Stresses due to vertical forces only ($H = 0$) are tabled from Table 15 to check the accuracy of the result graphical values are obtained from Plates V and VI and are given in Table 46. The influence lines of revised H are shown by full lines to the drawings.

The summations of positive and negative coefficients are given in Table 47 and the revised live load stresses in Table 48. For the detailed procedure see Article 42.

Table 44. Stresses Due to Revised Horizontal Thrust.

(a) Upper Chords.

Load at

Member	U_1	U_2	U_3	U_4	U_5	U_6	Member
I	.3333	.8334	1.5863	2.667	3.8889	4.5000	T
II′	.1426	.3565	.6796	1.1412	1.6639	1.9253	X′
I′	.0722	.1805	.3440	.5776	.8424	.9747	XI′
0	.0018	.0045	.0085	.0143	.0209	.0241	XII
I	.0810	.2025	.3860	.6481	.9452	1.0937	XI
II	.1567	.3919	.7468	1.2539	1.8287	2.1160	X
III	.2273	.5683	1.0831	1.8185	2.6521	3.0687	XI
IV	.2890	.7225	1.3770	2.3120	3.3717	3.9015	VIII

V	.3324	.8309	1.5833	2.6589	3.8775	4.4868	VII
VI	.3501	.8753	1.6682	2.8009	4.0846	4.7264	VI

All stress − for loads on cantilever arms.
All stress − for loads inside the supports.

(b) Lower Chords.

Load at

Member	L_0	L_1	L_2	L_3	L_4	L_5	Member
T	1.1982	1.5153	1.9885	2.7022	3.7255	4.8977	T
II′	.5127	.6483	.8508	1.1562	1.594	2.0955	X′
I′	.2595	.3282	.4307	.5853	.8070	1.0609	XI′
0	.0064	.0081	0.0107	.0145	.0200	.0262	XII
I	.2912	.3683	.4833	.6567	.9054	1.1903	XI
II	.5634	.7126	.9351	1.2707	1.7519	2.3030	X
III	.8171	1.0334	1.3561	1.8427	2.5407	3.3400	IX
IV	1.0388	1.3138	1.7240	2.3428	3.2301	4.2463	VIII
V	1.1947	1.5109	1.9827	2.6943	3.7147	4.8833	VII
VI	1.2585	1.5916	2.0886	2.8382	3.9130	5.1441	VI

All stress − for loads on cantilever arms.
All stress − for loads inside the supports.

(c) Diagonals.

Load at

Member	D_1	D_2	D_3	D_4	D_5	D_6	Member
T	.7394	.8766	1.0783	1.3289	1.3891	.6779	T
II'	.3164	.3750	.4614	.5686	.5943	.2900	X'
I'	.1602	.1899	.2336	.2879	.3009	.1468	XI'
0	.0040	.0047	.0058	.0071	.0075	.0036	XII
I	.1797	.2130	.2621	.3230	.3376	.1647	XI
II	.3477	.4122	.5071	.6249	.6532	.3188	X
III	.5042	.5978	.7354	.9062	.9473	.4623	IX
IV	.6408	.7600	.9349	1.1522	1.2043	.5877	VIII
V	.7372	.8740	1.0752	1.3250	1.3850	.6759	VII
VI	.7766	.9207	1.1326	1.3958	1.4590	.7120	VI

All stress – for loads on cantilever arms.

All stress – for loads inside the supports.

(d) Verticals.

Load at

Member	V_0	V_1	V_2	V_3	V_4	V_5	V_6	Member
T	.66	.7200	.7700	.7765	.6600	.2933	0	T
II'	.2824	.3081	.3294	.3322	.2824	.1255	0	X'
I'	.1430	.1560	.1668	.1682	.1430	.0636	0	XI'
0	.0035	.0039	.0041	.0042	.0035	.0016	0	XII
I	.1604	.1750	.1871	.1887	.1604	.0713	0	XI
II	..3104	.3386	.3621	.3651	.3104	.1379	0	X

III	.4501	.4910	.5251	.5295	.4501	.2001	0	IX
IV	.5722	.6242	.6676	.6732	.5722	.2543	0	VIII
V	.6581	.7179	.7677	.7742	.6581	.2925	0	VII
VI	.6932	.7562	.8087	.8155	.6932	.3081	0	VI

All stress − for loads on cantilever arms.

All stress − for loads inside the supports.

Table 45. Revised Stress Coefficients.

(a) Upper Chords.

Load at	U_2'	U_1'	U_1	U_{12}
II′	0	+ .5050	+ .7834	+ .8009
I′	0	0	+ .3908	+ .3982
0	0	0	+ .0018	+ .0045
I	0	0	− .3820	− .3762
II	0	0	− .2642	− .7655
III	0	0	− .1515	− .4734
IV	0	0	− .0477	− .2034
V	0	0	+ .0378	+ .0207
VI	0	0	+ .0976	+ .1809
VII	0	0	+ .1219	+ .2522
VIII	0	0	+ .1256	+ .2595
IX	0	0	+ .1010	+ .2211
X	0	0	+ .0725	+ .1604

XI	0	0	+ .0389	+ .0868
XII	0	0	+ .0018	+ .0045
XI′	0	0	− .0301	− .0347
XII′	0	0	− .0384	− .1250
	U_3	U_4	U_5	U_6
II′	+ .7910	+ .7106	+ .4966	+ .1580
I′	+ .3913	+ .3483	+ .2379	+ .0670
0	+ .0085	+ .0143	+ .0209	+ .0241
I	− .3493	− .2778	− .1351	+ .0520
II	− .7238	− .5979	− .3318	+ .0327
III	− 1.1228	− .9592	− .5887	− .0563
IV	− .5838	− 1.3918	− .9494	− .2652
V	− .1324	− .5819	− 1.5238	− .7216
VI	+ .1976	+ .0331	− .5451	− 1.5238
VII	+ .3578	+ .3440	+ .0194	− .7217
VIII	+ .3966	+ .4601	+ .2652	− .2653
IX	+ .3479	+ .4296	+ .3372	− .0564
X	+ .2566	+ .3280	+ .2855	+ .0327
XI	+ .1409	+ .1851	+ .1736	+ .0520
XII	+ .0085	+ .0143	+ .0209	+ .0241
XI′	− .0989	− .1146	− .0708	+ .0670
XII′	− .1894	− .2153	− .1207	− .1581

(b) Lower Chords.

Load at	L_1'	L_0'	L_0	L_1
II′	−.5740	−.9064	−.3937	−.4039
I′	0	−.4532	−.1937	−.1979
0	0	0	−.0064	−.0081
I	0	0	−.2912	+.1578
II	0	0	−.5634	−.2343
III	0	0	−.8171	−.6029
IV	0	0	−1.0388	−.9312
V	0	0	−1.1947	−1.1761
VI	0	0	−1.2585	−1.3046
VII	0	0	−1.1947	−1.2717
VIII	0	0	−1.0388	−1.1225
IX	0	0	−.8171	−.9795
X	0	0	−.6634	−.6169
XI	0	0	−.2912	−.3205
XII	0	0	−.0064	−.0081
XI′	0	0	+.2595	+.2804
X′	0	0	+.5127	+.5526
	L_2	L_3	L_4	L_5
II′	−.4046	−.3792	−.2876	−.0689

I′	− .1970	− .1824	− .1338	− .0213
0	− .0107	− .0145	− .0200	− .0262
I	+ .1444	+ .1110	+ .0354	+ .1081
II	+ .3203	+ .2647	+ .1297	+ .1386
III	− .2262	+ .4604	+ .2817	+ .0933
IV	− .7197	− .2956	+ .5331	+ .0826
V	− 1.1039	− .9030	− .4219	+ .5277
VI	− 1.3354	− 1.3028	− 1.0906	− .5061
VII	− 1.3550	− 1.4148	− 1.3627	− 1.0183
VIII	− 1.2218	− 1.3192	1.3485	− 1.1543
IX	− .9795	− 1.0750	− 1.1295	− 1.0210
X	− .6840	− .7589	− .8111	− .7570
XI	− .3578	− .4008	− .4350	− .4173
XII	− .0107	− .0145	− .0200	− .0262
XI′	+ .3052	+ .3294	+ .3366	+ .2879
X′	+ .5997	+ .6444	+ .6532	+ .5495

(c) Diagonals.

Load at	D_2'	D_1'	D_1	D_2
II′	+ .8855	+ .5602	− .0566	− .0308
I′	0	+ .8402	− .0263	− .0130
0	0	0	− .0040	− .0047

I	0	0	+.8473	−.0101
II	0	0	+.5859	+.8792
III	0	0	+.3360	+.5645
IV	0	0	+.1061	+.2731
V	0	0	−.0837	+.0300
VI	0	0	−.2164	−.1459
VII	0	0	−.2704	−.2283
VIII	0	0	−.2674	−.2434
IX	0	0	−.2241	−.2104
X	0	0	−.1610	−.1539
XI	0	0	−.0863	−.0839
XII	0	0	−.0040	−.0047
XI′	0	0	+.0668	+.0608
X′	0	0	+.1297	+.1167

	D_3	D_4	D_5	D_6
II′	+.0140	+.0968	−.2435	−.3440
I′	+.0099	+.0530	−.1255	−.1738
0	−.0058	−.0071	−.0075	−.0036
I	−.0384	−.0881	−.1622	−.1917
II	−.0597	−.1551	−.3024	−.3728
III	+.8276	−.2015	−.4211	−.5433
IV	+.5433	+.9956	−.5027	−.6957

V	+.4830	+.5543	+.0705	−.8109
VI	−.0239	+.2150	+.6457	+1.0804
VII	−.1513	+.0174	+.3689	+.8220
VIII	−.1958	−.0783	+.1988	+.6106
IX	−.1811	−.1008	+.1050	+.4364
X	−.1375	−.0880	+.0484	+.2803
XI	−.0773	−.0545	+.0138	+.1349
XII	−.0058	−.0071	−.0075	−.0036
XI′	+.0488	+.0194	−.0499	−.1528
XI	+.0918	+.0317	−.1073	−.3091

(d) Verticals.

Load at	V_2'	V_1'	V_0	V_1
II′	−1.00	−.7273	−.4491	+.0253
I′	0	−1.000	−.8097	+.0107
0	0	0	+.9965	+.0039
I	0	0	−.7563	−.9918
II	0	0	−.5229	−.7281
III	0	0	−.2999	−.4636
IV	0	0	−.0945	−.2249
V	0	0	+.0748	−.0246
VI	0	0	+.1932	+.1198

VII	0	0	+.2414	+.1875
VIII	0	0	+.2389	+.1999
IX	0	0	+.2001	+.1728
X	0	0	+.1437	+.1265
XI	0	0	+.0771	+.0689
XII	0	0	+.0035	+.0039
XI′	0	0	−.0597	−.0499
X′	0	0	−.1137	−.0960

	V_2	V_3	V_4	V_5	V_6
II′	−.0100	−.0577	−.1157	−.1625	0
I′	−.0071	−.0309	−.0597	−.0821	0
0	+.0041	+.0042	+.0035	+.0016	0
I	+.0274	+.0514	+.0771	+.0898	0
II	−.9574	+.0906	+.1437	+.1749	0
III	−.6625	−.8822	+.2001	+.2557	0
IV	−.3880	−.5817	−.7511	+.3284	0
V	−.1560	−.3238	−5086	−.6048	0
VI	+.0170	−.1257	−.3068	−.4696	−1.00
VII	+.1079	−.0101	−.1752	−.3556	0
VIII	+.1398	+.0458	−.0945	−.2642	0
IX	+.1292	+.0589	−.0499	−.1888	0
X	+.0982	+.0514	−.0229	−.1213	0

XI	+.0331	+.0318	−.0063	−.0583	0
XII	+.0041	+.0042	+.0035	+.0016	0
XI′	−.0348	−.0113	+.0237	+.0660	0
X′	−.0655	−.0186	+.0509	+.1337	0

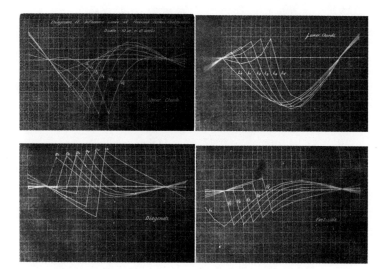

Table 46. Revised Stress Coefficients by Graphic Method.

(a) Upper Chords.

Load at	U_1	U_2	U_3	U_4	U_5	U_6
II′	+.780	+.800	+.780	+.700	+.301	+.160
I′	+.386	+.400	+.390	+.350	+.272	+.060
0	+.000	+.005	+.006	+.010	+.021	+.020

I	− .381	− .359	− 1.350	− .293	− .130	+ .052
II	− .263	− .762	− .716	− .587	− .311	+ .040
III	− .150	− .475	− 1.130	− .947	− .583	− .045
IV	− .050	− .209	− .588	− 1.410	− .990	− .315
V	+ .0366	+ .021	− .127	− .560	− .152	− .720
VI	+ .0971	+ .1831	+ .206	+ .027	− .545	− 1.510
VII	+ .121	+ .250	+ .358	+ .347	+ .020	− .720
VIII	− .125	+ .258	+ .397	+ .453	+ .272	− .315
IX	− .100	+ .225	+ .350	+ .427	+ .350	− .045
X	− .071	+ .158	+ .254	+ .320	+ .272	+ .040
XI	− .0381	+ .0381	+ .111	+ .123	+ .155	+ .052
XII	− .000	− .005	+ .006	+ .0101	+ .021	+ .020
XI′	− .300	− .065	− .100	− .115	− .071	+ .060
X′	− .058	− .120	− .173	− .220	− .121	+ .160

(b) Lower Chords.

Load at	L_1	L_2	L_3	L_4	L_5
II′	− .405	− .405	− .379	− .288	− .069
I′	− .201	− .190	− .182	− .134	− .021
0	− .009	− .010	− .015	− .020	− .026
I	+ .159	+ .139	+ .081	+ .037	+ .108
II	− .230	+ .319	+ .256	− .112	− .147

III	$-.607$	$-.219$	$+.446$	$+.261$	$-.098$
IV	$-.925$	$-.716$	$-.297$	$+.54$	$+.083$
V	-1.180	-1.095	$-.891$	$-.410$	$+.528$
VI	-1.300	-1.332	-1.295	-1.083	$-.506$
VII	-1.272	-1.350	-1.400	-1.345	-1.003
VIII	-1.125	-1.215	-1.321	-1.346	-1.174
IX	$-.828$	$-.975$	-1.083	-1.129	-1.034
X	$-.569$	$-.677$	$-.756$	$-.810$	$-.757$
XI	$-.310$	$-.338$	$-.378$	$-.415$	$-.399$
XII	$-.008$	$-.011$	$-.015$	$-.030$	$-.026$
XI$'$	$+.281$	$+.305$	$+.329$	$+.337$	$+.288$
X$'$	$+.553$	$+.600$	$+.644$	$+.653$	$+.550$

(c) Diagonals.

Load at	D_1	D_2	D_3	D_4	D_5	D_6
II$'$	$-.057$	$-.031$	$+.014$	$+.100$	$+.244$	$+.344$
I$'$	$-.026$	$-.013$	$+.010$	$+.053$	$+.126$	$+.174$
0	$-.004$	$-.005$	$-.006$	$-.007$	$-.008$	$-.004$
I	$+.850$	$-.000$	$-.022$	$-.066$	$-.138$	$-.192$
II	$+.584$	$+.832$	$-.054$	$-.144$	$-.285$	$-.373$
III	$+.333$	$+.561$	$+.928$	$-.202$	$-.417$	$-.587$
IV	$+.103$	$+.265$	$+.550$	$+.988$	$-.501$	$-.760$

V	$-.081$	$+.0261$	$+.221$	$+.56$	$+1.070$	$-.892$
VI	$-.214$	$-.149$	$-.022$	$+.213$	$+.638$	$+1.093$
VII	$-.259$	$-.219$	$-.140$	$-.0261$	$+.381$	$+.828$
VIII	$-.259$	$-.237$	$-.1881$	$-.067$	$+.208$	$+.611$
IX	$-.222$	$-.210$	$-.183$	$-.106$	$+.099$	$+.428$
X	$-.155$	$-.149$	$-.129$	$-.088$	$+.048$	$+.279$
XI	$-.0778$	$-.070$	$-.065$	$-.040$	$+.013$	$-.136$
XII	$-.002$	$-.005$	$-.008$	$-.007$	$-.008$	$-.004$
XI′	$+.067$	$-.061$	$+.049$	$-.019$	$-.050$	$-.153$
X′	$+.130$	$+.117$	$+.092$	$-.032$	$-.107$	$-.310$

(d) Verticals.

Load at	V_1	V_2	V_3	V_4	V_5
II′	$+.025$	$-.010$	$-.058$	$-.117$	$-.163$
I′	$+.011$	$-.007$	$-.031$	$-.060$	$-.082$
0	$+.004$	$+.004$	$+.004$	$+.004$	$+.002$
I	$-.985$	$+.077$	$+.053$	$+.066$	$+.088$
II	$-.7200$	$-.938$	$+.086$	$+.135$	$+.173$
III	$-.461$	$-.658$	$-.882$	$+.198$	$+.253$
IV	$-.223$	$-.393$	$-.574$	$-.707$	$+.324$
V	$-.022$	$-.159$	$-.326$	$-.508$	$-.619$
VI	$+.122$	$+.015$	$-.124$	$-.304$	$-.468$

VII	+.183	−.103	−.006	−.182	−.356
VIII	+.194	+.135	+.039	−.095	−.264
IX	+.173	+.131	+.059	−.046	−.185
X	+.122	+.093	+.052	−.023	−.120
XI	+.058	+.046	+.038	−.006	−.059
XII	−.004	+.004	+.004	+.004	+.002
XI′	−.049	−.035	−.011	+.024	+.070
X′	−.098	−.066	−.019	+.051	+.134

Table 47. Summation of Positive And Negative Stress Coefficients. (Stress due to loads at ends of cantilever arms not included.)

Member	Positive	Negative	Member	Positive	Negative
U_2'	0	0	D_2'	0	0
U_1'	0	0	D_1'	.8402	0
U_1	.9897	.8755	D_1	1.9421	1.3436
U_2	1.5888	1.8832	D_2	1.8076	1.0983
U_3	2.1057	3.0110	D_3	1.6479	.8766
U_4	2.1568	3.9232	D_4	1.8547	.7805
U_5	1.3606	4.1447	D_5	2.5160	1.4533
U_6	.3616	3.6103	D_6	3.5384	2.7744
L_1'	0	0	V_2'	0	0

L_0'	0	$-.4532$	V_1'	0	1.000
L_0	.2595	9.2754	V_0	1.1762	2.5430
L_1	.4382	8.6847	V_1	.8939	2.4764
L_2	.7699	8.2017	V_2	.5828	2.2058
L_3	1.1655	7.6815	V_3	.3383	1.9657
L_4	1.3165	6.7731	V_4	.5025	1.9850
L_5	.8982	5.2877	V_5	.9180	2.1447
			V_6	0	1.000

Table 48. Revised Live Load Stresses. (in kips)

(a) Positive Stresses.

Member	Coefficient	1st term	Load at II', X'	2nd term	Stress
U_2'	0	0	0	0	0
U_1'	0	0	.5050	225.3	225.3
U_1	.9897	139.9	.7834	349.2	489.1
U_2	1.5888	225.1	.8009	357.0	582.1
U_3	2.1057	299.9	.7910	352.5	652.4
U_4	2.1568	305.3	.7106	317.3	622.6
U_5	1.3606	185.2	.4966	221.4	406.6
U_6	.3516	49.8	.3160	141.0	217.9
L_1'	0	0	0	0	0
L_0'	0	0	0	0	0

L_0	.2595	36.8	.5127	229.0	265.8
L_1	.4382	62.0	.5526	247.2	309.2
L_2	.7699	108.9	.5997	267.8	376.7
L_3	1.1655	165.1	.6444	287.5	452.7
L_4	1.3165	186.3	.6532	291.5	478.8
L_5	.8982	127.1	.5495	245.1	372.2

Member	Coefficient	1st term	Load at II', X'	2nd term	Stress
D_2'	0	0	.8855	395.0	395.
D_1'	.8402	119.0	.5602	250.0	369.
D_1	1.9421	275.1	.1297	57.8	332.9
D_2	1.8076	256.2	.1167	52.1	308.3
D_3	1.6479	233.4	.1058	47.2	280.6
D_4	1.8547	262.8	.1305	58.3	321.1
D_5	2.5760	364.8	.2435	108.6	473.4
D_6	3.5384	501.3	.3440	153.5	654.8
V_2'	0	0	0	0	0
V_1'	0	0	0	0	0
V_0	1.1762	166.5	0	0	166.5
V_1	.8939	126.4	.0253	11.3	137.7
V_2	.5828	82.5	0	0	82.5
V_3	.3383	47.9	0	0	47.9

V_4	.5025	71.1	.0509	22.7	93.8
V_5	.9180	130.0	.1337	59.7	189.7
V_6	0	0	0	0	0

(b) Negative Stresses.

Member	Coefficient	1st term	Load at II',X'	2nd term	Stress
U_2'	0	0	0	0	0
U_1'	0	0	0	0	0
U_1	.8755	124.2	.0584	26.1	150.3
U_2	1.8832	266.5	.1250	55.8	322.3
U_3	3.0110	427.0	.1894	84.6	511.6
U_4	3.9232	557.2	.2153	96.2	653.4
U_5	4.1447	586.6	.1207	53.8	640.4
U_6	3.6103	511.2	0	0	511.2
L_1'	0	0	.5740	.2562	256.2
L_0'	.4532	64.3	.9064	.4041	468.4
L_0	9.2754	1312.1	.3937	175.5	1487.6
L_1	8.6847	1229.8	.4039	180.5	1410.3
L_2	8.2017	1160.1	.4046	180.7	1340.8
L_3	7.6815	1088.2	.3792	169.2	1257.4
L_4	6.773	953.5	.2876	128.3	1081.8
L_5	5.2877	748.2	.0689	30.7	778.9

Member	Coefficient	1st term	Load at II′,X′	2nd term	Stress
D_2'	0	0	0	0	0
D_1'	0	0	0	0	0
D_1	1.3436	189.7	.0566	25.2	214.9
D_2	1.0983	155.5	.0308	13.8	169.3
D_3	.8766	124.0	0	0	124.
D_4	.7805	110.5	0	0	110.5
D_5	1.4533	206.0	.1073	47.9	253.9
D_6	2.7744	392.3	.3091	137.8	530.1
V_2'	0	0	1.000	445.8	445.8
V_1'	1.000	141.6	.7273	324.0	465.6
V_0	2.5430	361.4	.5648	251.9	613.3
V_1	2.4764	350.1	.0960	42.8	392.9
V_2	2.2058	312.4	.0755	33.7	346.1
V_3	1.9657	278.7	.0762	34.0	312.7
V_4	1.9850	279.8	.1157	51.7	331.5
V_5	2.1447	304.3	.1625	72.5	376.8
V_6	1.0000	141.6	0	0	141.6

CHAPTER XXVI. REVISED IMPACT STRESSES

The revised impact stresses are given in Table 49. The

loaded lengths are measured from influence lines in Diagram 17 and the live load stresses taken from Table 48.

Table 49. Revised Impact Stresses. (in kips)

Member	Positive				Negative			
	1	Ratio	L. L.	Stress	1	Ratio	L. L.	Stress
U_2'	—	—	—	—	—	—	—	—
U_1'	250	.546	225.3	122.7	—	—	—	—
U_1	670	.309	489.1	154.3	530	.362	150.3	54.6
U_2	654	.315	582.1	183.1	546	.357	322.3	115.1
U_3	632	.322	652.4	211.1	568	.345	511.6	176.3
U_4	604	.332	622.6	208.5	596	.335	653.4	219.2
U_5	550	.353	406.6	143.2	650	.317	640.4	203.3
U_6	600	.333	217.9	72.6	600	.333	511.2	170.1
L_1'	—	—	—	—	250	.547	256.2	140.2
L_0'	—	—	—	—	300	.500	468.4	234.2
L_0	300	.500	265.8	132.9	900	.250	1487.6	371.0
L_1	372	.443	309.2	137.2	828	.266	1410.3	376.4
L_2	430	.411	376.7	150.9	770	.28	1340.8	376.2
L_3	480	.383	452.7	173.2	720	.294	1257.4	369.1
L_4	528	.362	478.8	173.8	672	.309	1081.8	334.3
L_5	574	.343	372.2	127.8	626	.324	778.9	252.3

Member	Positive				Negative			
	1	Ratio	L. L.	Stress	1	Ratio	L. L.	Stress
D_2'	250	.546	395.	215.9	—	—	—	—
D_1'	300	.5	369.	184.5	—	—	—	—
D_1	530	.361	332.9	120.3	630	.323	214.9	69.4
D_2	568	.345	308.3	106.2	632	.322	169.3	54.6
D_3	796	.274	280.6	76.9	404	.427	124	52.9
D_4	820	.268	321.1	86.1	380	.441	110.5	48.7
D_5	690	.303	473.4	143.7	510	.370	253.9	94.1
D_6	612	.329	654.8	215.4	588	.337	530.1	179.0
V_2'	—	—	—	—	250	.545	445.8	243.2
V_1'	—	—	—	—	300	.500	465.6	232.8
V_0	600	.333	166.5	55.4	600	.333	613.3	204.3
V_1	614	.489	137.7	67.3	586	.339	392.9	133.2
V_2	306	.494	82.5	40.7	804	.272	346.1	94.2
V_3	250	.545	47.9	26.1	950	.24	312.7	75.2
V_4	458	.395	93.8	37.1	742	.288	331.5	95.9
V_5	520	.366	189.7	69.5	680	.306	376.8	115.2
V_6	—	—	—	—	1200	.200	141.6	73.2

CHAPTER XXVII. REVISED DEAD LOAD STRESSES

From the preliminary estimate of weights the revised dead panel loads were computed in Table 39 and shown graphically in Diagram 12. Following the method given in Article 40 the revised dead load stresses are obtained and are given in Table 50. The graphical values as obtained from Plate I are given in Table 51.

The dead load stresses have been obtained by considering the arch as having an additional hinge at the centre of span. It will be interesting to see what these stresses would be if the arch were not erected as a combination type but to act as a pure two hinged truss. These stresses are given in Table 52 obtained by multiplying the stress coefficients of Table 45 by the proper dead panel loads. Since the stress coefficients are computed for loads on the upper chords there is a slight error induced in the stresses thus obtained, but as the effect is small, all the panel loads will be assumed to be acting on the upper chords. The stresses in verticals are afterwards corrected by the lower panel loads.

The comparison of stresses for the arch having two and two-hinges is given in Table 53 and Diagram 18. The symbols $2H$, and $3H$ are there used to denote the number of hinges. It will be seen that the $3H$ type is favorable for web members while the $2H$ type for chord members, the total stresses being much less in the latter than those in the former type. This is sufficient to show, for the case at hand at least, that combination type of arched trusses is slightly heavier than the pure two hinged arches. [1]

Table 50. Revise Dead Load Stresses. (in kips)

(a) Table of Negative Moments for Chord Members.

Member	M	139.3	87.1	121.3	96.0	82.9	76.1	74.6	72.3	
$D_1'L_1'U_1'$	3482.5	3482.5	2177.5	3032.5	2400	2072.5	1902.5	1865.	1807.5	25
$D_1L_3'L_0$	9142.5	6965.	4355	6065	4800	4145	3805	3730		50
$V_1D_2L_1U_1$	17835.	10447.5	6532.5	9097.5	7200	6217.5	5707.5			75
$V_2D_3L_2U_2$	28927.5	13930.	8710.	12130.0	9690	8290				100
$V_3D_4L_3U_3$	42092.5	17412.5	10887.5	15162.5	12000					125
$V_4D_5L_4U_4$	57160.0	20895.	13065.0	18195.						150
$V_5D_6L_5U_5$	74092.5	24377.5	15242.5							175
U_6	92832.5	27860.0								200
	325565.	125370	60970	63682.5	36000	20725	11415	5595	1807.5	325565

(b) Table of Positive Moments of V for Chord Members.

---

[1] Bridge Engineering. Waddell. P. 626.

$V_1 = 784.4$ kips.

Member	U_1, L_1	U_2, L_2	U_3, L_3	U_4, L_4	U_5, L_5	U_6
Distance	25	50	75	100	125	150
Moment	19510	39220	58830	78440	98050	117660

(c) Table of Negative Moments of H for Chord Members.

$H = 459.77$ kips

Member	U_1	U_2	$\cdot U_3$	U_4	U_5	U_6	L_s'
Ordinate	16.5	30	40.5	48	52.5	54	66
Moment	7586.4	13793.	18620.7	22069	24137.9	24827.6	30344.8

(d) Stresses in Upper Chords.

Member	M of Loads	M of Reaction		Total M	Lever Arm	Stress
	(−)	+	−			
U_2'	0	0	0	0	36	0
U_1'	3482.5	0	0	− 3482.5	49.5	+ 70.4
U_1	17835.0	19610	7586.4	− 5811.4	49.5	+ 117.3
U_2	28927.5	39220	13793.0	− 3500.5	36	+ 97.2
U_3	42092.5	58830	18620.7	− 1883.2	25.5	+ 73.8
U_4	57160.0	78440	22069.0	− 789.0	18	+ 43.8
U_5	74092.5	98050	24137.9	− 180.4	13.5	+ 13.3
U_6	92832.5	117660	24827.6	0	12	0

(e) Stress in Lower Chords.

Member	M of Load	M of Reactions		Total M	Lever Arm	Stress
	(–)	+	–			
L_1'	3482.5	0	0	– 3482.5	43.554	– 79.9
L_0'	9142.5	0	0	– 9142.5	55.084	– 166.1
L_0	9142.5	0	30344.8	– 39487.3	35.084	– 718.2
L_1	17835.0	19610	30344.8	– 28569.8	43.554	– 655.5
L_2	28927.5	39220	30344.8	– 20052.3	33.192	– 602.5
L_3	42092.5	58830	30344.8	– 13607.3	24.424	– 557.6
L_4	57160.0	78440	30344.8	– 9064.8	17.715	– 511.3
L_5	74092.5	98050	30344.8	– 6387.3	13.476	– 474.7

(f) Table of Negative Moments for Web Members.

Member	Load	Loads	Distance	1st Term	2nd Term	M
V_1', D_2'	139.3	139.3	66.668	9289.6	0	– 9289.6
D_1'	87.1	226.4	50	11320	3482.5	– 14802.5
D_1	121.3	347.7	100	34770	9142.5	43912.5
V_1, D_2	96.0	443.7	91.668	40673.6	17835.0	58508.6
V_2, D_3	82.9	526.6	85.714	45136.7	28927.5	74064.2
V_3, D_4	76.1	602.7	85	51230.	42092.5	93322.5
V_4, D_5	74.6	677.3	100	67730.	57160.0	124890.
V_5, D_6	72.3	749.3	225	168657.7	74092.5	242750.2

(g) Table of Positive Moments of V for Web Members.

$V = 784.4$ kips

Member	D_1	D_2, V_1	D_3, V_2	D_4, V_3	D_5, V_4	D_6, V_5
Distance	100	116.668	135.714	160	200	350
Moment	78440	91525	106453.7	125505.7	186880	274543.7

Negative Moment of H for all Web Members　459.77 × 66　30344.8

(h) Stresses in Diagonals.

Member	M of Loads	M of Reactions		Total M	Lever Arm	Stress
	(-)	+	-			
D_2'	-9289.6	0	0	-9289.6	75.291	+123.2
D_1'	-14802.5	0	0	-14802.5	89.26	+165.7
D_1	43912.5	78440	30344.8	-41827.0	89.26	+47.1
D_2	58508.6	91525	30344.8	-2671.6	75.291	+35.5
D_3	74064.2	106453.7	30344.8	-2044.7	61.206	+33.5
D_4	93322.5	125505.7	30344.8	-1838.4	49.665	+37.1
D_5	124890.	156880.	30344.8	-1645.2	47.512	+34.6
D_6	242750.2	274543.7	30344.8	-1448.9	97.362	+14.9

(i) Stress in Verticals.

Member	M of Loads		M of Reactions		Total M	Lever Arm	Stress
	-	+	+	-			
V_2'	—	—	0	0		85.714	-120.3
V_1'	-9289.6	-5106	0	0	-14395.6	91.668	-156.8
V_0	—	—	—	—	—	—	-252.0

V_1	−58508.6	4766.7	91525	30344.8	−7437.3	91.668	−81.1
V_2	74064.2	4037.1	106453.7	30344.8	−6081.8	85.714	−71.1
V_3	93322.5	3910.	125505.7	30344.8	−5748.4	85	−67.7
V_4	124890	4750.	156880.	30344.8	−6395.2	100	−64.0
V_5	242750	10642.4	274543.7	30344.8	−12091.3	225	−53.7
V_6	—	—	—	—	—	—	−41.8

Table 51. Revised Dead Load Stresses by Graphic Method. (in kips)

Member	Stress	Member	Stress
U_2'	+0	D_2'	+124.0
U_1'	+70.0	D_1'	+163.0
U_1	+115.0	D_1	+49.0
U_2	+96.0	D_2	+36.0
U_3	+71.0	D_3	+34.0
U_4	+42.0	D_4	+38.0
U_5	+12.0	D_5	+35.0
U_6	+0	D_6	+15.0
L_1'	−79.0	V_2'	−120.0
L_0'	−164.0	V_1'	−153.0
L_0	−716.0	V_0	−250.0
L_1	−655.0	V_1	−282.0

L_2	-601.0	V_2	-72.0
L_3	-556.0	V_3	-68.0
L_4	-510.0	V_4	-63.0
L_5	-473.0	V_5	-54.0
		V_6	-42.0

Table 52. Revised Dead Load Stresses for the Pure Two-hinged Arch.

(a) Upper Chords.

Load	U_2'	U_1'	U_1	U_2	U_3	U_4	U_5	U_6
139.3	0	-70.3	109.1	111.2	110.0	99.0	69.1	22.0
87.1	0	0	34.1	34.8	34.1	30.3	20.8	5.8
121.3	0	0	.2	.5	1.0	1.76	2.53	2.9
96.0	0	0	35.7	36.1	33.5	26.7	13.0	.5
82.9	0	0	21.9	63.3	60.0	49.5	27.5	2.7
76.1	0	0	11.5	36.0	85.6	72.9	44.8	4.3
74.6	0	0	3.6	15.1	43.6	104.1	70.9	19.8
72.3	0	0	2.7	1.5	9.6	42.1	110.7	52.2
69.6	0	0	6.8	12.6	13.7	2.3	37.9	106.2
72.3	0	0	8.8	18.3	25.9	24.9	1.4	52.5
74.6	0	0	9.3	19.3	29.6	34.4	19.8	19.8
76.1	0	0	7.7	16.8	26.5	32.7	24.9	4.3
82.9	0	0	6.0	13.2	21.5	27.2	23.7	2.7

96.0	0	0	3.7	8.3	13.4	17.8	16.6	5.0
121.3	0	0	.2	.5	1.0	1.76	2.53	2.9
87.1	0	0	2.6	5.6	8.6	10.0	6.1	5.8
139.3	0	0	8.2	17.4	26.4	29.9	16.8	22.0
Total	0	+70.3	+105.1	+63.5	+9.4	-63.0	-146.4	-186.5

(b) Lower Chords.

Load	L_1'	L_0'	L_0	L_1	L_2	L_3	L_4	L_5
139.3	79.9	126.1	54.8	56.2	56.5	52.8	40.2	9.6
87.1	0	39.5	16.9	17.3	17.1	15.9	11.7	1.9
121.3	0	0	.7	.9	1.3	1.7	2.4	2.9
96.0	0	0	27.9	15.1	13.9	10.7	3.4	10.4
82.9	0	0	46.6	19.5	26.5	21.9	10.7	11.5
76.1	0	0	62.1	45.7	17.2	35.0	21.5	7.1
74.6	0	0	77.5	69.5	53.7	22.1	39.8	6.2
72.3	0	0	86.5	85.2	80.0	65.2	30.5	38.1
69.6	0	0	87.8	90.7	92.8	90.7	75.9	35.0
72.3	0	0	86.5	92.2	98.2	102.5	96.0	73.3
74.6	0	0	77.5	83.8	91.2	98.3	100.1	86.5
76.1	0	0	62.1	67.6	74.4	81.9	86.	77.8
82.9	0	0	46.7	51.1	56.7	62.9	67.1	62.7
96.0	0	0	27.9	30.8	34.4	38.5	41.7	40.1
121.3	0	0	.7	.9	1.3	1.7	2.4	2.9

87.1	0	0	22.6	24.4	26.6	28.7	29.4	25.1
139.3	0	0	71.3	76.8	83.5	90.0	91.1	76.5
Total	−79.9	−165.6	−668.3	−595.7	−524.3	−447.5	−358.1	−275.8

（c）Diagonals.

Load	D_2'	D_1'	D_1	D_2	D_3	D_4	D_5	D_6
139.3	123.2	92.5	7.9	4.3	1.9	13.7	33.9	48.1
87.1	0	73.2	2.2	1.1	.9	4.6	10.9	15.2
121.3	0	0	.5	.6	.7	.9	.9	.4
96.0	0	0	81.1	1.0	5.7	8.5	15.6	18.4
82.9	0	0	48.6	72.8	4.9	12.9	25.1	30.9
76.1	0	0	25.6	43.0	62.8	15.3	32.1	41.3
74.6	0	0	7.9	20.4	40.6	74.3	37.4	52.0
72.3	0	0	6.1	2.2	15.6	40.1	76.7	58.8
69.6	0	0	15.1	10.1	1.6	15.0	45.0	75.2
72.3	0	0	19.5	16.5	10.9	1.3	26.7	59.6
74.6	0	0	19.9	18.1	14.6	5.8	14.8	45.7
76.1	0	0	17.0	15.9	13.8	7.67	8.0	33.2
82.9	0	0	13.2	12.7	11.4	7.3	4.0	23.2
96.0	0	0	9.3	8.0	7.4	5.2	1.2	13.0
121.3	0	0	.5	.6	.7	.9	.9	.4
87.1	0	0	5.9	5.3	4.3	1.7	4.3	13.3

139.3	0	0	18.1	16.2	12.8	4.4	14.9	42.2
Total	+123.2	+165.7	+76.0	+71.0	+67.4	+90.6	+90.0	+65.5

(d) Verticals.

Load	V_2'	V_1'	V_0	V_1	V_2	V_3	V_4	V_5	V_6
139.3	139.3	101.5	62.6	3.5	1.4	9.4	16.1	22.7	0
87.1	0	87.1	70.5	.9	.6	2.7	4.2	7.1	0
121.3	0	0	.4	.5	.5	.5	.4	.2	0
96.0	0	0	72.6	95.1	2.6	4.9	7.4	8.6	0
82.9	0	0	43.3	59.9	79.2	7.5	11.9	14.5	0
76.1	0	0	22.7	35.4	50.5	6.7	15.2	19.5	0
74.6	0	0	7.1	16.8	29.0	43.5	56.8	24.2	0
72.3	0	0	5.3	1.8	11.3	23.4	36.9	43.8	0
69.6	0	0	13.4	8.3	1.2	8.8	21.4	32.7	69.6
72.3	0	0	17.5	13.5	7.8	7.3	12.7	25.8	0
74.6	0	0	17.8	14.9	10.4	3.4	7.1	19.7	0
76.1	0	0	15.2	13.1	9.9	5.2	3.8	14.3	0
82.9	0	0	11.9	10.5	8.1	4.3	1.9	10.1	0
96.0	0	0	7.4	6.6	5.3	3.1	.6	5.6	0
121.3	0	0	14.0	.5	.5	.5	.4	.2	0
87.1	0	0	5.2	4.4	3.0	1.0	2.1	5.8	0
139.3	0	0	16.1	13.4	9.1	2.6	6.9	18.7	0
Total	−120.3	−157.6	−152.0	−111.1	−102.0	−106.2	−92.2	−65.1	−41.8

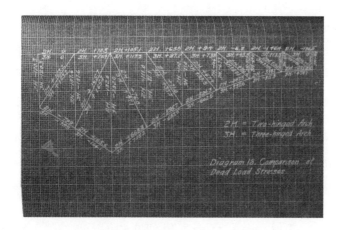

Table 53. Comparison of Dead Load Stresses.

(a) Upper Chords.

Member	U_2'	U_1'	U_1	U_2	U_3	U_4	U_5	U_6
2H	0	−70.3	−105.1	−53.5	−9.4	−63.	−146.4	−186.5
3H	0	−70.4	−117.3	−97.2	−73.8	−43.8	−13.3	0
3H − 2H	0	0	−12.2	−35.7	−64.4	−19.2	−133.1	−186.5

Total Stresses in favor of 2H = 2268 kips.

	L_1'	L_0'	L_0	L_1	L_2	L_3	L_4	L_5
2H	−79.9	−165.6	−668.3	−595.7	−524.3	−447.5	−358.1	−257.8
3H	−79.9	−166.1	−718.2	−655.5	−602.5	−557.6	−511.3	−474.7
3H − 2H	0	0	−49.9	−59.8	−38.2	−110.1	−163.2	−198.9

Total Stresses in favor of 2H = 620.1 kips.

Member	D_2'	D_1'	D_1	D_2	D_3	D_4	D_5	D_6
2H	−123.2	−165.7	−76.0	−71.0	−67.4	−90.6	−90.0	−65.5
3H	−123.2	−165.7	−47.1	−35.5	−33.5	−37.1	−34.6	−14.9
2H−3H	0	0	−28.9	−35.5	−33.9	53.5	55.4	50.6

Total Stresses in favor of 3H = 257.8 kips.

	V_2'	V_1'	V_0	V_1	V_2	V_3	V_4	V_5	V_6
2H	−120.3	−157.6	−152.	−111.1	−102.	−106.2	−92.5	−65.1	−41.8
3H	−120.3	−156.8	−252.	−81.1	−71.	−67.7	−64.0	−53.7	−41.8
2H−3H	0	0	101	30.0	31.	38.5	28.5	11.4	0

Total Stresses in favor of 3H = 240.4 kips

Grand total of Stresses in favor of 2H = 348.7 kips.

CHAPTER XXIII. REVISED WIND STRESSES

The stresses in chord members due to direct action of winds remain the same as for preliminary stresses. For stresses due to overturning effect the equivalent loads of page may now be treated as the variable dead panel loads of Table 52 and a more exact result obtained than by the use of mean loads. The revised wind stresses due to overturning effect only are given in Table 54.

Table 54. Revised Wind Stresses. (in kips)

For S. Wind

(a) Upper Chords.

Load	U_2'	U_1'	U_1	U_2	U_3	U_4	U_5	U_6
117.8	0	59.5	92.3	94.2	93.1	83.8	58.4	18.6
90.9	0	0	35.5	36.2	35.7	31.7	21.6	6.1
50.2	0	0	.1	.2	.4	.7	1.0	1.2
120.3	0	0	46.0	45.2	42.0	33.4	16.2	6.3
83.9	0	0	22.2	65.3	60.7	50.1	27.8	2.7
54.9	0	0	8.3	26.1	62.0	52.7	32.3	3.1
33.5	0	0	1.6	6.8	19.2	46.6	31.8	8.9
19.6	0	0	.7	.4	2.6	11.4	29.9	14.1
13.8	0	0	1.3	2.5	2.7	.4	7.5	21.1
19.6	0	0	2.2	4.9	7.0	6.8	.4	14.1
33.5	0	0	4.2	8.7	13.3	15.4	8.9	8.9
54.9	0	0	5.5	12.1	19.1	23.6	18.5	3.1
83.9	0	0	6.1	13.5	21.6	27.5	23.9	2.8
120.3	0	0	4.7	9.7	16.9	22.3	20.9	6.2
50.2	0	0	.1	.2	.4	.7	1.0	1.2
90.9	0	0	2.7	5.9	9.0	10.4	6.4	6.1
117.8	0	0	6.9	14.7	22.3	25.3	14.2	18.6
+	0	59.5	152.7	182.6	210.2	212.9	154.6	58.8
−	0	0	87.7	163.0	155.8	229.8	156.1	38.1
Total	0	+59.5	+65.0	+19.6	+54.4	−16.9	−11.5	−20.7

(b) Lower Chords.

Load	L_1'	L_0'	L_0	L_1	L_2	L_3	L_4	L_5
117.8	67.5	105.8	46.3	47.6	47.7	44.6	33.9	8.1
90.9	0	41.2	17.6	18.0	18.0	16.6	12.2	1.9
50.2	0	0	.3	.4	.5	.7	1.0	1.3
120.5	0	0	35	19.0	17.4	13.3	4.2	13.1
83.9	0	0	47.3	19.7	26.9	22.2	10.9	11.6
54.9	0	0	44.8	34	12.3	25.5	15.4	5.1
33.5	0	0	33.8	31.2	24.1	9.9	17.9	2.8
19.6	0	0	23.5	23.1	21.6	17.7	8.2	10.3
13.8	0	0	17.3	18.5	18.4	18.0	15.1	7.0
19.6	0	0	23.5	25.0	26.6	27.7	26.8	19.9
33.5	0	0	33.8	37.6	41.1	44.0	45.2	38.7
54.9	0	0	44.9	48.8	53.7	59.0	62.0	56.1
83.9	0	0	47.3	51.7	57.4	63.7	68.2	63.5
120.3	0	0	35.0	38.5	43.1	48.1	52.3	50.1
50.2	0	0	.3	.4	.5	.7	1.0	1.3
90.9	0	0	23.6	25.4	27.7	29.9	30.6	26.2
117.8	0	0	60.5	65.2	70.5	75.8	76.9	64.7
+	0	0	84.1	106.9	142.5	166.7	155.9	104.0
−	67.5	148.0	450.7	394.5	365.0	350.7	325.0	277.7
Total	−67.5	−148.	366.6	−107.6	−222.5	−184.0	−169.1	−173.7

(c) Diagonals.

Load	D_2'	D_1'	D_1	D_2	D_3	D_4	D_5	D_6
117.8	103.4	-66.1	6.7	3.6	1.6	11.6	28.7	40.5
90.9	0	76.3	2.4	1.2	-.9	4.8	11.4	15.8
80.2	0	0	.1	.2	-.5	.3	.8	.2
120.3	0	0	101.6	1.2	-4.6	16.1	19.5	23.1
83.9	0	0	49.4	73.6	-5.0	13.1	25.4	31.3
54.9	0	0	18.5	31.0	45.5	11.1	23.1	29.9
33.5	0	0	3.5	9.1	18.2	33.3	16.8	23.3
19.6	0	0	1.6	.6	4.3	10.8	21.0	15.9
13.8	0	0	3.0	2.0	.3	2.9	8.9	14.9
19.6	0	0	5.3	4.5	2.9	.3	7.2	16.1
33.5	0	0	8.9	8.1	6.5	2.6	6.6	2.0
54.9	0	0	12.3	11.5	9.9	5.5	57.6	24.0
83.9	0	0	13.3	12.9	11.56	7.4	4.1	23.6
120.3	0	0	10.3	10.1	9.3	6.5	1.6	16.0
50.2	0	0	.3	.2	.5	.3	.4	.2
90.9	0	0	6.1	5.5	4.4	1.8	4.5	13.9
117.8	0	0	15.3	13.7	10.8	3.7	12.6	36.4
+	103.4	142.4	194.9	133.5	85.7	69.2	147.1	152.9
-	0	0	63.4	53.7	51.1	65.6	102.7	174.2
Total	+103.4	+142.4	+131.5	+79.8	+34.6	+3.6	+44.4	-21.3

(d) Verticals.

Load	V_2'	V_1'	V_0	V_1	V_2	V_3	V_4	V_5	V_6
117.8	117.8	85.6	52.9	2.9	.2	6.7	13.6	19.1	0
90.8	0	90.8	73.4	.9	-.6	2.8	5.4	7.4	0
50.2	0	0	.2	.2	.2	.2	.2	.1	0
120.3	0	0	90.9	90.1	3.3	6.2	17.3	19.5	0
83.9	0	0	43.8	60.5	80.3	7.6	12.1	14.7	0
54.9	0	0	16.5	25.4	36.4	48.4	11.0	14.1	0
33.5	0	0	3.2	7.5	13.0	19.5	25.5	11.0	0
19.6	0	0	1.4	.5	3.0	6.3	10.0	11.8	0
13.8	0	0	2.6	1.6	.2	1.7	4.2	6.4	13.8
19.6	0	0	4.7	3.7	2.1	.2	3.4	7.0	0
33.5	0	0	8.0	6.6	4.7	1.5	3.2	3.9	0
54.9	0	0	11.0	9.5	7.1	3.2	1.3	10.3	0
83.9	0	0	12.1	10.6	8.3	4.3	1.9	10.2	0
120.3	0	0	9.2	8.3	6.6	3.8	.7	6.9	0
50.2	0	0	.2	.2	.2	.2	.2	.1	0
90.9	0	0	5.4	4.5	3.2	1.0	2.2	6.0	
117.8	0	0	13.6	11.3	7.7	2.2	6.0	15.8	0
+	0	0	49.4	44.5	32.7	27.0	49.0	81.3	0
-	117.8	176.4	328.5	200.1	144.4	88.8	69.2	88.0	13.8
Total	-87.4	-124.2	-279.1	-73.9	-57.0	-24.7	-3.3	-.6	-12.4

CHAPTER XXIX. REVISED TEMPERATURE STRESSES

As found in Chapter XIX the change of span due to variation of temperature is 1.755 inches. The value of displacement δ' was there found by assuming the area of U_6 as the weighted mean of all the members of the truss and the preliminary temperature stresses were obtained entirely based on this assumption. It is now proposed to check the accuracy of this method and to compute more exact values of the temperature stresses. It is found in Table 42 the summation of $\dfrac{T^2 L}{A}$ for all the members of the truss is equal to 2×28.66705 or 57.3341 units in which L is expressed in feet, A in square inches and H assumed 1 kip per square inch. The actual displacement for H 1 kip will be equal to this quantity divided by E 29000 and multiplied by 12, or equal to 0.023724 inch.

By direct proportion the value of H resulted from a displacement of 1.755 inches is found to be $\dfrac{1.755}{0.02344}$ or 73.976 kips which being the thrust produced by the maximum effect of temperature. The value of H obtained for preliminary stresses was

67.68 kips, only 6. 3 kips in error or 8. 5% of the revised value. This serves to indicate that the approximate method is accurate enough for all practical purposes.

The revised temperature stresses are given in Table 55.

Table 55. Revised Temperature Stresses. (in kips)

For fall in temperature

Member	T	Stress	Member	T	Stress
U_1	$-.3333$	-24.6	D_1	$+.7394$	$+54.6$
U_2	$-.8334$	-61.7	D_2	$+.8766$	$+64.7$
U_3	-1.5883	-117.4	D_3	$+1.0783$	$+79.7$
U_4	-2.6667	-197.2	D_4	$+1.3289$	$+98.2$
U_5	-3.8889	-288.0	D_5	$+1.3891$	$+102.7$
U_6	-4.500	-332.7	D_6	$+0.6779$	$+50.2$
L_0	$+1.1982$	$+88.6$	V_0	-0.66	-48.8
L_1	$+1.5153$	$+112.1$	V_1	-0.7200	-53.2
L_2	$+1.9885$	$+147.1$	V_2	-0.7700	-56.9
L_3	$+2.7022$	$+199.7$	V_3	-0.7765	-57.4
L_4	$+3.7255$	$+275.6$	V_4	0.6600	-48.8
L_5	$+4.8977$	$+362.1$	V_5	-0.293333	-21.6
			V_6	0	0

CHAPTER XXX. SUMMATION OF REVISED STRESSES

The summation of revised stresses is given in Table 56 in which the dead load stresses are taken from table 50, live load stresses from Table 48, impact stresses from Table 49, temperature stresses from Table 55 and wind stresses from Table 54. It will be seen that the stresses thus obtained differ only slightly from those for preliminary stresses. This is more clearly seen in the following diagram in which the stresses given are those which control the design.

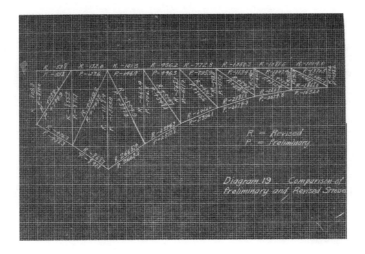

Table 56. Summation of Revised Stresses.

(a) Upper Chords.

Member	U_2'	U_1'	U_1	U_2	U_3	U_4	U_5	U_6
(Positive)								
D. L.	0	70.4	117.3	97.2	73.8	43.8	13.3	0
L. L.	0	225.3	489.1	582.1	652.4	622.6	406.6	217.9
Imp.	0	122.7	154.3	183.1	211.1	208.5	143.2	72.6
Temp. Rise	0	0	24.6	61.7	117.4	197.2	288.0	332.7
S. Wind Direct	53.8	−178.9	−11.8	80.8	156.0	213.7	253.9	276.5
S. Wind Over.	0	59.5	65.0	19.6	54.4	0	0	20.7
N. Wind Over.	0	0	0	0	0	16.9	11.5	0
(Negative)								
L. L.	0	0	150.3	322.3	511.6	653.4	640.4	511.2
Imp.	0	0	54.6	115.1	176.3	219.2	203.3	170.1
Temp Fall	0	0	24.6	61.7	117.4	197.2	288.0	332.7
N. Wind Direct	+53.8	+178.9	11.8	80.8	156.0	213.7	253.9	276.5
N. Wind Over.	0	59.5	6.5	19.6	54.4	0	0	20.7
S. Wind Over.	0	0	0	0	0	16.9	11.5	0
(Without Wind)								
Maximum	0	+418.4	+785.3	+924.1	+1054.7	+1072.1	+851.1	+623.2
Minimum	0	+46.9	−151.3	−456.2	−772.8	−1050.3	−1081.6	−1000.4
(With Wind)								

Maximum	+53.8	+597.3	+797.1	+1024.3	+1265.1	+1285.8	+1105.	+920.4
Minimum	-53.8	-132.0	-163.0	-556.6	-983.2	-1264.0	-1335.5	-1297.6

(b) Lower Chords.

Member	L_1'	L_0'	L_0	L_1	L_2	L_3	L_4	L_5
(Positive)								
L. L.	0	0	265.8	309.2	376.7	452.7	478.8	372.2
Imp.	0	0	132.9	137.2	150.9	173.2	173.8	127.8
Temp. Fall	0	0	88.6	112.1	147.1	199.7	275.6	362.1
S. Wind Direct	5.9	-22.7	16.5	44.1	63.8	77.0	85.2	89.0
N. Wind Over.	67.5	148.0	366.6	107.6	222.5	184.0	169.1	173.7
(Negative)								
D. L.	79.9	166.1	718.2	655.5	602.5	557.6	511.3	474.7
L. L.	256.2	468.4	1487.6	1410.3	1340.8	1257.4	1081.8	778.9
Imp.	140.2	234.2	371.0	376.4	376.2	369.1	334.3	252.3
Temp. Rise	0	0	88.6	112.1	147.1	199.7	275.6	362.1
N. Wind Direct	+5.9	+22.7	16.5	44.1	63.8	77.0	85.2	89.0
S. Wind Over.	67.5	148.0	366.6	107.6	222.5	184.0	169.1	173.7
(Without Wind)								
Maximum	-53.2	-111.0	+9.3	+121.5	+272.7	+454.6	+587.2	+184.0
Minimum	-476.3	-868.7	-2665.4	-2554.3	-2466.6	-2383.8	-2203.0	-1868.0
(With Wind)								
Maximum	+20.2	+59.7	+375.9	+229.1	+495.2	+638.6	+756.4	+357.7

(c) Diagonals.

Member	D'_2	D'_1	D_1	D_2	D_3	D_4	D_5	D_6
(Positive)								
D. L.	123.2	165.7	47.1	35.5	33.5	37.1	34.6	14.9
L. L.	395.0	369.0	322.9	308.3	280.6	321.1	473.4	654.8
Imp.	215.9	184.5	120.3	106.2	76.9	86.1	143.7	215.4
Temp. Fall	0	0	54.6	64.7	79.7	98.2	102.7	50.2
S. Wind Direct	103.4	142.4	131.5	79.8	34.6	3.6	44.4	0
(Negative)								
L. L.	0	0	214.9	189.3	124.0	110.5	253.9	530.1
Temp.	0	0	69.4	54.6	52.9	48.7	94.1	179.0
Temp. Rise	0	0	54.6	64.7	79.7	98.2	102.7	50.2
S. Wind Over.	103.4	142.4	131.5	79.8	34.66	3.6	44.4	0
(Without Wind)								
Maximum	+734.1	+719.2	+564.9	+514.7	+470.7	+542.5	+754.4	+935.3
Minimum	+82.2	+110.2	−307.5	−264.9	−234.3	−230.7	−427.6	−749.4
(With Wind)								
Maximum	+837.5	+861.6	+686.4	+594.5	+505.3	+546.1	+798.8	+956.6
Minimum	−21.2	−32.2	−439.0	−344.7	−268.9	−234.3	−472.0	−770.7

(d) Verticals.

Member	V_2'	V_1'	V_0	V_1	V_2	V_3	V_4	V_5	V_6
(Positive)									
L. L.	0	0	166.5	137.7	82.5	47.9	93.8	189.7	0
Imp.	0	0	55.4	67.3	40.7	26.1	37.1	69.5	0
Temp. Rise	0	0	48.8	53.2	56.9	57.4	48.8	21.6	0
N. Wind Over.	87.4	124.2	279.1	73.9	57.0	24.7	3.3	.6	12.4
(Negative)									
D. L.	120.3	156.8	252.0	81.1	71.1	67.7	64.	53.7	41.8
L. L.	445.8	465.6	613.3	392.9	346.1	312.7	331.5	376.8	141.6
Imp.	243.2	232.8	204.3	133.2	94.2	75.2	95.9	115.2	73.2
Temp. Fall	0	0	48.8	53.2	56.9	57.4	48.8	21.6	0
S. Wind Over.	87.6	124.2	279.1	73.9	57.	24.7	3.3	.6	12.4
(Without Wind)									
Maximum	−80.3	−104.2	+102.7	+204.2	+132.9	+86.3	+137.0	+244.9	+27.9
Minimum	−809.3	−855.2	−1118.4	−660.4	−568.3	−513.0	−54.0	−567.3	−256.6
(With Wind)									
Maximum	+7.4	+20.2	+381.8	+278.1	+189.9	+111.0	+140.3	+245.5	+159.0
Minimum	−896.9	−979.4	−1397.5	−734.3	−625.3	−537.7	−543.3	−567.9	−269.0

CHAPTR XXXI. REVISED DESIGN

Since the revised stresses are very nearly the same as the

preliminary values, the preliminary design may be considered as final if theoretical correctness is not required. To obtain more economical sections, however, several changes should be made both to save material and to give more close agreement to the computed stresses. The composition of sections of members is kept the same as before, modifications being made only in the dimensions.

Article 62. Verticals.

Member V_2'

Greatest Stresses: – Without Wind = – 809. 3 kips

With Wind = – 856. 5 kips

Same section used as in the preliminary design.

Area supplied 65. 3 sq. in.

Unit Stress 12. 57 kips per sq. in.

Area required: –

Without Wind stresses = 64. 5 sq. in.

With Wind stresses = 57. 1 sq. in.

Member V_1'

Greatest Stresses: – Without wind = – 855. 2 kips

With wind = – 979. 4 kips

Same section used as in the preliminary design.

Area supplied 77. 8 sq. in.

Unit Stress 11. 45 kips per sq. in.

Area required: −

Without wind stresses = 74. 7 sq. in.

With wind stresses = 68. 5 sq. in.

Member V_0

Greatest Stresses: − Without wind = − 1118. 4 kips

Wind wind = − 1446. 9 kips

$$2 \text{ Webs } 25'' \times \frac{7}{8}'' = 43. 7$$

$$4 \lfloor S \ 6'' \times 4'' \times \frac{7}{8}'' = 32. 0$$

$$2 \text{ Fills } 13'' \times \frac{7}{8}'' = 22. 7$$

$$8 \text{ Flats } 5 \times \frac{1}{2} \quad = \underline{20.}$$

Area supplied   118. 4 sq. in.

Least r about axis A = 8. 7 in.

Length = 792 in.

Unit Stress 9. 63 kips per sq. in.

Area required: −

Without wind stresses = 115. 6 sq. in.

With wind stresses   = 115. 5 sq. in.

Member V_1

Greatest Stresses: − Without wind = 660. 4 kips

With wind = 778. 6 kips

Same section used as in the preliminary design.

Area supplied 65. 9 sq. in.

Unit Stress 11. 39 kips per sq. in.

Area required: −

Without wind stresses = 58. 0 sq. in.

With wind stresses = 55. 0 sq. in.

Member V_2

Greatest Stresses: − Without wind = − 568. 3 kips

With wind = − 658. 0 kips

Same section used as V_3 of preliminary design.

Area supplied 44 sq. in.

Unit Stress 13. 0 kips per sq. in.

Area required: −

Without wind stresses = 44. 0 sq. in.

With wind stresses = 40. 8 sq. in.

Member V_3

Greatest Stresses: − Without wind = − 513. 0 kips

With wind = − 573. 1 kips

Same section used as V_3 of preliminary design.

Area supplied 44 sq. in.

Unit stress 13.5 kips per sq. in.

Area required: –

Without wind stresses = 38.0 sq. in.

With wind stresses   = 33.4 sq. in.

Member V_4

Greatest Stresses: –   Without wind = −540.0 kips

With wind   = −606.1 kips

Same section used as V_3 of preliminary design.

Area supplied 44 sq. in.

Unit stress 14 kips per sq. in.

Area required: –

Without wind stresses = 40 sq. in.

With wind stresses   = 36.2 sq. in.

Member V_5

Greatest Stresses: –   Without wind = −567.3 kips

With wind   = −654.7 kips

Same section used as V_3 of preliminary design.

Area supplied 44 sq. in.

Unit stress 14 kips per sq. in.

Area required: –

  Without wind stresses $= 40.5$ sq. in.

  With wind stresses $= 37.3$ sq. in.

Member V_6

Greatest Stresses: – Without wind $= -256.6$ kips

         With wind $= -269.0$ kips

 Same section used as in the preliminary design.

Area supplied 29.5 sq. in.

Unit stress 14 kips per sq. in.

Area required: –

  Without wind stresses $= 18.2$ sq. in.

  With wind stresses $= 15.3$ sq. in.

Article 63. Diagonals.

The sections of diagonals used in the preliminary design will be adopted here without modification. The maximum revised stresses are all smaller than the preliminary values showing that no investigation of tensile strength is necessary. The compressive revised stresses, however, are little greater than the preliminary stresses but as the sections are more than amply provided for compression, it is found that the increased stresses are not sufficient to take the difference between the area supplied and that required for compression in the preliminary design. For these reasons the detailed investigations will not be reproduced.

Article 64. Lower Chords.

Member L_1'

Greatest Stresses: – Without wind = −476.3 kips

With wind　　= −549.7 kips

Same section used as in the preliminary design.

Area supplied 49.0 sq. in.

Unit stress 13.57 kips per sq. in.

Area required: –

Without wind stresses = 35.1 sq. in.

With wind stresses　 = 32.4 sq. in.

Member L_0'

Greatest Stresses: – Without wind = −868.7 kips

With wind　　= −1016.7 kips

Same section used as in the preliminary design.

Area supplied 73.4 sq. in.

Unit stress 13.5 kips per sq. in.

Area required: –

Without wind stresses = 64. 7 sq. in.

With wind stresses = 60. 0 sq. in.

Member L_0

Greatest Stresses: – Without wind = – 2665. 4 kips

With wind = – 3116. 1 kips

Same section used as in the preliminary design.

Area supplied 195. 8 sq. in.

Unit stress 13. 65 kips per sq. in.

Area required: –

Without wind stresses = 195. 5 sq. in.

With wind stress = 182. 0 sq. in.

Member L_1

Greatest Stresses: – Without wind = – 2554. 3 kips

With wind = – 2948. 8 kips

Same section used as in the preliminary design.

Area supplied 185. 8 sq. in.

Unit stress 13. 7 kips per sq. in.

Area required: –

Without wind stresses = 185. 6 sq. in.

With wind stresses = 172. 0 sq. in.

Member L_2

Greatest Stresses: – Without wind = – 2466. 6 kips

With wind   = – 2831. 6 kips

Same section used as L_1 of preliminary design.

Area supplied 185. 8 sq. in.

Unit stress 13. 8 kips per sq. in.

Area required: –

Without wind stresses = 180 sq. in.

With wind stresses   = 161 sq. in.

Member L_3

Greatest Stresses: – Without wind = – 2383. 8 kips

With wind   = – 2734. 5 kips

Same section used as L_2 of preliminary design.

Area supplied 175. 8 sq. in.

Unit stress 13. 8 kips per sq. in.

Area required: –

Without wind stresses = 173. 2 sq. in.

With wind stresses   = 160. Sq. in.

Member L_4

Greatest Stresses: – Without wind = – 2203. 0 kips

With wind   = -2528.0 kips

Same section as L_2 of preliminary design.

Area supplied 175.8 sq. in.

Unit stress 14 kips per sq. in.

Area required: -

Without wind stresses = 157.0 sq. in.

With wind stresses   = 144.0 sq. in.

Member L_5

Greatest Stresses: - Without wind = -1868.0 kips

With wind   = -2145.7 kips

Same section used as L_3 of preliminary design.

Area supplied 153.4 sq. in.

Unit stress 14 kips per sq. in.

Area required: -

Without wind stresses = 134.0 sq. in.

With wind stresses   = 122.4 sq. in.

Article 65. Upper Chords.

Since the revised stresses which control the design are all smaller than the preliminary stresses, the sections for upper chords will be taken from the preliminary design without modification. For several members the tensile stress is very great but it is found not to endanger the section. The detailed investigations will be omitted.

CHAPTER XXXII. FINAL STRESSES

We have seen that large part of the preliminary design has been adopted for revised stresses without necessitating any important modifications. Nevertheless, the changes made, no matter how slight, overturn the theoretical accuracy of the revised stresses and require a third design if very exact values are to be sought. This, while is theoretically correct, is not justifiable in practice. As the areas are increased in some members but are decreased in others their effects on the values of H tend to balance each other, at least to such a degree as to preserve the practical correctness of the revised stresses. To repeat the calculations for another set of stresses is not warrantable inasmuch as the object of the present work is to show the process rather than the actual result. For this reason the revised design will be considered as final and is shown in the detailed drawings. The distribution of areas and composition of sections are shown in

the accompanying diagrams. Table 57 gives the principal
constants of the different members, I being the moment of inertia
of the member about an axis perpendicular to the plane of truss
and C the distance from extreme fibres to the neutral axis of the
section, measured in the plane of truss.

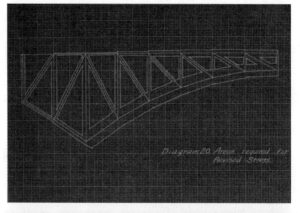

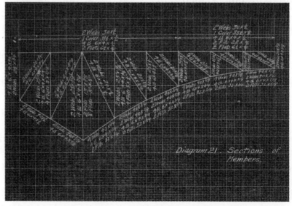

Table 57. Constants of Members.

Member	L	A	Least r	I	$2C$
	feet	sq. in.	in.	in.^{4}	in.
U_2'	25	77.2	12.90	12155	T 28.10 B 34.40
U_1'	25	77.2	12.90	12155	T 28.10 B 34.40
U_1	25	77.2	12.90	12155	T 28.10 B 34.40
U_2	25	77.2	12.90	12155	T 28.10 B 34.40
U_3	25	77.2	12.90	12155	T 28.10 B 34.40
U_4	25	99.2	10.00	13165	T 28.74 B 33.76
U_5	25	99.2	10.00	13165	T 28.74 B 33.76
U_6	25	99.2	10.00	13165	T 28.74 B 33.76
L_1'	28.413	49.0	9.73	4686	40
L_0'	29.954	73.4	10.10	7500	40

Member	L	A	Least r	I	$2C$
	feet	sq. in.	in.	in.^{4}	in.
L_0	29.954	195.8	10.30	21080	40
L_1	28.413	185.8	10.30	19760	40
L_2	27.116	185.8	10.30	19760	40
L_3	26.101	175.8	10.25	12410	40
L_4	26.402	175.8	10.25	18410	40
L_5	25.045	153.0	10.30	16260	40

Member	L	A	Least r	I	$2C$
	feet	sq. in.	in.	in.^{4}	in.
D_2'	43.830	31.37	6.95	2485	20
D_1'	55.456	51.37	6.95	2485	20
D_1	55.456	38.130	7.11	1932	20
D_2	43.830	35.630	7.19	1849	20
D_3	35.711	35.630	7.19	1849	20
D_4	30.806	38.130	7.11	1932	20
D_5	28.413	53.86	6.87	2602	20
D_6	27.731	63.05	6.78	2903	20
V_2'	35.0	65.3	8.80	5070	25.0
V_1'	49.5	77.8	9.13	6785	26.25
V_0	66.0	118.4	8.70	10825	27.0
V_1	49.5	65.9	9.02	5872	26.0
V_2	36.0	44.0	8.57	3295	25.0

V_3	25.5	44.0	8.57	3295	25.0
V_4	18.0	44.0	8.57	3295	25.0
V_5	13.5	44.0	8.57	3295	25.0
V_6	12.0	29.5	5.95	1495	20.0

CHAPTER XXXIII. DETAILING

The detail drawings of the bridge are found in Plates XII and XIII in which all the necessary details are clearly indicated. The design of the details that are particular to two hinged arches is given in the following, for the rest of details which being the same as for simple trusses the design will not be reproduced.

The end hinges will be first considered. They are governed by shear as the moment is small.

Maximum reaction occurs when the arch span is fully loaded.

Vertical reaction due to dead load = 784.4 kips.

Horizontal reaction due to dead load = 459.77 kips.

Vertical reaction due to live load = $141.6 \times (0.00536 + 0.24304 + 0.47023 + 0.68195 + 86700 + 0.99708) \times 2 + 141.6 \times 1.05032 = 1075$ kips.

Total vertical reactions = 1634.4 kips.

Total horizontal reactions = 1534. 8 kips.

Resultant reaction　　　 = 2245. 0 kips.

Inclined at on angle of 46°48′ from horizontal. (The top of pedestal is bevelled at 45° from horizontal).

Admissible shearing stress of pins = 12 kips per sq. in.

Area of pin required $= \frac{1}{2} \cdot \frac{2245}{12} = 93. 5$ sq. in.

Diameter required = 11. 5 inches.

12″ pins will be used for the end hinges.

Bearing value of pins = 24 kips per sq. in.

　　　　　　　　　 = 288 kips per linear inch.

Required thickness of pin packing on each side of centre line $= \frac{2245}{2} \div 288 = 3. 9$ inches.

The hinges at ends of cantilever arms are next considered. The greatest shearing force is the maximum reaction of the suspended span. This is found as follows.

Reaction for dead load　 = 95 kips.

Reaction for live load = Maximum end shear = 375 kips.

Total reaction = 470 kips.

Area of pin required $= \frac{470}{2} \div 12 = 19. 6$ sq. in.

Diameter required = 5. 12 inches.

6″ pins will be used.

Bearing value of pins = 24 kips per sq. in.

$$= 144 \text{ kips per linear inch.}$$

Required thickness of pin packing on each side of centre

line $= \dfrac{470}{2} \div 144 = 1. 63$ inches.

In designing the crown hinge inserted during section, the only stress needs consideration is that of L_5 due to dead load only and is equal to 474. 7 kips. The shearing force is therefore 237. 4 kips requiring a 6″ pin. Bearing thickness of pin packing should be at least 1. 65 inches.

The bearing area required on the masonry $= \dfrac{2245000}{600} =$ 3740 sq. in.

PART Ⅴ. SECONDARY STRESSES

CHAPTER XXIV. INTRODUCTION

In the previous analyses of stresses there are various assumptions which are not realised in practice. The one which diverges most apparently from the theory is the assumption of the frictionless joint. In a riveted truss the connections of the members meeting at any joint are made rigid in the plane of the truss, it being impossible to move individual member without affecting the other. As a result when the structure is deformed under load it is no longer composed of straight bars, but the members are bent and twisted so as to keep the constancy of the angles at the joints. Fibre stresses are produced forming the principal source of secondary stresses. This kind of secondary stresses (due to rigidity of joints) will be considered in the following paragraphs. To give a clear presentation of the subject influence lines of those stresses will be constructed, and conclusions drawn therefrom.

CHAPTER XXXV. METHODS OF CALCULATING SECONDARY STRESSES

Through a critical study of the different methods for computing secondary stresses, Mr. Pans in his thesis [1] mentioned four exact methods and three approximate ones. The exact ones are methods of Manderla, Muller-Breslau, Ritter and Mohr; while the approximate ones are methods of Muller-Breslau, Rngesear and Landsberg, and his own. By applying the different methods (exact) to the same numerical example (a six panel Warren truss) he found that only the methods of Manderla and Mohr are suitable for practical purposes.

The method of Manderla is largely advocated by Mr. Turnesure[2](called by him the method of Winkler) while that of

---

[1]   Thesis No. $75^{\#}$ (1913).

[2]   Engineering News. Vol. 68. P. 438.

Mohr is recommended by Mr. Kunz. [1] The amount of work involved in these two methods is about the same while the method of Mohr possesses the following advantages: [2]

(1) The moment is expressed in terms of ϕ which is constant for any joint of the truss while in Manderla, it is in terms of τ which is different for each member at any joint.

(2) The values of ψ and $K\psi$ are more rapidly calculated than the values $\Sigma \Delta\delta$, $K \Sigma \Delta\delta$ and $\Sigma (K \Sigma \Delta\delta)$ of Manderla.

(3) The absolute terms of method of Mohr can be found more directly than in that of Manderla.

(4) The values of the unknown angle ϕ of Mohr's method have the same signs at each side of a certain point while the signs of the reference deflection angle τ of Manderla's method do not follow any regularity to serve as a check for the result. (The point at each side of which the values of ϕ have the same sign was said by Mr. Pass to be the load point, but is not true for all cases. The position of this point depends on a number of factors: loads, type of truss, construction of displacement diagram, etc.)

---

[1]   Engineering News. Vol. 56. P. 379.
[2]   Thesis No. 75#. P. 123.

(5) The value of ϕ can be actually determined by a sensitive level.

Another advantage of Mohr's method is that the construction of displacement diagrams in finding the change of slope of members, serves at the same time the purpose of constructing influence lines for vertical deflections of the various panel points of the truss.

For the reasons enumerated above, the method of Mohr will be used in the following pages.

CHAPTER XXXVI. METHOD OF MOHR

When a truss is loaded in any manner and deflects, any member 12 owing to

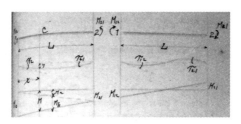

the rigid connection of its ends, is subjected to two moments M_{12} and M_{21} acting on its ends besides the axial force S. The moment of $S = Sy$ may be neglected. The moments of M_{12}, M_{21} cause bonding moment in the member which vary along the member in straight line ratio between the values M_{12} and M_{21} at the ends.

The greatest bending or secondary stresses occur therefore at the ends and are found by the well known formula

$$f = \frac{MC}{I}.$$

The main problem is, therefore, to find the moments M_{12} and M_{21}. This can be done by applying the equation of elastic line of the member to determine the relation of M and τ, the deflection angle, and afterwards determine the values of τ by the assumption that the total moments around any joint is zero provided there be no eccentric connections. In two hinged arches the moments about the end hinges are also zero so that the above method applies without modification.

The differential equation of the elastic line of the member is

$$\frac{\mathrm{d}^2 y}{\mathrm{d}x^2} = -\frac{M}{E1}.$$

Where M is the bending moment at any point C having the coordinates x and y. This moment can be expressed as follows:

$$M = M_a + M_b = M_{12}\frac{L-x}{L} - M_{21}\frac{x}{L},$$

$$\therefore \frac{\mathrm{d}^2 y}{\mathrm{d}x^2} = -\frac{1}{E1}\left(M_{12}\frac{L-x}{L} - M_{21}\frac{x}{L}\right).$$

Integrating, there is obtained an expression for $\dfrac{\mathrm{d}y}{\mathrm{d}x} = \tan\ \tau$.

Where τ is the deflection angle. Since τ is very small, we can put τ in arc measure for $\tan\ \tau$. The constant of integration is obtained by integrating once more and considering that for $x = 0$, and $x = L$, $y = 0$. By introducing the values of $x = 0$, $x = 2$ respectively in the expression for τ we get

$$M_{12} = \frac{2E1}{L}(2\tau_{12} + \tau_{21})\,,$$

$$M_{21} = \frac{2E1}{L}(2\tau_{21} + \tau_{12})\,.$$

In finding M, Mohr makes τ_{12} and τ_{21} functions of the angle ψ_{12} which the straight axis $1-2$ forms with the original location of this axis, and angles ϕ_1 and ϕ_2 which the original straight axis forms with the tangents to the elastic curve, as follows:

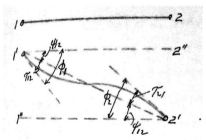

Line $1 - 2$ is the original axis of the member. $1'$ and $2'$, the displaced ends of 1 and 2.

$1'2'$ is the straight axis of $1'$, $2'$.

Lines $1'2'$ and $1''2'$ are drawn parallel to $1-2$.

From the figure,

$$\tau_{12} = \phi_1 - \psi_{12},$$

$$\tau_{21} = \phi_2 - \psi_{12},$$

$$M_{12} = \frac{2E1}{L}(2\phi_1 - 2\psi_{12} + \phi_2 - \psi_{12})$$

i. e. , $M_{12} = \dfrac{2E1}{L}(2\phi_1 + \phi_2 - 3\psi_{12})$.

Similarly $M_{21} = \dfrac{2E1}{L}(\phi_1 + 2\phi_2 - 3\psi_{12})$

or $M_{21} = \dfrac{2E1}{L}(2\phi_2 + \phi_1 - 3\psi_{12})$.

Now the moment is expressed in terms of the change of slope ψ and the angle of rotation ϕ of the joint in consideration. The values of ψ are geometrical functions of the axial deformations of the truss and can be easily found either by a displacement diagram or by equating the internal and external works of the truss due to the rotation of the members by an amount of ψ due to a unit moment applied at the joint to which the members are connected. The graphic method is most convenient for the case at hand and will be explained briefly. Refer to Plates VII to IX. The displacement of G with reference to H consists of the

556

movements from H for an amount of change in length in HG in the direction of HG and a movement at right angles thereto. Hence the line perpendicular to HG is the arc described by the end G of the member, and that line divided by the length of HG is therefore the angle ψ of that member. Correspondingly, the line drawn at right angles to the axial deformation of any other member represents the value ψL of that member.

These should be scaled from the diagram with the same scale in which the axial deformations are drawn. The signs of ψ are determined as follows: For member HG assume H as fixed (H is the point on the diagram obtained before G is constructed) the displacement diagram will show that G moves upwards or clockwise about H and ψ is therefore positive, if G would move counter clockwise about H, ψ will be negative. The same result is obtained by assuming G as fixed. To secure uniformity, the movement must always be considered from the arbitrary immovable point towards the movable point.

After ψ is known, M may be expressed in terms of ϕ alone and the problem is solved if ϕ can be determined. Note that ϕ is constant for all members at any joint, this being due to the fact the joints are rigid and the members must turn through the same angles should any rotation occur. To find ϕ, an equation of equilibrium of moments is introduced for every joint of the truss, i. e. , $\sum_1^n M = M_{12} + M_{13} + M_{14} + \cdots + M_{1n} = 0$.

provided there be no eccentric connection.

Calling $\dfrac{I}{L} = K$ a constant,

$$M_{12} = 2E(2K_{12}\phi_1 + K_{12}\phi_2 - 3K_{12}\psi_{12})$$

$$M_{13} = 2E(2K_{13}\phi_1 + K_{13}\phi_3 - 3K_{13}\psi_{13})$$

$$M_{14} = 2E(2K_{14}\phi_1 + K_{14}\phi_4 - 3K_{14}\psi_{14})$$

...

$$M_{1n} = 2E(2K_{1n}\phi_1 + K_{1n}\phi_n - 3K_{1n}\psi_{1n})$$

$$\sum M_{1n} = 2E(2\phi_1 \sum K_{1n} + \sum K_{1n}\phi_n - 3\sum K_{1n}\psi_{1n}) = 0$$

or $\quad 2\phi_1 \sum K_{1n} + \sum K_{1n}\phi_n - 3\sum K_{1n}\psi_{1n} = 0$,

where the subscript n represents the members of the adjacent joints.

Since there are M joints and for each joint there is an unknown ϕ it is possible to form M equations to contain the M unknowns. Solving simultaneously, M values of ϕ are obtained and substituted in the expression

$$M_{1n} = 2E(2\phi_1 K_{1n} + K_{1n}\phi_n - 3K_{1n}\psi_{1n}).$$

The value of M is obtained from which the secondary stress may be determined.

$$f = \frac{2C}{L}(2\phi_a + \phi_n - 3\psi_{an})E.$$

There ϕ_a = value of ϕ for the near end of the member.

ϕ_n = value of ϕ for the far end of the member.

ψ_{an} = change of slope of that member.

CHAPTER XXXVII. DETAILS OF PROCEDURE

The secondary stresses will be found separately for the vertical and horizontal forces. Each will be considered as if the arch were a simple truss and the method of Mohr is followed strictly in each case. By properly combining the resulting stresses, secondary stresses for the arch will be obtained.

To construct influence lines of stresses it is necessary to consider either of the two cases:

(1) the stresses in members of one half of the truss for a unit load at every panel point along the span.

(2) the stresses in members of the whole truss for a unit load at every panel point of one half of the truss.

The second case is more suitable for our purposes and will

be adopted.

Since the secondary stresses are directly proportional to the external loads so long as the positions of the loads are fixed it will save labor if the values of the horizontal and vertical loads are so chosen as to shorten the calculations. For horizontal thrust the value will be taken as equal to the load or unity in this case for constructing influence lines. This is the first case (I) to be considered in the following computations. The actual stresses are found by multiplying the stresses thus obtained by the value of the thrust corresponding to the position of the vertical load. For vertical loads, there are nine cases requiring consideration, i. e. for a unit load at each panel point of one half of the truss. The amount of the loads will be so chosen as to give the unit left hand reactions for all the positions of the loads. [1] The following cases are considered.

Case II, load 1 kips at 0; Case III, load 2 kips at VI; Case IV, load 2. 4 kips at VII; Case V, load 3 kips at VIII; Case VI, load 4 kips at IX; Case VII, load 6 kips at X; Case VIII, load 12

---

① Modern Framed Structures. Part II. P. 442.

kips upward at XI; Case IX, load 12 kips upwards at XI'; and Case X, load 6 kips upwards at X'. The actual stresses for unit load are found by dividing the stresses of the above cases by the values of the corresponding loads.

Article 66. Computation of Primary Stresses.

For Case I, the primary stresses are the same as τ. For Case II, only one member is stressed, i. e. the post over the support at 0, and its value is equal to the load, or unity. For cases III to X since the left hand reaction is the same irrespective of the position of the load, the stresses for members on the left of the load will be the same for all cases and are those given on p. 139. For members on the right of the load the stresses are obtained by multiplying the stresses for unit right hand reaction by the value of the reaction for the position of the load. The values of the right reactions are: for Case III, 1 kip; Case IV, 1.4 kips; Case V, 2 kips; Case VI, 3 kips; Case VII, 5 kips; Case VIII, 11 kips; Case IX, 13 kips, and Case X, 7 kips.

The primary stresses for Cases III to X are given on Table 58. All numbers above the step lines are the same as those immediately on the right of the step lines and those below the step lines are equal to the stresses in the symmetrical members of the left half of truss multiplied by the corresponding right hand reactions.

Article 67. Construction of Displacement Diagrams.

As described before the changes of slope ψ of members of the truss are found by constructing displacement diagrams. In Table 59 are given the axial deformations in members due to the primary stresses given on Table 58. For members on the left half of arch the deformations are the same for cases III to X, while for those on the right half the deformations for Cases I and III are the same as those of the left half. Only one member is deformed in Case II, i. e. , V_0, whose change in length $= -6.68$. The value of E is assumed to be equal to 1, so that the deformations thus computed are E times greater than the real values.

The displacement diagram for the ten cases are given on Plates VII to IX, in all of which the point H and the direction Hh are considered as fixed. On account of symmetry of the truss it is necessary to construct only one half of the diagrams for Cases I and III. For the remaining cases it is sufficient to construct the diagrams for that part of the truss which is on the right of the

Table 58. Primary Stresses for Cases III to X.

Member	Cases III to X	Member	Case III	Case IV	Case V	Case VI	Case VII	Case VIII	Case IX	Case X
U_2'	0	U_7	−12.50	−12.50	−12.50	−12.50	−12.50	−12.50	−12.50	−12.50
U_1'	0	U_8	−9.26	−12.96	−12.96	−12.95	−12.96	−12.96	−12.96	−12.96
U_1	−.51	U_9	−5.56	−7.79	−11.12	−11.12	−11.12	−11.12	−11.12	−11.12
U_2	−1.39	U_{10}	−2.94	−4.11	−5.82	−8.82	−8.82	−8.88	−8.82	−8.82
U_3	−2.94	U_{11}	−1.39	−1.95	−2.78	−4.17	−6.95	−6.95	−6.95	−6.95
U_4	5.56	U_{12}	−.51	−.71	−1.02	−1.53	−2.55	−5.56	−5.56	−5.56
U_5	−9.26	U_{12}'	0	0	0	0	0	0	0	0
U_6	−12.50	U_{11}'	0	0	0	0	0	0	0	0
L_1'	0	L_6	9.28	12.99	12.99	12.99	12.99	12.99	12.99	12.99
L_0'	0	L_7	5.64	7.91	11.29	11.29	11.29	11.29	11.29	11.29
L_0	0	L_8	3.07	4.30	6.14	9.21	9.21	9.21	9.21	9.21

Member	Cases III to X	Member	Case III	Case IV	Case V	Case VI	Case VII	Case VIII	Case IX	Case X
L_1	.57	L_9	1.51	2.11	3.02	4.53	7.53	7.53	7.53	7.53
L_2	1.51	L_{10}	.57	.80	1.14	1.71	2.85	6.31	6.31	6.31
L_3	3.07	L_{11}	0	0	0	0	0	0	5.45	5.45
L_4	5.64	L_{11}	0	0	0	0	0	0	5.45	5.45
L_5	9.28	L_{10}'	0	0	0	0	0	0	0	3.45

Member	Cases III to X	Member	Case III	Case IV	Case V	Case VI	Case VII	Case VIII	Case IX	Case X
D_2'	0	D_7	3.59	-.51	-.51	-.51	-.51	-.51	-.51	-.51
D_1'	0	D_8	4.21	5.88	2.10	2.10	2.10	2.10	2.10	2.10
D_1	1.12	D_9	3.22	4.51	6.44	2.82	2.82	2.82	2.82	2.82
D_2	1.55	D_{10}	2.22	3.10	4.44	6.66	2.68	2.68	2.68	2.68
D_3	2.22	D_{11}	1.55	2.17	3.10	4.65	7.75	2.43	2.43	2.43
D_4	3.22	D_{12}	1.12	1.57	2.24	3.36	5.50	12.30	2.41	2.41
D_5	4.21	D_{12}'	0	0	0	0	0	0	-10.08	-3.36

	D_6		D_{11}'							
D_6	3.59	D_{11}'	0	0	0	0	0	0	0	-5.32
V_2'	0	V_6	-2.00	0	0	0	0	0	0	0
V_1'	0	V_7	-1.56	-2.18	.22	.22	.22	.22	.22	.22
V_0	-1.00	V_8	-2.00	-2.80	-4.00	-1.00	-1.00	-1.00	-1.00	-1.00
V_1	1.27	V_9	-1.88	-2.64	-3.76	-5.65	-1.65	-1.65	-1.65	-1.65
V_2	1.58	V_{10}	1.58	-2.21	-3.16	-4.74	-7.92	-1.92	-1.92	-1.92
V_3	1.88	V_{11}	1.27	-1.78	-2.54	-3.81	-6.35	-14.00	-2.00	-1.92
V_4	-2.00	V_{12}	-1.00	-1.40	-2.00	-3.00	-5.00	-11.00	-13.00	-2.00
V_5	-1.56	V_{11}'	0	0	0	0	0	0	12.00	7.00
		V_{10}'	0	0	0	0	0	0	0	4.36
			0	0	0	0	0	0	0	6.00

Table 59. Alterations in Length of Members for Cases I to X.

(a) For Members of Left Half of Arch.

Member	L	A	L/A	Case I	Cases III to X
U_2'	300	77.2	3.89	0	0
U_1'	300	77.2	3.89	0	0
U_1	300	77.2	3.89	1.29	−1.98
U_2	300	77.2	3.89	3.24	−5.41
U_3	300	77.2	3.89	6.18	−11.43
U_4	300	99.2	3.89	8.08	−16.82
U_5	300	99.2	3.89	11.79	−28.05
U_6	300	99.2	3.89	13.62	−37.90
L_1'	341	49.0	6.97	0	0
L_0'	360	73.4	4.91	0	0
L_0	360	195.8	1.84	−2.21	0
L_1	341	185.8	1.83	−2.78	1.64
L_2	325	185.8	1.75	−3.48	2.64
L_3	313	175.8	1.78	−4.80	5.47
L_4	305	175.8	1.73	−6.45	9.75
L_5	301	153.4	1.96	−9.60	18.19

Member	L	A	L/A	Case I	Cases III to X
D_2'	526	51.37	10.22	0	0

D_1'	565	51.37	12.95	0	0
D_1	665	38.13	17.45	-12.89	19.52
D_2	526	35.63	14.75	-12.93	22.84
D_3	428	35.63	12.0	-12.95	26.65
D_4	370	38.13	9.70	-12.90	32.24
D_5	341	53.86	6.33	-8.80	26.68
D_6	332	63.05	5.27	-3.58	18.90
V_2'	432	65.3	6.61	0	0
V_1'	594	77.8	7.63	0	0
V_0	792	118.4	6.68	4.41	-6.68
V_1	594	65.9	9.02	6.48	-11.44
V_2	432	44.0	9.81	7.55	-15.50
V_3	306	44.0	6.95	5.42	-13.07
V_4	216	44.0	4.91	3.24	-9.82
V_5	162	44.0	3.68	1.07	-5.74
V_6	144	29.5	4.88	0	-9.76*

For Case III only.

For Case II, change in length of V_0 -6.68.

(b) For Members of Right Half of Arch.

Member	L/A	Case III	Case IV	Case V	Case VI	Case VII	Case VIII	Case IX	Case X
U_7	3.03	-37.90	-37.90	-37.90	-37.90	-37.90	-37.90	-37.90	-37.90

Member	L/A	Case III	Case IV	Case V	Case VI	Case VII	Case VIII	Case IX	Case X
U_8	3.03	-28.05	-39.20	-39.20	-39.20	-39.20	-39.20	-39.20	-39.20
U_9	3.03	-16.82	-23.55	-33.70	-33.70	-33.70	-33.70	-33.70	-33.70
U_{10}	3.89	-11.43	-16.00	-22.90	-34.30	-34.30	-34.30	-34.30	-34.30
U_{11}	3.89	-5.41	-7.58	-10.81	-16.21	-27.05	-27.05	-27.05	-27.05
U_{12}	3.89	-1.98	-2.76	-3.96	-5.95	-9.91	-21.62	-21.62	-21.62
U_{12}'	3.89	0	0	0	0	0	0	0	-11.78
U_{11}'	3.89	0	0	0	0	0	0	0	0
L_6	1.96	18.19	25.43	25.43	25.43	25.43	25.43	25.43	25.43
L_7	1.73	9.75	13.69	19.52	19.52	19.52	19.52	19.52	19.52
L_8	1.78	5.47	7.65	10.82	16.40	16.40	16.40	16.40	16.40
L_9	1.75	2.64	3.69	5.28	7.94	13.18	13.18	13.18	13.18
L_{10}	1.83	1.04	1.46	2.09	3.13	5.21	11.54	11.54	11.54
L_{11}	1.84	0	0	0	0	0	0	10.01	10.01
L_{11}'	4.91	0	0	0	0	0	0	26.75	26.75
L_{10}'	6.97	0	0	0	0	0	0	0	24.03

Member	L/A	Case III	Case IV	Case V	Case VI	Case VII	Case VIII	Case IX	Case X
D_7	5.27	18.90	-2.68	-2.68	-2.68	-2.68	-2.68	-2.68	-2.68
D_8	6.33	26.68	37.31	13.30	13.30	13.30	13.30	13.30	13.30
D_9	9.70	32.24	44.76	62.48	27.36	27.36	27.36	27.36	27.36
D_{10}	12.00	26.65	37.21	53.32	80.05	32.15	32.15	32.15	32.15
D_{11}	14.75	22.84	32.02	45.75	68.60	114.35	35.85	35.85	35.85

D_{12}	17.45	19.52	27.40	39.06	58.68	97.62	214.52	42.10	42.10
D_{12}'	12.95	0	0	0	0	0	0	-130.50	-43.55
D_{11}'	10.22	0	0	0	0	0	0	0	-54.4
V_6	4.88	-9.76	0	0	0	0	0	0	0
V_7	3.68	-5.74	-8.02	.81	.81	.81	.81	.81	.81
V_8	4.91	-9.82	-13.74	-19.62	-4.91	-4.91	-4.91	-4.91	-4.91
V_9	6.95	-13.07	-18.35	-26.15	-39.28	-11.48	-11.48	-11.48	-11.48
V_{10}	9.81	-15.50	-21.72	-31.00	-46.51	-77.68	-18.83	-18.83	-18.83
V_{11}	9.02	-11.44	-16.05	-22.90	-34.34	-57.25	-126.15	-18.04	-18.04
V_{12}	6.68	-6.68	-9.34	-13.36	-20.04	-33.40	-73.48	86.93	46.82
V_{11}'	7.63	0	0	0	0	0	0	91.51	33.30
V_{10}'	6.61	0	0	0	0	0	0	0	39.65

load. The members cut by a section passed through the vertical post immediately on the left of the load should also be included. The points from which the diagrams are started are traced from the diagram for the preceding case. This is plainly shown in Plate IX.

After the diagrams are constructed the values of ζ (the distance which the end of the member has swung to reach its deformed position) are measured and are recorded in the Plates VII to IX. These are also given in Table 60 in which a division is made for the members on the left and right halves of the truss. For members on the right half, the values of ζ for Cases I and III are the same as those for the left half and for Case II, are equal to zero. The values of ζ, as the axial deformations are E times greater than their real values.

The system of numbering the panel points used in Table 60 and shown in Diagram 22 may require explanation. These are so labelled as to facilitate the solution of equations. Arabic numbers are used, the joints being membered in such a manner as to follow order used in constructing the displacement diagrams if the end vertical V_2' or $1-2$ is considered as fixed. The joints of the right half of the truss will then have a number which when added to the symmetrical joint of the left half of truss will give a sum of 35 or one greater than the number of joints. These notations are important, as will be seen later, the arrangement of the table for facilitating the solution of equations, proposed by the writer, is largely based on this fact.

The values of ψ are then computed by taking the ratio of ζ and L, the length of member, both being in inches. The values of ψ thus obtained are for a left reaction $= 1$ kip and are E times greater than their real values. The product $K\psi$ are formed for every member of the truss and are given on Table 60. For members on right half of truss, the values of ψ and $K\psi$ are the same as those for the left half for Cases I and III, while for Case II, the values are equal to zero. The values of $K\psi$ for the ten cases are marked on the truss diagrams in Diagram 22 and those of K in Diagram 23, the latter being the same for all cases. Summations of $K\psi$ and K are then formed for every joint of the truss and are recorded nearby the joints enclosed by a parenthesis.

Table 60. Values of ψ and $K\psi$.

(a) For Members of Left Half of Arch.

	Member	1 – 3	3 – 5	6 – 8	8 – 10	10 – 12	12 – 14	14 – 16	16 – 18
	I	12155	12155	12155	12155	12155	13165	13165	13165
	L	300	300	300	300	300	300	300	300
	K	40.52	40.52	40.52	40.52	40.52	43.88	43.88	43.88
for Case Except 16 – 18	I	– 143	– 140	– 158	150	– 137	– 123	– 89	– 35
	II	3.3	3.2	0	0	0	0	0	0
	III to X	297	295	325	314	308	285	230	133

for Case Except 16−18									
	I	−.477	−.466	−.526	−.500	−.457	−.410	−.297	−.117
	II	.011	−.011	0	0	0	0	0	0
	III to X	.990	.982	1.083	1.046	1.027	.950	.766	.443

for Case Except 16−18									
	I	−19.3	−18.9	−21.3	−29.2	−18.5	−18.0	−13.0	−5.1
	II	.5	.4	0	0	0	0	0	0
	III to X	40.1	39.8	43.8	42.5	41.6	41.7	33.6	19.5

Member	2−4	4−6	6−7	7−9	9−11	11−13	13−15	15−17
I	4686	7500	21080	19760	19760	18410	18410	16260
L	341	360	360	341	325	313	305	301
K	13.76	20.85	38.70	57.82	60.75	58.82	60.35	54.10

for Case									
	I	−162	−163	−186	−159	−153	−130	−95	−37
	II	3.8	−4.1	8.0	0	0	0	0	0
	III to X	335	348	383	358	334	304	241	130

for Case									
	I	−.475	−.453	−.517	−.495	−.471	−.415	−.312	−.123
	II	.011	−.011	.022	0	0	0	0	0
	III to X	.982	.967	1.062	1.049	1.027	.971	.790	.432

for Case									
	I	−6.5	−9.5	−30.3	−28.6	−28.6	−24.4	−18.8	−6.7
	II	.2	.2	1.3	0	0	0	0	0
	III to X	13.5	20.2	62.3	60.7	62.4	57.1	47.7	23.4

Member	2−3	4−6	6−7	8−9	10−11	12−13	14−15	16−17
I	2485	2465	1932	1849	1849	1849	1932	2003

L	526	655	665	526	428	370	341	332
K	4.72	3.74	2.91	3.51	4.32	5.22	7.63	8.75

for Case								
I	-248	-315	-307	-230	-174	-136	-94	-37
II	5.7	7.5	0	0	0	0	0	0
III to X	523	661	651	498	393	315	241	127

for Case								
I	-.472	-.475	-.462	-.437	-.407	-.367	-.276	-.111
II	.011	.011	0	0	0	0	0	0
III to X	.995	.995	.950	.947	.918	.851	.707	.383

for Case								
I	-2.2	-1.8	-1.3	-1.5	-1.8	-1.9	-2.1	-1.0
II	.1	0	0	0	0	0	0	0
III to X	4.7	3.7	2.9	3.3	4.0	4.4	5.4	3.4

Member	1 - 2	3 - 4	5 - 6	8 - 7	10 - 9	12 - 11	14 - 13	16 - 15	18 - 17
I	5070	6785	10825	5872	3295	3295	3295	3295	3295
L	432	594	792	594	432	306	216	162	144
K	11.73	11.42	13.69	9.88	7.62	10.75	15.25	20.30	10.38

for Case Except 18 - 17									
I	-204	-281	-371	-265	-178	-108	-60	-24	0
II	4.9	6.8	4.4	0	0	0	0	0	0
III to X	431	590	783	570	394	250	142	61	0

for Case Except 18 - 17									
I	-.472	-.473	-.468	-.446	-413	-353	-277	-.148	0
II	.011	.011	.003	0	0	0	0	0	0
III to X	.998	.992	.990	.960	.912	.817	.657	.376	0

for Case Except 18 – 17	I	– 5.5	– 5.4	– 6.4	– 4.4	3.1	– 3.8	– 4.2	– 3.0	0
	II	.1	.1	.1	0	0	0	0	0	0
	III to X	11.7	11.4	13.6	9.5	6.9	8.8	10.0	7.6	0

（b）For Members of Right Half of Arch.

	Member	16 – 18	18 – 19	19 – 21	21 – 23	23 – 25	25 – 27	27 – 30	30 – 32	32 – 34
	I	13165	13165	13165	13165	12155	12155	12155	12155	12155
	L	300	300	300	300	300	300	300	300	300
	K	43.88	43.88	43.88	43.88	40.52	40.52	40.52	40.52	40.52
for Case	IV	115	– 67	– 283	– 372	– 397	– 421	– 426	– 386	– 394
	V	115	– 67	– 221	– 401	– 456	– 478	– 490	– 429	– 430
	VI	115	– 67	– 221	– 322	– 492	– 532	– 536	– 457	– 460
	VII	115	– 67	– 221	– 322	– 384	– 614	– 625	– 497	– 500
	VIII	115	– 67	– 221	– 322	– 384	– 433	– 808	– 533	– 532
	IX	115	– 67	– 221	– 322	– 384	– 433	– 471	– 600	– 332
	X	115	– 67	– 221	– 332	– 384	– 433	– 471	– 483	– 494
for Case	IV	.383	– .223	– 942	– 1.239	– 1.322	– 1.400	– 1.420	– 1.321	– 1.312
	V	.383	– .223	– .737	– 1.335	– 1.519	– 1.592	– 1.630	– 1.429	– 1.430
	VI	.383	– .223	– .737	– 1.073	– 1.640	– 1.772	– 1.792	– 1.520	– 1.539
	VII	.383	– .223	– .737	– 1.073	– 1.280	– 2.051	– 2.081	– 1.663	– 1.672
	VIII	.383	– .223	– .737	– 1.073	– 1.280	– 1.445	– 2.691	– 1.780	– 1.772
	IX	.383	– .223	– .737	– 1.073	– 1.280	– 1.445	– 1.570	– 2.000	– 1.105
	X	.383	– .223	– .737	– 1.073	– 1.280	– 1.445	– 1.570	– 1.610	– 1.652

for Case									
IV	16.8	-9.8	-41.3	-54.4	-53.6	-56.8	-57.6	-53.5	-53.2
V	16.8	-9.8	-32.3	-58.7	-61.5	-64.5	-66.1	-57.9	-58.0
VI	16.8	-9.8	-32.3	-47.1	-66.5	-71.8	-72.5	-61.6	-62.0
VII	16.8	-9.8	-32.3	-47.1	-51.9	-83.1	-84.5	-67.3	-67.6
VIII	16.8	-9.8	-32.3	-47.1	-51.9	-58.6	-109.2	-72.1	-71.7
IX	16.8	-9.8	-32.3	-47.1	-51.9	-58.6	-63.6	-81.0	-44.8
X	16.8	-9.8	-32.3	-47.1	-51.9	-58.6	-63.6	-65.3	-66.7

Member	17-20	20-22	22-24	24-26	26-28	28-29	29-31	31-33
I	16260	18410	18410	19760	19760	21080	7500	4686
L	301	305	313	325	341	360	360	341
K	54.10	60.35	58.82	60.75	57.82	58.70	20.85	13.76
for Case IV	-100	-296	-395	-439	-473	-500	-454	-445
V	-70	-248	-429	-500	-538	-561	-498	-490
VI	-70	-236	-375	-545	-593	-620	-525	-527
VII	-70	-236	-345	-496	-678	-720	-563	562
VIII	-70	-236	-345	-427	-627	-898	-562	-603
IX	-70	-236	-345	-427	-500	-446	-696	-480
X	-70	-236	-345	-427	-500	-491	-580	-540

	IV	−.333	−.987	−1.261	−1.350	−1.388	−1.388	−1.260	−1.305
	V	−.233	−.813	−1.370	−1.538	−1.580	−1.555	−1.382	−1.430
	VI	−.233	−.773	−1.197	−1.680	−1.741	−1.723	−1.455	−1.545
for Case	VII	−.233	−.773	−1.105	−1.525	−1.992	−2.001	−1.564	−1.653
	VIII	−.233	−.773	−1.105	−1.311	−1.842	−2.491	−1.561	−1.773
	IX	−.233	−.773	−1.105	−1.311	−1.471	−1.242	−1.931	−1.405
	X	−.233	−.773	−1.105	−1.311	−1.471	−1.362	−1.610	−1.584
	IV	−18.0	−59.6	−74.2	−82.0	−80.3	−81.4	−26.3	−18.0
	V	−12.6	−49.1	−80.7	−93.5	−91.4	−89.8	−28.8	−19.8
	VI	−12.6	−46.6	−70.4	−102.0	−100.7	−101.1	−30.3	−21.2
for Case	VII	−12.6	−46.6	−64.9	−92.8	−115.1	−117.4	−32.6	−22.7
	VIII	−12.6	−46.6	−64.9	−79.7	−106.2	−146.5	−32.5	−24.3
	IX	−12.6	−46.6	−64.9	−79.7	−84.8	−72.6	−40.3	−19.4
	X	−12.6	−46.6	−64.9	−79.7	−84.8	−80.0	−33.6	−21.8

Member	17 − 19	20 − 21	22 − 23	24 − 25	26 − 27	28 − 30	30 − 31	32 − 33
I	2903	2602	1932	1849	1849	1932	2485	2485
L	332	341	370	428	526	665	665	526
K	8.75	7.63	5.22	4.32	3.51	2.91	3.74	4.72

for Case								
IV	−75	−294	−408	−504	−653	−851	−866	−681
V	75	−243	−430	−559	−717	−947	−953	−754
VI	−75	−243	−371	−573	−756	−1005	−1028	−817
VII	−75	−243	−371	−505	−778	−1067	−1109	−876
VIII	−75	−243	−371	−505	−687	−1089	−1187	−939
IX	−75	−243	−371	−505	−687	−931	−855	−580
X	−75	−243	−371	−505	−687	−931	−910	−726

for Case								
IV	−.226	−.862	−1.102	−1.178	−1.240	−1.280	−1.300	−1.295
V	−.226	−.712	−1.161	−1.300	−1.361	−1.424	−1.431	−1.432
VI	−.226	−.712	−1.001	−1.338	−1.442	−1.514	−1.545	−1.550
VII	−.226	−.712	−1.001	−1.182	−1.483	−1.602	−1.665	−1.665
VIII	−.226	−.712	−1.001	−1.182	−1.305	−1.640	−1.780	−1.791
IX	−.226	−.712	−1.001	−1.182	−1.305	−1.400	−1.285	−1.101
X	−.226	−.712	−1.001	−1.182	−1.305	−1.400	−1.372	−1.381

for Case								
IV	−2.0	−6.6	−5.8	−5.1	−4.4	−3.7	−4.9	−6.1
V	−2.0	−5.4	−6.1	−5.6	−4.8	−4.2	−5.4	−6.8
VI	−2.0	−5.4	−5.2	−5.8	−5.1	−4.4	−5.8	−7.3
VII	−2.0	−5.4	−5.2	−5.1	−5.2	−4.7	−6.2	−7.9
VIII	−2.0	−5.4	−5.2	−5.1	−4.6	−4.8	−6.7	−8.4
IX	−2.0	−5.4	−5.2	−5.1	−4.6	−4.1	−4.8	−5.2
X	−2.0	−5.4	−5.2	−5.1	−4.6	−4.1	−5.1	−6.5

Member		18 – 17	19 – 20	21 – 22	23 – 24	25 – 26	27 – 28	30 – 29	32 – 31	34 – 33
I		1995	3295	3295	3295	3295	5872	10825	6785	5070
L		144	162	216	306	432	594	792	594	432
K		10.38	20.30	15.25	10.75	7.62	9.88	13.69	11.42	11.73
for Case	IV	0	– 69	– 175	– 318	– 510	– 744	– 1025	– 772	– 560
	V	0	– 67	– 166	– 332	– 559	– 825	– 1129	– 851	– 619
	VI	0	– 67	– 162	– 318	– 563	– 864	– 1209	– 921	– 674
	VII	0	– 67	– 162	– 308	– 545	– 895	– 1303	– 990	– 722
	VIII	0	– 67	– 162	– 308	– 517	– 851	– 1370	– 1058	– 771
	IX	0	– 67	– 162	– 308	– 517	– 790	– 1069	– 708	– 476
	X	0	– 67	– 162	– 308	– 517	– 790	– 1094	– 807	– 570
for Case	IV	0	– .425	– .811	– 1.040	– 1.183	– 1.251	– 1.295	– 1.302	– 1.295
	V	0	– .413	– .768	– 1.084	– 1.295	– 1.391	– 1.425	– 1.430	– 1.431
	VI	0	– 413	– .785	– 1.040	– 1.305	– 1.451	– 1.525	– 1.551	– 1.560
	VII	0	– .413	– .785	– 1.005	– 1.263	– 1.525	– 1.645	– 1.665	– 1.674
	VIII	0	– .413	– .785	– 1.005	– 1.195	– 1.431	– 1.742	– 1.782	– 1.791
	IX	0	– .413	– .785	– 1.005	– 1.195	1.331	– 1.352	– 1.192	– 1.105
	X	0	– .413	– .785	– 1.005	– 1.195	– 1.331	– 1.382	– 1.361	– 1.321

	IV	0	−8.7	−12.4	−11.2	−9.0	−12.4	−17.7	−14.9	−15.2
	V	0	−8.4	−11.7	−11.7	−9.9	−13.8	−19.5	−16.4	−16.8
	VI	0	−8.4	−12.0	−11.2	−10.0	−14.4	−20.9	17.7	−18.4
for Case	VII	0	−8.4	−12.0	−10.8	−9.6	−14.9	−22.5	−19.1	−19.2
	VIII	0	−8.4	−12.0	−10.8	−9.1	−14.2	−23.7	−20.4	−21.0
	IX	0	−8.4	−12.0	−10.8	−9.1	−13.2	−18.5	−13.6	−12.9
	X	0	−8.4	−12.0	−10.8	−9.1	−13.2	−18.9	−15.5	−15.5

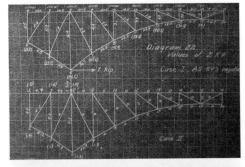

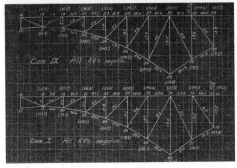

（excerpt）

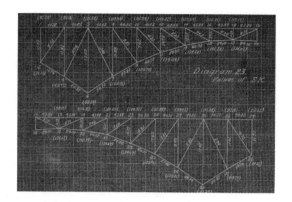

Article 68. Formulation of Equations.

It is now ready to form the equation:

$$2\phi_1 \Sigma K_{1n} + \Sigma \phi_n K_{1n} - 3 \Sigma K_{1n} \psi_{1n} = 0$$

for every joint of the truss for all the cases of loading. It will be seen that these equations are composed of two parts one including the unknown ϕ and the other not. The second part is therefore known and is found by taking thrice the value of $K\psi$ given in Diagram 22. These will be called the absolute terms of equations and are given in Table 61. The values of those terms for Cases I and III for joints of right half of truss are the same as those for the left half, and for Case II are equal to zero.

The first part of the equations contains the unknowns ϕ and their coefficients K and ΣK, which are given in Diagram 23.

To arrange the unknowns and coefficients in a systematic manner the following method will be found convenient. Refer to Plate X. Construct a table with the top column denoting the subscripts of ϕ and the left hand one the number of joints. There will then be as many vertical columns as there are horizontal lines each being equal to the total number of joints, i. e. , thirty four in this case. Opposite the number of joint along any horizontal line write down twice the value of ΣK for that joint under the vertical column of ϕ which has the same number of subscript. Repeating the process for all joints of the truss there is obtained a diagonal line of coefficients as shown in Plate X under the dotted line AB. Now complete the table by filling in the horizontal lines the values of K of those members which meet at the corresponding joints.

Table 61. Absolute Terms of Equations.

Joint	Case I	Case II	Case III	Case IV	Cases V to X
	+	−	−	−	−
1	74.4	1.5	155.4		
2	42.6	1.2	89.7		
3	137.4	3.0	288.0		
4	69.6	.3	146.4		
5	149.1	1.2	311.4		
6	138.6	3.6	288.3		
7	193.8	3.9	406.2	Same as	Same as
8	142.2	0	297.3	Case III	Case IV
9	185.4	0	399.9		
10	130.8	0	285.		
11	175.8	0	396.9		
12	126.6	0	289.5		
13	147.9	0	357.6		
14	111.9	0	272.1		
15	91.8	0	252.3	252.3	
16	66.3	0	192.3	184.2	
17	0	0	0	20.4	36.6
18	0	0	0	21.0	21.0
19	185.4	157.5	157.5	157.5	157.5	157.5	157.5
20	278.7	226.5	219.0	219.0	219.0	219.0	219.0
21	344.1	324.3	290.4	290.4	290.4	290.4	290.4
22	456.0	442.8	402.6	386.1	386.1	386.1	386.1
23	375.0	413.7	390.0	345.0	345.0	345.0	345.0

24	514.5	574.2	568.2	520.8	481.5	481.5	481.5
25	365.1	424.5	462.3	449.1	374.1	374.1	374.1
26	521.7	598.8	653.4	668.1	598.8	534.6	534.6
27	393.6	447.6	691.4	563.1	559.8	420.	420.
28	533.4	597.6	661.8	756.3	815.1	824.1	546.3
29	376.2	414.6	456.9	517.5	608.1	394.2	397.8
30	412.2	459.6	495.6	555.6	649.5	516.	471.3
31	192.0	211.2	225.0	241.8	251.7	234.3	228.
32	383.1	417.3	445.8	485.7	517.8	433.8	462.
33	117.6	130.2	140.7	150.6	160.8	112.5	131.4
34	205.2	224.4	241.2	261.6	277.8	173.1	246.6

It will be seen from Plate X the set of thirty four simultaneous equations thus obtained are symmetrical about two axes. One of those is the diagonal line AB which was discovered by Mr. Paez enabling him to apply the method of Gauss to the solution of the equations. Another symmetry, found by the writer, is about the horizontal columns of joints 17 and 18 as marked by two crossing dotted lines. This is more plainly seen by adding two more columns along the right hand and bottom sides of the tables; one with the number of subscript of ϕ and the other the number of joints. The order of the numbers is just the reverse of that of the left and top columns. It is now evident that the same coefficients of ϕ are obtained by referring to either of the sets of columns, left and top or right and bottom. By waiving the repeated portions of the table the same results will be obtained by using the first eighteen columns with reference to both sets of remark columns as those when the whole table is used with reference to but one set of remark columns. See Plate XI, where the space required for the coefficients of ϕ is reduced to one quarter its size of Plate X.

It must be admitted that the symmetry about horizontal axis is possible only when the joints are numbered in a proper manner such that the sum of the numbers of symmetrical joints is always equal to one greater than the number of joints, i. e. , 35 in this case. Failure to secure this result has to account for the fact that it is unknown for a long time.

Article 69. Solution of Equations.

By the use of method of Gauss, as proposed by Mr. Paez, the solution of the equations can be made very mechanical and checked at every step during the process. The following is a detailed description of the process in which the arrangement of equations suggested by the writer is incorporated.

In Plate XI are constructed a table in which the coefficients of ϕ (refer to two sets of headings) and absolute terms are recorded. A column of check term is also added which being the sum of coefficients of ϕ and the absolute term of the equation. It is to be treated just as an absolute term in the solution so that at every step during the progress the check term must be satisfied by being equal to the sum of coefficients and the absolute term after some multiplication or addition having been done.

The horizontal columns bounded by heavy lines are now divided into longitudinal rectangles the number of which is to be determined as follows: For equations (1) or (34), 1 line; equation (2) or (33), 2 lines; equation (3) or (32) and all the remaining equations 4 lines; except in those whose first unknown is the same as that of the previous one, in which case 5 lines should be allowed. The coefficients of equation (Plate X), absolute terms, (Table 61) and check terms, marked by numbers without primes are now entered into those cells which are next to the bottom of each horizontal column. The step lines, starting from the first cell are next drawn as shown in Plate XI. Numbers immediately on the right of the step lines will be called the first column of that equation, the next one, the second column, the third one, the third, etc.

To begin the solution, multiply (1) by the ratio of $\left(-\dfrac{2\text{nd column}}{1\text{st column}} \right)$ of (1), calling it (1″) and write it down in the first line of the next cell below, i. e. , that of equation (2). Multiply (1) by $\left(\dfrac{3\text{rd column}}{1\text{st column}} \right)$ of (1), calling it (1″) and write it down in the first line of the third cell below, i. e. , that of equation (3). If there is a fourth column in (1), multiply (1) by $\left(\dfrac{4\text{th column}}{1\text{st column}} \right)$ of (1), calling it (I^{IV}) and write it down in the first line of the fourth cell, i. e. , that of equation (4). The sign of (1″) is always negative, while that of (1″) is not fixed jet. In the second cell there will be only one line left which is to be filled by the sum of (2) and (1″). This line will be designated by the letter (II). It is to be noticed that the first unknown of (2) is now eliminated in (II).

Multiply (II) by the ratio of $\left(\dfrac{2\text{nd column}}{1\text{st column}} \right)$ of (II), calling it (II″) and write it down in the second line of third cell, i. e. , that of equation (3), in which the first line has already been

586 |

occupied by (1″). Multiply (II) by $\left(\dfrac{\text{3rd column}}{\text{1st column}}\right)$ of (II),

calling it (II″) and write it down in the first line of the fourth

cell, i. e. , that of equation (4). If there is a fourth column in

(II), multiply (II) by $\left(\dfrac{\text{4th column}}{\text{1st column}}\right)$ of (II), calling it (II^{IV})

and write it down in the first line of the fifth cell, i. e. , that of

equation (5).

Now in the third cell all the lines are filled except the last

one which is to be occupied by the sum of (1″), (II″) and

(3). Now fix the sign of (1″) so that its first term will cancel

that of equation (3). The last line thus obtained will not contain

the first two unknowns of (3) and will be designated by the letter

(III). Equation (III) is to be treated in the same manner as

equation (II) has described above.

It should be noted that the above statements apply equally

well to equations (34) (XXXIII), etc.

The check term should be satisfied at every step of the

process so as to ensure correctness at every point in the solution.

This is exceedingly important in avoiding the accumulation of

errors which are hard to correct.

Repeating the same process as described the bottom line is eventually arrived when the thirty four equations will be reduced to six equations involving six unknown, i. e. , equations (XV) , (XVI) , (XVII) , (XVIII) , (XIX) and (XX) , involving ϕ_{15} , ϕ_{16} , ϕ_{17} , ϕ_{18} , ϕ_{19} , and ϕ_{20}. The solution of these six equations which is not troublesome, gives the values of ϕ_{15} , ϕ_{16} , etc, and the latter when substituted successively in equations marked by Roman capitals, will give all the unknowns required. The values of ϕ thus obtained are for a left reaction equal to 1 kip and are E times greater than their real values, i. e. , we have obtained $E\phi$, instead of ϕ. Those are given in Table 62, in which for joints of left half of arch the values of ϕ for cases VI to X are so nearly equal that they will be made equal and for joints of right half of arch, the values of ϕ for cases I and III are the same as those on the left half and for case II are equal to zero.

By arranging the equations as shown in Plate XI the following advantages are found in contrasting with the method suggested by Mr. Paez:

(1)In regard to the coefficients of the unknowns, it saves 50% of labor by reducing 18 equations instead of 34.

Table 62. Values of ϕ.

Joint	Case I	Case II	Case III	Case IV	Case V	Cases VI to X
	−	+	+	+	+	+
1	.478	.007	1.001			
2	.471	.017	.993			
3	.466	.013	.978			
4	.462	0	.975	Same as	Same as	Same as
5	.490	0	1.028	Case III	Case IV	Case V.
6	.482	.016	.997			
7	.504	.011	1.056			
8	.513	0	1.060			
9	.477	0	1.028			
10	.468	0	1.022	1.023		
11	.441	0	.995	.997		
12	.427	0	.976	.975		
13	.362	0	.870	.861	.860	
14	.349	0	.830	.841	.841	
15	.223	0	.637	.654	.661	
16	.191	0	.612	.555	.558	.562
17	0	0	0	.091	.115	.127
18	0	0	0	.102	.080	.085
19	.550	.463	.484	.480	.482	.481	.484
20	.697	.528	.532	.530	.530	.531	.530
21	1.090	.995	.872	.914	.895	.905	.902
22	1.128	1.080	.953	.930	.948	.930	.946

23	1.270	1.429	1.303	1.099	.188	1.146	1.160
24	1.290	1.462	1.431	1.262	1.180	1.262	1.195
25	1.292	1.521	1.715	1.615	1.210	1.385	1.342
26	1.339	1.550	1.722	1.752	1.469	1.115	1.404
27	1.415	1.607	1.760	2.108	2.079	1.371	1.502
28	1.390	1.552	1.720	2.015	2.210	2.649	1.490
29	1.293	1.418	1.565	1.718	2.160	.907	1.328
30	1.358	1.520	1.635	1.815	2.293	1.805	1.593
31	1.279	1.408	1.482	1.572	1.545	1.760	1.582
32	1.305	1.410	1.505	1.635	1.675	1.528	1.585
33	1.300	1.445	1.570	1.680	1.830	1.161	1.381
34	1.312	1.439	1.548	1.680	1.802	.934	1.590

(2) It requires a table of only one quarter as large as the ordinary method. This refers to coefficients only.

(3) It saves work in substitution in the finding of the remaining unknowns, after some of those are obtained as 15, 16, etc. One substitution gives the coefficients of two unknowns the sum of whose number of subscripts is equal to 35.

(4) It has less chance of catching errors because it deals with only one half the numbers that usually have to be considered. This refers to coefficients only.

(5) It gives a more accurate result by avoiding a long, successive substitution in finding the values of the first few

unknowns.

(6) It saves time in verifying the check terms.

Article 70. Calculation of Secondary Stresses.

After the values of ϕ are obtained the secondary stresses for the ten cases may be obtained from the formula

$$f = \frac{2C}{L}(2\phi_a + \phi_n - 3\psi_{an})E.$$

Table 63. Secondary Stresses for Cases I to X. (In pounds)

Joint	Member	$\dfrac{2C}{L}$	$\dfrac{2C}{L}$	$(2\phi_a + \phi_n - 3\psi_{an})$		
			Case I	Case I	Case II	Case III to X
	1 – 2	.0578	– .63	– .16	.05
1	1 – 3 T	.0936	.84	– .56	.93
	B	.1145	1.03	.68	1.14
	2 – 1	.0578	– .23	.41	– .40
2	2 – 3	.0381	.30	.53	– .80
	2 – 4	.1172	2.46	.11	1.76
	3 – 1 T	.0936	1.96	– 1.21	– 1.21
	B	.1145	2.41	1.49	1.49
3	3 – 2	.0381	.49	.84	– 1.37
	3 – 4	.0442	.97	.22	– 1.99
	3 – 5 T	.0936	– 2.25	.56	3.56
	B	.1145	2.76	.68	4.36

	4 – 2	.1172	3.52	– 1.88	– .35
4	4 – 3	.0442	1.15	– .93	– 2.12
	4 – 5	.0301	.33	– .98	– .21
	4 – 6	.1111	5.56	– 1.89	5.12
	T	.0936	– 4.48	– 1.87	8.22
	5 – 3 B	.1145	5.50	2.29	10.08
	5 – 4	.0301	– .51	– .99	1.39
5	5 – 6	.0341	– 1.98	– .06	2.83
	5 – 7	.0301	– 2.95	.33	5.18
	5 – 8 T	.0936	7.95	0	– 12.44
	B	.1145	8.73	0	– 15.21
	6 – 4	.1111	– 7.77	– .11	7.55
6	6 – 5	.0341	– 1.70	.47	1.77
	6 – 7	.1111	9.22	– 3.56	– 15.13
	7 – 6	.1111	6.78	– 3.11	6.03
7	7 – 5	.0301	– 3.37	.66	12.80
	7 – 8	.0438	– 8.04	.96	– .82
	7 – 9	.1172	0	2.58	

Joint	Member	$\dfrac{2C}{L}$	$\dfrac{2C}{L}$	$(2\phi_a + \phi_a - 3\psi_{an})$				
				Case I	**Case II**	**Case III**	**Case IV**	**Cases V to X**
	8 – 5 T	.0936	5.801		0	– 9.45		
	B	.1145	7.10		0	11.54		
8	8 – 7	.0438	– 8.41		.48	13.01		
	8 – 9	.0381	– 7.22		0	11.69		
	8 – 10 T	.0936	3.61		0	.37		
	B	.1145	6.86		0	.45		
	9 – 7	.1172	3.16		1.29	– 4.10		
9	9 – 8	.0381	– 5.95		0	11.47		
	9 – 10	.0578	– 10.58		0	19.80		
	9 – 11	.1230	2.22		0	– 3.69	Same as	Same as
	10 – 8 T	.0936	4.77		0	– 3.18	Case Ⅲ	Case Ⅳ
	B	.1145	5.84		0	3.90		
10	10 – 9	.0578	– 10.05		0	19.45		
	10 – 11	.0467	– 7.28		0	12.81		
	10 – 12 T	.0936	.75		0	– 5.70		
	B	.1145	.91		0	6.97		
	11 – 9	.1230	6.65		0	– 7.74		
11	11 – 10	.0467	– 6.02		0	12.04		
	11 – 12	.0817	– 20.41		0	41.76		
	11 – 13	.1278	.12		0	– 6.78		

Joint	Member	$\frac{2C}{L}$	Case I	Case III	Case IV	Case V	Cases VI to X
	12 – 10 T						
	B	.0936	4.58	0	– 10.01	– 9.91	
	12 – 11	.1145	5.61	0	12.24	12.12	
12	12 – 13	.0817	– 19.31	0	40.50	40.68	
	12 – 14 T	.0541	– 6.21	0	14.54	14.04	
	B	.0956	2.57	0	– 6.51	– 5.45	
		.1122	3.02	0	7.62	6.43	
	13 – 11	.1278	10.21	0	– 22.74	– 24.81	– 24.81
13	13 – 12	.0541	– 2.70	0	8.71	7.78	7.78
	13 – 14	.1156	– 28.05	0	69.18	68.53	68.5
	13 – 15	.1310	– 1.44	0	.92	.79	1.70

$$\frac{2C}{L} \left(2\phi_a + \phi_n - 3\psi_{an} \right)$$

Joint	Member	$\frac{2C}{L}$	Case I	Case III	Case IV	Case V	Cases VI to X
	14 – 12 T	.0956	10.02	– 20.45	– 18.45	– 18.44	– 18.44
	B	.1122	12.84	26.21	21.66	21.66	21.66
14	14 – 13	.1156	– 26.53	64.68	66.19	66.09	66.09
	14 – 15	.0587	– 5.44	10.31	12.61	13.02	13.02
	14 – 16 T	.0956	.19	– 2.48	– 5.83	– 5.56	– 5.15
	B	.1122	.22	2.92	6.84	6.51	6.05
	15 – 13	.1310	16.79	– 29.54	– 26.30	– 24.59	– 24.59
15	15 – 14	.0587	1.938	1.04	1.63	2.45	2.45
	15 – 16	.1542	– 29.76	116.95	113.21	115.87	116.69
	15 – 17	.1328	– 10.21	– 2.92	13.68	18.72	20.32

Joint	Member	$\frac{2C}{L}$				
16	16 – 14 T	.0956	15.29	– 24.29	– 33.21	– 32.59	– 31.90
	B	.1122	17.90	28.51	39.01	38.31	37.40
	16 – 15	.1542	– 24.75	114.49	98.14	100.20	101.18
	16 – 17	.0602	– 2.94	4.51	3.13	4.92	6.13
	16 – 18 T	.0956	– 2.95	– 10.02	– 11.21	– 12.71	– 11.44
	B	.1122	3.37	11.78	13.10	14.89	13.44
17	17 – 15	.1328	19.39	– 87.39	– 61.09	– 53.69	– 50.90
	17 – 16	.0602	8.54	– 32.31	– 24.78	– 21.69	– 20.09
	17 – 18	.1388	0	0	39.37	43.09	47.12
	17 – 19	.0602	– 8.53	32.29	18.57	26.79	27.04
	17 – 20	.1328	19.38	– 87.39	64.86	53.18	56.10
18	18 – 16 T	.0956	15.27	– 68.47	– 54.45	– 58.47	– 57.10
	B	.1122	17.87	80.38	63.98	68.52	67.01
	18 – 17	.1388	0	0	28.39	22.18	23.52
	18 – 19 T	.0956	– 15.27	– 68.47	30.37	35.14	33.87
	B	.1122	17.87	80.38	36.29	41.81	39.79

$$\frac{2C}{L}\left(2\phi_a + \phi_n - 3\phi_{an}\right)$$

Joint	Member	$\frac{2C}{L}$	Case IV	Case V	Case VI	Case VII	Case VIII	Case IX	Case X
19	19 – 18 T	.0956	– 31.42	– 16.87	– 20.41	– 19.76	– 20.09	– 19.87	– 20.39
	B	.1122	36.87	19.89	24.03	23.21	23.59	23.32	24.10
	19 – 17	.0602	– 19.87	– 8.01	– 9.80	– 9.31	– 9.58	– 9.45	– 9.81
	19 – 20	.1542	– 80.43	– 33.54	– 40.65	– 39.18	– 39.76	– 39.59	– 40.38
	19 – 21 T	.0956	60.79	27.69	35.53	32.19	34.56	32.87	32.61
	B	.1122	71.22	32.43	41.59	37.76	39.52	38.59	38.29

	20 – 17	.1328	– 40. 38	– 32. 18	– 31. 59	– 31. 10	– 31. 10	– 31. 39	– 31. 10
	20 – 19	.1542	– 103. 21	– 44. 69	– 48. 19	– 46. 78	– 47. 20	– 47. 32	– 47. 49
20	20 – 21	.0587	5. 98	4. 97	11. 87	9. 67	10. 85	10. 07	10. 37
	20 – 22	.1310	57. 74	39. 69	39. 52	43. 10	40. 65	42. 79	41. 02
	21 – 19 T	.0956	9. 16	– 23. 18	– 1. 63	– 9. 26	– 5. 82	– 7. 64	– 7. 36
	B	.1122	10. 77	29. 21	1. 91	10. 88	6. 85	9. 03	8. 65
	21 – 20	.0587	– 17. 04	– 22. 39	– 8. 05	– 12. 84	– 10. 61	– 11. 84	– 11. 43
21	21 – 22	.1156	– 101. 12	– 88. 62	– 39. 59	– 46. 69	– 44. 31	– 44. 59	– 45. 69
	21 – 23 T	.0956	25. 45	56. 09	16. 47	27. 89	23. 01	25. 10	24. 39
	B	.1122	30. 21	65. 69	19. 29	32. 79	27. 09	29. 45	28. 58
	22 – 20	.1310	1. 04	– 32. 58	– 15. 59	– 9. 32	– 13. 69	– 9. 43	– 13. 49
	22 – 21	.1156	– 105. 59	– 98. 47	– 49. 02	– 48. 54	– 50. 47	– 47. 52	– 50. 78
22	22 – 23	.0541	– 11. 87	– 5. 72	– 11. 13	– 2. 22	– 4. 53	– . 31	– 2. 91
	22 – 24	.1278	30. 24	62. 38	32. 38	22. 69	28. 59	22. 69	27. 21
	23 – 21 T	.0956	8. 31	14. 51	– 24. 79	10. 21	– 4. 96	2. 10	– . 29
	B	.1122	9. 74	17. 04	29. 10	12. 00	5. 84	2. 46	. 34
	23 – 22	.0541	– 19. 57	– 24. 57	– 30. 09	– 6. 92	– 17. 49	– 12. 01	– 14. 37
23	23 – 24	.0817	– 58. 41	– 88. 01	– 74. 79	– 36. 43	– 44. 18	– 44. 13	– 40. 76
	23 – 25 T	.0936	12. 52	16. 65	56. 01	25. 31	34. 76	15. 25	16. 65
	B	.1145	15. 36	20. 39	68. 69	31. 41	29. 16	18. 69	20. 39

$$\frac{2C}{L}\left(\phi_a + \phi_n - 3\psi_{an}\right)$$

Joint	Member	$\frac{2C}{L}$	Case IV	Case V	Case VI	Case VII	Case VIII	Case IX	Case X
	24 – 22	.1278	9.59	13.51	38.59	– 19.82	– 1.01	– 19.76	– 4.59
	24 – 23	.0817	– 59.47	– 89.79	– 85.52	– 49.72	– 43.52	– 53.47	– 43.69
24	24 – 25	.0467	– 12.94	– 25.42	– 21.64	– 27.91	– 16.75	– 17.46	– 9.23
	24 – 26	.1230	16.095	17.19	56.10	36.76	12.36	35.71	16.66
	25 – 23 T	.0935	10.52	8.04	17.53	45.69	21.69	7.11	.36
	B	.1145	12.76	9.84	21.39	56.01	26.59	8.76	.45
	25 – 24	.0467	– 13.04	– 28.21	– 34.78	– 44.39	– 3.07	– 23.32	– 16.12
25	25 – 26	.0578	– 22.21	– 40.76	– 71.51	– 69.45	– 17.59	– 17.39	– 29.09
	26 – 27 T	.0936	18.78	11.87	11.21	76.01	15.32	18.15	13.94
	B	.1145	23.10	14.52	13.76	92.86	18.76	22.19	17.01
	26 – 24	.1230	10.08	6.34	20.32	– 23.49	– 23.10	– 54.10	– 9.01
26	26 – 25	.0578	– 24.92	– 42.59	– 72.10	– 77.52	– 32.59	– 1.73	– 32.69
	26 – 27	.0381	– 14.19	– 46.72	– 33.60	– 44.71	– 42.10	11.94	– 15.03
	26 – 28	.1172	11.23	10.30	6.57	82.90	43.69	– 55.01	13.15
	27 – 25 T	.0936	7.29	3.83	7.01	29.81	– 96.82	19.53	– 1.03
	B	.1145	8.92	4.70	8.60	36.52	118.21	23.76	1.26
	27 – 26	.0381	– 17.10	– 48.91	– 35.10	– 38.29	– 65.32	– 2.21	– 18.85
27	27 – 28	– .0438	– 20.59	– 26.21	– 39.40	– 75.31	– 91.21	– 61.42	– 22.10
	27 – 30 T	.0936	6.67	14.59	20.11	19.47	151.01	15.19	10.61
	B	.1145	8.25	17.87	24.65	23.92	185.20	18.59	12.93

Joint	Member	$\frac{2C}{L}$	Case IV	Case V	Case VI	Case VII	Case VIII	Case IX	Case X
28	28 – 26	.1172	5.27	10.10	6.82	22.30	-43.32	-.35	3.05
	28 – 27	.0438	-19.54	-23.83	-37.29	-71.15	-96.89	-117.49	-21.54
	28 – 30	.0301	-8.96	-10.94	-16.38	-31.42	-54.10	-87.32	-11.21
	28 – 29	.1111	10.10	7.02	17.21	22.36	98.87	390.01	58.03

$$\frac{2C}{L}\left(2\phi_a + \phi_n - 3\phi_{an}\right)$$

Joint	Member	$\frac{2C}{L}$	Case IV	Case V	Case VI	Case VII	Case VIII	Case IX	Case X
29	29 – 28	.1111	20.92	30.79	34.47	50.01	104.72	82.76	76.30
	29 – 30	.0341	-2.00	-2.75	-8.52	-14.18	-48.62	14.71	-3.71
	29 – 31	.1111	-9.44	-10.81	-27.60	-47.58	-131.83	246.45	65.75
	30 – 27 T	.0936	12.12	22.76	31.70	47.10	129.50	-25.43	2.05
	30 – 27 B	.1145	14.92	27.82	39.01	57.52	158.80	31.12	2.52
	30 – 28	.0301	-8.01	10.01	-13.81	-25.42	-56.54	-62.00	-14.32
30	30 – 29	.0341	-4.23	-6.23	-8.86	-15.76	-53.12	-15.92	-12.69
	30 – 31	.0301	-2.96	-3.75	-3.51	-6.23	-23.43	-45.57	-19.81
	30 – 32 T	.0936	-5.71	-15.25	-20.13	-26.69	-86.12	-80.69	-5.52
	B	.1145	7.30	18.72	24.63	32.50	105.31	99.70	6.75
	31 – 29	.1111	-7.89	-9.81	-18.21	-25.81	-63.31	151.51	38.69
31	31 – 30	.0301	-.48	-1.38	1.08	1.08	-1.29	-44.31	-19.38
	31 – 32	.0442	1.63	2.81	8.01	9.56	25.41	-65.31	-29.51
	31 – 33	.1172	6.73	3.41	11.85	14.81	45.76	54.71	22.91

		T	.0936	−.75	−4.95	−7.95	−9.85	−28.39	−106.51	−6.27
	32−30	B	.1145	.91	6.07	9.73	12.01	34.69	130.40	7.67
32	32−31		.0442	.48	2.74	5.12	6.75	19.68	−55.01	−29.71
	32−33		.0381	−.95	.95	4.08	1.71	−7.25	−2.93	−47.73
	32−34	T	.0936	1.12	2.90	9.68	5.61	14.83	89.82	−135.30
		B	.1145	1.37	3.55	11.82	6.87	18.13	110.01	165.51
	33−31		.1172	4.22	−.94	1.52	2.11	12.31	15.62	46.30
33	33−32		.0381	−.76	−.38	.19	0	1.33	11.03	−39.76
	33−34		.0578	−1.56	−2.26	−.46	−1.74	−5.33	−2.54	−22.73
		T	.0936	.09	.18	−1.03	1.40	2.91	33.21	−135.60
34	34−32	B	.1145	.11	.22	1.26	1.72	3.56	40.52	166.30
	34−33		.0578	−2.26	−1.91	.81	−1.73	−3.70	15.62	−34.82

Since the values of ϕ and ψ given in Tables 5 and 3 are E times greater than their real values and are for loads expressed in kips, f in pounds $= 1000 \dfrac{2C}{L}(2\phi_a + \phi_n - 3\psi_{an})$, in which C is the distance of the extreme fibre of the member from the neutral axis and is in the plane of truss.

The values of f for the cases I to X are given in Table 63. The signs given there are those of the moment and not those of the fibre stress. For the joint of right half of arch, the stresses in members for cases I and III are the same as those for left half of arch and for case II, they are equal to zero. The signs for cases

IX and X are reversed so as to correct the use of an upward load on the cantilever arm.

In table 64 are given the secondary stresses in all the members of one half the truss for a unit load at the different panel points. For members of the other half of arch, the stresses may be obtained from the symmetry of truss. In those tables the letters V, H and S have been used to denote the secondary stresses in pounds due to vertical and horizontal forces and their algebraic sum. The stresses referred are for the top and bottom fibres of the upper chord; the top fibre of the bottom chord; the upper fibre of diagonals, and those fibres of the vertical posts that are facing the centre of the truss. The values of V are obtained by dividing the stresses given in Table 63 by the amount of load of the different cases, i. e. , II to X; of H, by multiplying the value of f in Table 63 for case I by the value of H corresponding to the position of the load; and of S by the algebraic sum of V and H. Revised primary stresses are also given in those tables and percentages of secondary stresses in respect to them are added in the fifth and ninth columns.

The influence lines of the secondary stresses are plotted in

Diagram 24. In those curves the full line represents the stress in that end of the member which is first named in Table 64 and the dotted line for the stress in the other end. Primary stresses are indicated by dots and dashes. Since it is the object in constructing those diagrams to show the distribution of stresses rather than their numerical quantities, two sets of scales are used for the primary and secondary stresses that for the former is fifty times smaller than that for the latter.

Table 64. Secondary Stresses for a Load of One Kip. (In pounds)

Member $U_2'(1-3)$

Load at	Primary	END 1				END 3			
	Stress	V	H	S	%	V	H	S	%
II'	T	22.60	.36	22.94		-22.55	-.84	-23.39	
	B	-24.40	-.44	-24.84		27.70	1.03	28.73	
I'	T	-2.76	.18	-2.28		7.38	-.42	6.96	
	B	3.37	-.22	3.15		-9.17	-.52	-8.65	
0	T	.56	-0	.56		-1.22	.01	-1.21	
	B	-.69	.01	-.68		1.49	-.01	1.48	
I	T	.24	-.20	.04		-1.22	.48	-.74	
	B	-.30	.25	-.05		1.51	-.59	.92	
	T	.23	-.40	-.17		-.93	.92	-.01	

II	B	− .29	.48	.19	1.14	− 1.13	.01
	T	− .26	− .57	− .31	− 2.42	1.33	− 1.09
III	B	.32	.70	1.02	2.95	− 1.64	1.31
	T	.06	− .73	− .67	− .97	1.70	.73
IV	B	− .07	.89	.82	1.18	− 2.09	− 1.91
	T	.04	− .83	− .79	− .46	1.96	1.50
V	B	− .05	1.02	1.97	.57	− 2.40	− 1.83
	T	− .47	− .88	− 1.35	− .61	2.07	1.46
VI	B	.57	1.08	1.65	.75	− 2.54	− 1.79
	T	− .39	− .83	− 1.22	− .51	1.96	1.45
VII	B	.48	1.02	1.50	.62	− 2.40	− 1.78
	T	− .31	− .73	− 1.04	− .40	1.70	1.30
VIII	B	.38	.89	1.27	.50	− 2.09	− 1.59
	T	− .23	− .57	− .80	− .30	1.33	1.03
IX	B	.29	.70	.99	.37	− 1.64	− 1.27
	T	− .16	− .40	− .56	− .20	− .92	.72
X	B	.19	.48	.67	.25	− 1.13	− .88
	T	− .08	− .20	− .28	− .10	.48	.38
XI	B	.09	.25	.34	.12	− .59	− .47
	T	− 0	− 0	0	− 0	.01	.01
XII	B	0	.01	.01	0	− .01	− .01
	T	.08	.18	.86	.10	− .42	− .32
XIII	B	− .09	− .22	− .31	− .12	.52	.40
	T	.16	.36	.52	.20	− .84	− .64
X′	B	− .19	− .44	− .63	− .25	1.03	.78

Member U_1' $(3-5)$

Load at	Primary Stress		END 3				END 5			
			V	H	S	%	V	H	S	%
II′		T	1.04	−.96	.08	.016	−.92	1.92	1.00	.198
	505.5	B	−1.28	1.18	.10	.020	1.13	−2.35	−2.22	.440
I′		T	8.87	−.49	8.30		−6.71	.96	−5.75	
		B	−10.87	.60	−10.27		8.31	−1.19	7.12	
0		T	−.56	.01	−.55		1.87	−.02	1.85	
		B	.69	−.01	.68		−2.29	.03	−2.26	
I		T	−2.36	.55	−1.81		7.17	−1.09	6.08	
		B	2.89	−.57	2.32		−8.78	1.33	−7.45	
II		T	−1.64	1.05	.60		4.28	−2.11	2.17	
		B	2.01	−1.30	.71		−5.43	2.68	−2.85	
III		T	−1.98	1.53	−.45		5.03	−3.06	1.97	
		B	2.43	−1.88	.55		−6.15	3.75	−2.40	
IV		T	−1.56	1.95	.39		5.09	−3.89	1.20	
		B	2.02	−2.39	−.37		−6.24	4.77	−1.47	
V		T	−.31	2.24	1.93		2.38	−4.47	−2.09	
		B	.38	−2.75	−2.37		−2.91	5.49	2.58	
VI		T	−1.78	2.36	.58		4.11	−4.72	−.61	
		B	2.18	−2.90	−.72		−5.04	5.78	.74	
VII		T	−1.48	2.24	.76		3.42	−4.47	−1.05	
		B	1.81	−2.75	−.94		−4.20	5.49	1.29	
VIII		T	−1.19	1.95	.76		2.74	−3.89	−1.15	
		B	1.45	−2.39	−.94		−3.36	4.77	1.41	

IX	T	-.89	1.53	.64	2.06	-3.06	-1.00	
	B	1.09	-1.88	-.79	-2.52	3.75	1.23	
X	T	-.59	1.05	.46	1.37	-2.11	-.74	
	B	.73	-1.30	-.57	-1.68	2.58	.90	
XI	T	-.30	.55	.25	.69	-1.04	-.35	
	B	.36	-.57	-.21	-.84	1.33	.49	
XII	T	-0	-.01	-.01	0	-.02	.02	
	B	0	-.01	-.01	-0	.03	.03	
'XI'	T	.30	-.49	-.19	-.69	.96	.30	
	B	-.36	.60	.24	.84	-1.19	.35	
'X'	T	.59	-.96	-.37	-1.37	1.92	.55	
	B	-.73	1.18	.35	1.68	-2.35	-.67	

Member $U_1(5-8)$

Load at	Primary Stress		END 5				END 8			
			V	H	S	%	V	H	S	%
		T	-3.50	3.41	.09	.01	2.50	-2.48	-.02	.00
II'	783.4	B	4.20	-3.74	.46	.58	-3.15	3.05	-.10	.01
		T	-2.11	1.72	-.39	.10	1.60	-1.49	.11	.03
I'	390.8	B	2.59	-1.89	.70	.18	-2.56	1.54	-1.02	.26
		T	0	-.04	-.04	2.22	-0	.03	.03	1.66
0	1.8	B	0	.05	.05	2.78	0	-.04	-.04	2.22
		T	11.79	-1.93	9.86	2.58	-12.59	1.39	-11.20	2.86
I	-382.0	B	-13.22	2.10	-11.12	2.84	15.42	-1.73	13.69	3.59
		T	7.83	-3.74	4.09	1.55	-3.25	2.72	-.53	.20

II	−264.2	B	−9.61	4.11	−5.50	2.08	3.99	−3.34	.65	.24
		T	7.95	−5.42	2.43	1.61	−5.03	3.90	−1.13	.75
III	−151.5	B	−9.75	5.96	−3.79	2.50	6.15	−4.84	1.31	.87
		T	7.61	−6.89	.72	1.51	−4.87	5.03	.16	.33
IV	−47.7	B	−9.27	7.57	−1.70	3.57	5.97	−6.16	−1.19	.40
		T	5.03	−7.92	−2.89	7.90	−2.79	5.78	2.99	7.91
V	37.8	B	−6.21	8.71	2.50	6.62	3.43	−7.08	−3.65	9.65
		T	6.22	−8.37	−2.15	2.20	−4.72	6.11	1.39	1.43
VI	97.6	B	−7.62	9.20	1.58	1.62	5.78	−7.47	−1.69	1.73
		T	5.18	−7.92	−2.74	2.24	−3.93	5.78	1.85	1.61
VII	121.9	B	−6.38	8.71	2.36	1.93	4.82	−7.08	−2.26	1.85
		T	4.15	−6.89	−2.74	2.18	−3.13	5.03	1.90	1.51
VIII	125.6	B	−5.07	7.57	2.50	1.99	3.85	−6.16	−2.31	1.84
		T	3.11	−5.42	−2.31	2.19	−2.36	3.90	1.54	1.52
IX	101.0	B	−3.81	5.96	2.15	2.13	2.89	−4.84	−1.95	1.93
		T	2.07	−3.74	−1.67	2.31	−1.57	2.72	1.15	1.54
X	72.5	B	−2.54	4.11	1.57	2.17	1.93	−3.34	−1.41	1.95
		T	1.04	−1.93	−.89	2.29	−.79	1.39	.60	1.54
XI	38.9	B	−1.27	2.10	1.83	2.13	1.96	−1.73	−.77	1.97
		T	0	−.04	−.04	2.22	−0	.03	.03	1.67
XII	1.8	B	−0	.05	.05	2.78	0	−.04	−.04	2.22
		T	−1.04	1.72	.58	2.26	.79	−1.49	−.70	2.33
'XI'	−30.1	B	1.27	−1.89	−.62	2.06	−1.96	1.54	.58	1.93
		T	−2.07	3.41	1.34	2.30	1.57	−2.48	−.91	1.55
'X'	−58.4	B	2.54	−3.74	−1.20	2.05	−1.93	3.05	1.12	1.91

Member $U_2(8-10)$

Load at	Primary Stress		END 8				END 10			
			V	H	S	%	V	H	S	%
II′	800.9	T	.17	2.40	2.57	.32	2.32	-2.04	.28	.04
		B	-.21	-2.93	-3.14	.39	-2.83	2.49	-.34	.04
I′	398.2	T	-1.61	1.21	-.40	.10	1.51	-1.03	.48	.12
		B	1.98	-1.48	.50	.12	-1.85	1.26	-.59	.15
0	4.5	T	-0	0	0	0	0	.03	.03	.67
		B	0	0	0	0	0	-.03	-.03	.67
I	-382.0	T	8.06	-1.36	6.70	1.78	1.27	1.16	.11	.02
		B	-9.86	1.67	-8.19	2.14	-1.57	-1.42	-.15	.04
II	-264.2	T	4.97	-2.64	2.33	.89	-12.65	2.25	-10.40	3.78
		B	-6.08	3.23	-2.85	1.08	15.45	-2.75	12.70	4.81
III	-161.5	T	1.75	-3.82	-2.07	1.36	-3.75	3.27	.46	.31
		B	-2.15	4.68	2.53	1.67	3.43	-3.99	-.56	.37
IV	-47.7	T	1.28	-4.87	-3.59	.75	-3.97	4.15	.18	.38
		B	-1.56	5.95	4.39	.92	4.83	-5.07	-.24	.50
V	37.8	T	3.04	5.59	-2.55	6.42	-.83	4.78	3.95	10.41
		B	-3.72	6.85	3.13	8.25	1.56	-5.83	-4.27	11.21
VI	97.6	T	-.19	-5.91	-7.10	7.28	-1.59	5.03	3.44	3.52
		B	23	7.22	7.45	7.61	1.92	-6.15	-4.20	4.28
VII	121.9	T	-.16	-5.59	-5.75	4.74	-1.32	4.78	3.46	2.84
		B	.19	6.85	7.04	5.81	1.62	-5.83	-4.21	3.46
VIII	125.6	T	-.12	-4.87	-4.99	3.95	-1.06	4.15	3.09	2.45
		B	.15	5.95	6.10	4.84	1.30	-5.07	-3.77	2.98

		T	V	H	S	%	V	H	S	%
		T	-.09	-3.82	-3.91	3.87	-.79	3.27	2.48	2.46
IX	101.0	B	.11	4.68	4.79	4.74	.97	-3.99	-3.02	2.98
		T	-.06	-2.64	-2.70	3.72	-.53	2.25	1.72	2.46
X	72.5	B	.08	3.23	3.31	4.56	.65	-2.75	-2.10	2.89
		T	-.03	-1.36	-1.39	3.56	-.25	1.16	.91	2.31
XI	38.9	B	.04	1.67	1.71	4.37	.32	-1.42	-1.10	2.81
		T	-0	-0	0	0	0	.03	.03	1.78
XII	1.8	B	0	0	0	0	0	-.03	-.03	1.78
		T	.03	1.21	1.24	4.13	.25	-1.03	-.78	2.60
'XI'	-30.1	B	-.04	-1.48	-1.52	5.07	-.32	1.26	.94	3.13
		T	.06	2.40	2.46	4.20	.53	-2.04	-1.51	2.58
'X'	-58.4	B	-.08	-2.93	-3.01	5.13	-.65	2.49	1.84	3.14

Member $U_3(10-12)$

Load at	Primary Stress		END 10				END 12			
			V	H	S	%	V	H	S	%
		T	-.06	.32	.26	.02	2.78	-1.96	.82	.10
II'	791.0	B	.07	-39	-.32	.04	-3.40	2.39	-1.01	.13
		T	-.59	.16	-.43	.11	1.28	-.99	.29	.07
I'	391.3	B	.73	-.20	.50	.13	-1.56	1.21	.35	.09
		T	0	-0	0	0	0	.02	.02	.23
0	8.5	B	0	0	0	0	0	-.03	-.03	.35
		T	1.80	-.18	1.62	.46	-2.90	1.12	-1.78	.51
I	-349.3	B	-2.22	.22	-2.00	.57	2.43	-1.36	1.07	.31
		T	7.61	-.35	7.26	1.01	-4.20	2.16	-2.04	.28

II	−723.8	B	−9.33	.43	−8.96	1.23	5.10	−2.34	2.46	.34
		T	4.37	−.51	3.86	.34	−14.00	3.13	−10.87	.96
III	−1122.8	B	−5.36	.62	−4.74	.42	17.15	−3.82	13.33	1.17
		T	2.62	−.65	1.97	.34	−5.57	3.98	−1.59	.27
IV	−583.8	B	−3.28	.79	−2.49	.43	6.80	−4.86	1.94	.33
		T	4.38	−.75	3.63	2.74	−5.22	4.57	−.65	.49
V	−132.4	B	−5.33	.91	−4.42	3.35	6.42	−5.58	.84	.63
		T	2.85	.79	2.06	1.02	−5.01	4.83	−.18	.09
VI	197.6	B	−3.49	.96	−2.53	1.27	6.12	−5.91	.21	.11
		T	2.38	−.75	1.63	.45	−4.17	4.57	.40	.11
VII	357.8	B	−2.91	.91	−2.00	.56	5.11	−5.58	−.47	.13
		T	1.90	−.65	1.25	.31	−3.34	3.98	.64	.16
VIII	396.6	B	−2.33	.79	1.54	.39	4.08	−4.86	−.78	.20
		T	1.42	−.51	.91	.26	−2.50	3.13	.63	.18
IX	347.9	B	−.175	.62	−1.13	.32	3.06	−3.82	−.78	.22
		T	.95	−.35	.60	.23	−1.67	2.16	.49	.19
X	256.6	B	−1.16	.43	−.73	.29	2.04	−2.64	−.50	.23
		T	.47	−.18	.29	.21	−.84	1.12	.28	.20
XI	140.9	B	−.58	.22	−.36	.25	1.02	−1.36	−.34	.24
		T	0	−0	0	0	−0	.02	.02	.24
XII	8.5	B	−0	0	0	0	0	−.03	−.03	.35
		T	−.47	.16	−.31	.31	.84	−.99	−.15	.15
'XI'	−98.9	B	.58	−.20	.38	.38	−1.02	1.21	.19	.19
		T	−.95	.32	−.63	.34	1.67	−1.96	−.29	.25
'X'	−189.4	B	1.16	−.39	.77	.40	−2.04	2.39	−.35	.19

Member $U_4(12-14)$

Load at	Primary Stress		END 12				END 14			
			V	H	S	%	V	H	S	%
II'	710.6	T	.05	1.10	1.15	.16	4.07	-4.28	-.21	.29
		B	-.06	-1.30	-1.36	.19	-4.77	5.48	.71	1.00
I'	348.3	T	-.17	.56	.39	.11	2.09	-2.70	-.61	.17
		B	.21	-.65	-.44	.13	-2.46	2.78	.32	.09
0	14.3	T	0	-0	0	0	0	.05	.05	.35
		B	0	.0	0	0	0	-.07	-.07	.49
I	-277.8	T	-.41	-.63	-1.04	.37	-1.91	2.44	.53	.19
		B	.48	.74	1.22	.44	2.26	-3.12	-.86	.31
II	-597.9	T	1.70	-1.21	.49	.82	-4.62	4.71	.09	.01
		B	-2.00	1.42	-.58	.97	5.47	-6.04	-.57	.95
III	-959.2	T	-6.20	-1.76	-7.96	.83	-4.12	6.82	2.70	.28
		B	7.28	2.07	9.35	.97	4.82	-8.75	-3.93	.41
IV	-1391.8	T	4.83	-2.24	2.59	.18	-18.70	8.67	-10.03	.72
		B	-5.68	2.63	-3.05	.22	21.90	-11.10	10.80	.78
V	-581.9	T	3.46	-2.57	.89	.15	-10.60	10.0	-.60	.10
		B	-4.06	3.02	-1.04	.18	12.50	-12.20	.30	.05
VI	33.1	T	3.25	-2.71	.54	1.63	-10.23	10.53	.30	.90
		B	-3.81	3.19	-.62	1.87	13.10	-13.50	-.40	1.21
VII	344.0	T	2.27	-2.57	-.30	.09	-7.68	10.0	2.32	.67
		B	-2.66	3.02	.36	.11	9.82	-12.80	-3.98	1.16
VIII	460.1	T	1.82	-2.24	-.42	.09	-6.14	8.67	2.53	.55
		B	-2.13	2.63	.50	.11	7.87	-11.10	-4.23	.92

			V	H	S	%	V	H	S	%
		T	1.36	-1.76	-.40	.09	-4.61	6.82	2.21	.51
IX	429.6	B	-1.60	2.07	.33	.08	5.90	-8.75	-2.85	.66
		T	.91	-1.21	-.30	.09	-3.07	4.71	1.64	.50
X	328.0	B	-1.06	1.42	.34	.10	3.93	-6.04	-2.11	.64
		T	.45	-.63	-.18	.10	-1.53	2.44	.91	.49
XI	185.1	B	-.53	.74	.21	.11	1.96	-3.12	-1.16	.63
		T	0	-0	0	0	-0	.05	.05	.35
XII	14.3	B	-0	0	0	0	0	-.07	-.07	.49
		T	-.45	.56	.11	.09	1.53	-2.70	-1.17	1.02
'XI'	-114.6	B	.53	-.65	-.12	.10	-1.96	2.78	.82	.71
		T	-.91	1.10	.09	.04	3.07	-4.28	-1.19	.55
'X'	-215.3	B	1.06	-1.30	-.24	.11	-3.93	5.48	1.55	.72

Member $U_5(14-16)$

Load at	Primary Stress		END 14				END 16			
			V	H	S	%	V	H	S	%
		T	1.21	.08	1.29	.26	5.43	-6.55	-1.12	.23
II'	496.6	B	-1.44	-.10	-1.54	.31	-6.38	7.66	1.28	.26
		T	.64	.04	.68	.28	2.74	-3.30	-.56	.23
I'	237.9	B	-.75	-.05	-.80	.33	-3.21	3.88	.67	.28
		T	0	-0	0	0	0	.08	.08	.38
0	20.9	B	0	0	0	0	0	-.10	-.10	.49
		T	-.49	-.05	-.54	.40	-2.88	3.72	.84	.62
I	-135.1	B	.57	.03	.62	.46	3.29	-4.36	-1.07	.79
		T	-1.55	-.05	-1.60	.42	-5.37	7.19	1.82	.55

II	−331.8	B	1.81	.11	1.92	.58	6.58	−8.44	−1.86	.56
		T	−.41	−.13	−.54	.09	−8.89	10.50	1.61	.27
III	−588.7	B	.48	.15	.63	.11	10.41	−12.30	−1.89	.32
		T	−7.72	−.17	−7.89	.83	−9.23	13.41	4.18	.44
IV	−949.4	B	9.73	.20	9.93	1.04	10.81	−15.55	−4.74	.50
		T	3.82	−.19	3.63	.24	−25.30	15.25	−10.05	.66
V	−1523.8	B	−4.49	.22	−4.27	.28	29.50	−17.89	11.61	.76
		T	1.24	.20	1.04	.19	−12.15	16.05	3.90	.71
VI	−545.1	B	−1.46	.24	−1.22	.22	14.25	−18.75	−4.50	.82
		T	2.43	−.19	2.24	11.60	−13.80	15.25	1.45	7.48
VII	19.4	B	−3.87	.22	−3.65	18.80	16.20	−17.89	−1.69	8.71
		T	1.85	−.17	1.68	.63	−10.85	13.41	2.56	.96
VIII	265.2	B	−2.16	.20	−1.96	.73	12.74	−15.55	−2.81	1.05
		T	1.29	−.13	1.16	.35	−7.97	10.50	2.53	.75
IX	337.2	B	−1.52	.15	−1.37	.41	9.35	−12.30	−2.95	.87
		T	.86	−.05	.81	.29	−5.30	7.19	1.89	.66
X	285.5	B	−1.01	.11	−.90	.31	6.23	−8.44	−2.21	.77
		T	.43	−.05	.38	.22	−2.65	3.72	1.07	.61
XI	173.6	B	−.51	.05	−.46	.26	−3.11	−4.36	−1.25	.72
		T	0	0	0	0	−0	.08	.08	.39
XIII	20.9	B	−0	0	0	0	0	−.10	−.10	.48
		T	−.43	.04	−.39	.55	2.65	−3.30	−.65	.92
'XI'	−70.8	B	.51	−.05	.46	.65	−3.11	3.88	.77	1.08
		T	−.86	−.10	−.96	.79	5.30	−6.55	−1.25	1.03
'X'	−120.7	B	1.01	.08	1.09	.90	−6.23	7.66	1.43	1.18

Member U_6 (16 – 18)

Load at	Primary Stress		END 16				END 18			
			V	H	S	%	V	H	S	%
II′	158.0	T	3.40	-1.26	2.14	1.34	5.55	-6.53	-.98	.62
		B	-4.00	1.44	-2.56	1.62	-6.64	7.65	1.01	.64
I′	67.0	T	1.66	-.64	1.02	1.52	2.78	-3.31	-.53	.79
		B	-1.94	.74	-1.20	1.79	-3.32	3.87	.55	.82
0	24.1	T	0	.01	.01	.04	0	.08	.08	.34
		B	0	-.01	-.01	.04	0	-.10	-.10	.42
I	52.0	T	-1.68	.71	-.97	1.86	-2.78	3.72	.94	1.81
		B	1.97	-.82	1.15	2.21	3.32	-4.35	-1.03	1.98
II	32.7	T	-3.30	1.39	-1.91	5.85	-5.55	7.19	1.64	5.01
		B	3.86	-1.59	2.27	6.95	6.64	-8.41	-1.77	5.70
III	-56.3	T	-5.10	2.02	-3.08	5.48	-8.73	10.50	1.77	3.14
		B	6.00	-2.30	3.70	6.55	9.94	-12.30	-2.36	4.16
IV	-265.2	T	-5.63	2.57	-3.06	1.13	-11.56	13.25	1.69	.63
		B	6.63	-2.93	3.70	1.38	13.68	-15.69	-2.01	.75
V	-721.6	T	-13.15	2.95	-10.20	1.42	-12.89	15.25	2.36	.33
		B	15.38	-3.37	12.01	1.67	15.10	-17.85	-2.75	.38
VI	-1523.8	T	5.01	3.11	8.12	.53	-34.25	16.05	-18.20	1.19
		B	-5.89	-3.55	-9.44	.62	40.20	-18.80	21.40	1.40
VII	-721.7	T	4.77	2.95	7.72	1.06	-22.70	15.25	-7.45	1.03
		B	-5.47	-3.37	-8.84	1.22	26.60	-17.85	8.75	1.21
VIII	-265.3	T	4.23	2.57	6.80	2.56	-19.48	13.25	-6.23	2.35
		B	-4.97	-2.93	-7.90	2.97	22.85	-15.69	7.16	2.70

			V	H	S	%	V	H	S	%
IX	-56.4	T	2.86	2.02	4.88	8.61	-14.27	10.50	-3.77	6.65
		B	-3.37	-2.30	-5.67	10.01	16.75	-12.30	4.45	7.87
X	32.7	T	1.91	1.39	3.30	10.10	-9.52	7.19	-2.33	7.15
		B	-2.24	-1.59	-3.83	11.60	11.18	-8.41	2.77	8.49
XI	52.0	T	.95	.71	1.66	3.17	-4.76	3.72	-.96	1.84
		B	-1.12	-.82	-1.94	3.71	5.58	-4.35	1.23	2.36
XII	24.1	T	0	.01	.01	.04	0	.08	.08	.33
		B	-0	-.01	-.01	.04	0	-.10	-.10	.41
'XI'	67.0	T	-.95	-.64	-1.59	2.37	4.76	-3.31	1.45	2.16
		B	1.12	.74	1.86	2.77	-5.58	3.87	-1.71	2.55
'X'	158.1	T	-1.91	-1.26	-3.17	2.00	9.52	-6.53	2.99	1.90
		B	2.24	.44	2.68	1.69	-11.18	7.65	-3.53	2.23

Member L_1' $(2-4)$

Load at	Primary Stress	END 2				END 4			
		V	H	S	%	V	H	S	%
II′	-574.0	-7.75	1.05	-6.70	1.17	3.82	-1.50	2.32	.41
I′		-1.30	.53	-.77		4.56	-.76	3.80	
0		-.12	-.01	-.13		-1.88	.02	-1.86	
I		1.02	-.60	.42		-3.80	.85	-2.95	
II		.35	-1.15	-.80		-2.47	1.65	-.82	
III		.38	-1.68	-1.30		-2.96	2.40	-.56	
IV		.31	-2.13	-1.82		-1.13	3.05	1.92	
V		1.75	-2.45	-.70		-2.79	3.51	.72	
VI		-.88	-2.58	-3.46		-.18	3.70	3.52	

VII	−.72	−2.45	−3.17	−.15	3.51	3.36
VIII	−.59	−2.13	−2.72	−.12	3.05	2.93
IX	−.43	−1.68	−2.11	−.09	2.40	2.31
X	−.29	−1.15	−1.44	−.06	1.65	.99
XI	−.14	−.60	−.74	−.03	.85	.82
XII	−0	−.01	−.01	−0	.02	.02
'XI'	.14	.53	.67	.03	−.76	−.73
'X'	.29	1.05	1.34	.06	−1.50	−1.44

Member L_0' (4 − 6)

Load at	Primary	END 5				END 6			
	Stress	V	H	S	%	V	H	S	%
II'	−906.4	−6.46	−2.37	−8.83	.97	10.95	3.31	14.26	1.57
I'	−453.2	−12.61	−1.20	−13.81	3.04	20.50	1.68	22.18	4.89
0		1.89	.03	1.92		.11	−.04	.07	
I		−5.27	1.35	−3.92		10.98	−1.88	9.10	
II		−4.30	2.61	−1.69		7.93	−3.64	4.29	
III		−4.56	3.78	−.88		6.87	−6.28	1.59	
IV		−3.26	4.82	1.56		3.63	−6.72	−3.09	
V		−3.29	5.54	2.25		3.94	−7.73	−3.79	
VI		−2.56	5.83	3.27		3.77	−8.15	−4.38	
VII		−2.13	5.54	3.41		3.14	−7.73	−4.59	
VIII		−1.71	4.82	3.11		2.51	−6.72	−4.21	
IX		−1.28	3.78	2.50		1.88	−5.28	−3.40	
X		−.85	2.61	1.76		1.26	−3.64	−2.38	

XI		−.43	1.35	.92	.63	−1.88	−1.25
XII		−0	.03	.03	0	−.04	−.04
'XI'		.43	−1.20	−.77	−.63	1.68	1.05
'X'		.85	−2.37	−1.52	−1.26	3.31	2.05

Member $L_0(6-7)$

Load at	Primary	END 6				END 7			
	Stress	V	H	S	%	V	H	S	%
II′	−393.7	−12.65	3.93	−8.72	2.21	9.67	−2.90	6.77	1.72
I′	−193.7	−6.89	1.99	−4.90	2.54	32.30	−1.47	30.83	16.00
0	−6.4	3.56	−.03	3.51	54.87	−3.11	.04	−3.07	47.80
I	−291.2	8.72	−2.24	6.48	2.22	−2.83	1.65	−1.18	.41
II	−563.4	8.32	−1.33	3.99	.71	−3.73	3.19	−.54	.09
III	−817.1	8.62	−6.28	2.34	.29	−4.30	4.63	4.33	.53
IV	−1038.8	10.26	−7.99	2.27	.22	−2.33	5.89	3.56	.34
V	−1194.7	8.71	−9.18	−.47	0	−4.11	6.77	2.66	.22
VI	−1258.5	7.57	−9.68	−2.11	.17	−4.28	7.15	2.87	.23
VII	−1194.7	6.31	−9.18	−2.87	.24	−3.55	6.77	3.22	.27
VIII	−1038.8	5.04	−7.99	−2.95	.28	−2.85	5.89	3.04	.29
IX	−817.1	3.78	−6.28	−2.50	.31	−2.14	4.63	2.49	.30
X	−563.4	2.52	−4.33	−1.81	.32	−1.43	3.19	1.76	.31
XI	−291.2	1.26	−2.24	−.98	.34	−.71	1.55	.94	.32
XII	−6.4	0	−.05	−.05	.78	−0	.04	.04	.62
'XI'	259.5	−1.26	1.99	.73	.28	.71	−1.47	−.76	.29
'X'	512.7	−2.52	3.93	1.41	.27	1.43	−2.90	−1.47	.30

Member $L_1 (7-9)$

Load at	Primary	END 7				END 9			
	Stress	V	H	S	%	V	H	S	%
II′	−403.9	−.51	0	−.51	.13	2.19	−1.35	.84	.21
I′	−197.9	−.03	0	−.03	.01	4.57	−.69	3.88	1.96
0	−8.1	−2.58	0	−2.58	31.80	1.29	.02	1.31	16.1
I	157.8	−3.62	0	−3.62	2.31	−3.64	.77	−2.87	1.82
II	−234.3	3.67	0	3.67	2.56	−8.72	1.49	−7.23	3.09
III	−602.9	1.70	0	1.70	.28	−2.64	2.16	−.48	.08
IV	−931.2	3.37	0	3.37	.36	−3.44	2.75	−.79	.09
V	−1176.1	2.20	0	2.20	.19	−4.68	3.16	−1.52	.13
VI	−1304.6	.41	0	.41	.03	−2.05	3.33	1.28	.09
VII	−1271.7	.34	0	.34	.03	−1.71	3.16	1.45	.11
VIII	−1122.5	.27	0	.27	.02	−1.37	2.75	1.38	.12
IX	−889.9	.20	0	.20	.02	−1.03	2.16	1.13	.13
X	−616.9	.14	0	.14	.02	−.69	1.49	.80	.13
XI	−320.5	.07	0	.07	.02	−.34	.77	.43	.13
XII	−8.1	0	0	0	0	−0	.02	.02	.25
'XI'	280.4	−.07	0	−.07	.02	.34	−.69	−.35	.13
'X'	552.6	−.14	0	−.14	.02	.69	−1.35	−.66	.12

Member $L_2 (9-11)$

Load at	Primary	END 9				END 11			
	Stress	V	H	S	%	V	H	S	%
II′	−404.6	1.50	.95	2.45	.61	2.79	−2.84	−.05	.01

Load at	Primary Stress	V	H	S	%	V	H	S	%
I'	-197.0	4.50	.48	4.98	2.57	2.98	-1.44	1.54	.78
0	-10.7	0	-.01	-.01	.09	0	.04	.04	.37
I	144.4	-1.93	-.54	-2.47	1.72	-1.03	1.62	.59	.41
II	320.3	-3.92	-1.04	-4.96	1.55	-6.13	3.12	-3.01	.94
III	-226.2	5.08	-1.51	3.57	1.58	-14.00	4.53	-9.47	4.18
IV	-719.7	2.13	-1.92	.21	.03	-5.73	5.76	.03	0
V	-1103.9	4.21	-2.21	2.00	.18	-0.60	6.63	.03	0
VI	-1335.4	1.85	-2.33	-.48	.04	-3.88	7.00	3.12	.24
VII	-1355.0	1.54	-2.21	-.67	.05	-3.23	6.63	3.40	.25
VIII	-1221.8	1.23	-1.92	-.69	.06	-2.59	5.76	3.17	.26
IX	-979.5	.92	-1.51	-.59	.06	-1.95	4.53	2.58	.26
X	-684.0	.62	-1.04	-.42	.06	-1.29	3.12	1.83	.27
XI	-357.8	.31	-.54	-.23	.06	-.64	1.62	.98	.27
XII	-10.7	0	-.01	-.01	.09	-0	.04	.04	.37
'XI'	305.2	-.31	.48	.17	.06	.64	-1.44	-.80	.26
'X'	599.7	-.62	.95	.33	.06	1.29	-2.88	-1.59	.27

Member L_3 (11 – 13)

Load at	Primary Stress	END 11				END 13			
		V	H	S	%	V	H	S	%
II'	-379.2	.77	.05	.82	.22	4.53	-4.37	.16	.04
I'	-182.4	1.65	.03	1.68	.92	2.89	-2.21	.68	.37
0	-14.5	0	-0	0	0	0	.05	.05	.37
I	111.0	-.08	-.03	-.11	1.00	-2.38	2.48	.10	.90
II	264.7	-3.30	-.06	-3.36	1.27	-3.78	4.80	1.02	.39

		V	H	S	%	V	H	S	%
III	460.4	−9.65	−.09	−9.74	2.12	−8.11	6.95	−1.16	.25
IV	−295.6	4.51	−.11	4.40	1.48	−20.80	8.85	−11.95	4.03
V	−903.0	4.00	−.13	3.87	.43	−12.60	10.18	−2.42	.27
VI	−1302.8	3.39	−.13	3.26	.25	−11.38	10.71	−.67	.05
VII	−1414.8	2.83	−.13	2.70	.19	−10.32	10.18	−.14	.01
VIII	−1319.2	2.25	−.11	2.15	.16	−8.26	8.85	.59	.05
IX	−1075.0	1.69	−.09	1.60	.15	−6.20	6.95	.75	.07
X	−758.9	1.13	−.06	1.07	.14	−4.13	4.80	.67	.09
XI	−400.8	.57	−.03	.54	.14	−2.07	2.48	.41	.10
XII	−14.5	0	−0	0	0	−0	.05	.05	.35
'XI'	329.4	−.57	.03	−.54	.16	2.07	−2.21	−.14	.04
'X'	644.4	−1.13	.05	−1.08	.17	4.13	−4.37	−.04	.06

Member $L_4(13-15)$

Load at	Primary Stress	END 13				END 15			
		V	H	S	%	V	H	S	%
II′	−287.6	2.25	−.06	2.19	.77	6.82	−7.15	−.33	.11
I′	−133.8	.79	−.03	.76	.56	3.57	−3.63	−.14	.10
0	−20.0	0	0	0	0	0	.09	.09	.45
I	33.4	−1.16	.03	−1.13	3.18	−3.39	4.08	.69	1.94
II	129.7	−1.55	.06	−1.49	1.14	−7.18	7.90	.72	.55
III	281.7	−3.90	.09	−3.81	1.36	−9.90	11.53	1.63	.58
IV	533.1	−10.85	.11	−10.74	2.01	−13.21	14.56	1.35	.25
V	−421.9	−.44	.13	−.31	.73	−23.90	16.76	−7.14	1.69
VI	−1090.6	−.46	.14	−.32	.03	−14.80	17.65	2.85	.26

Load at	Primary Stress	V	H	S	%	V	H	S	%
VII	-1362.7	-.33	.13	-.20	.01	-10.90	16.76	5.86	.43
VIII	-1348.5	-.57	.11	-.46	.03	-8.20	14.56	6.36	.45
IX	-1129.5	-.43	.09	-.34	.03	-6.14	11.53	5.39	.48
X	-811.1	-.28	.06	-.22	.03	-4.10	7.90	3.80	.47
XI	-435.0	-.14	.03	-.11	.03	-2.05	4.08	2.03	.47
XII	-20.0	-0	0	0	0	-0	.09	.09	.45
'XI'	336.6	.14	-.03	.11	.03	2.05	-3.63	-1.58	.47
'X'	653.2	.28	-.06	.22	.03	4.10	-7.15	-3.05	.47

Member L_5 (15 – 17)

Load at	Primary Stress	END 15				END 17			
		V	H	S	%	V	H	S	%
II′	-68.9	5.18	-4.36	.82	1.19	9.32	-8.30	1.02	2.50
I′	-21.3	2.59	-2.21	.38	1.76	4.66	-4.19	.47	2.18
0	-26.2	0	.05	.05	.19	0	.10	.10	.38
I	-103.1	-2.59	2.48	-.11	.10	-4.66	4.72	.06	.06
II	-138.6	-5.18	4.91	-.27	.19	-9.32	9.11	-.21	.15
III	-93.3	-7.90	6.97	-.93	1.00	-14.00	13.21	-.79	.85
IV	82.6	-10.75	8.85	-1.90	2.30	-17.75	16.80	-.95	1.14
V	527.7	-16.80	10.20	6.60	1.25	-27.00	19.32	-7.68	1.46
VI	-506.1	1.46	10.73	12.19	2.41	-43.7	20.4	-23.30	4.61
VII	-1018.3	-5.79	10.20	4.41	.43	-25.50	19.32	-6.18	.61
VIII	-1154.3	-6.23	8.85	2.62	.23	-17.90	16.80	-1.10	.10
IX	-1021.0	-5.07	6.97	1.90	.19	-12.60	13.21	.61	.06
X	-757.0	-3.38	4.91	1.53	.20	-8.43	9.11	.68	.09

XI	-417.3	-1.69	2.48	$.79$	$.19$	-4.21	4.72	$.51$	$.12$
XII	-26.2	-0	$.05$	$.05$	$.19$	-0	$.10$	$.10$	$.38$
'XI'	287.9	1.69	-2.21	$-.52$	$.18$	4.21	-4.19	$.03$	$.01$
'X'	549.5	3.38	-4.36	-1.02	$.19$	8.43	-8.30	$.13$	$.02$

Member D_2' $(2-3)$

Load at	Primary Stress	END 2				END 3			
		V	H	S	%	V	H	S	%
II'	885.5	6.63	.13	6.76	.76	-7.95	$-.22$	-8.17	.92
I'		$-.91$	.07	$-.84$		$-.24$	$-.11$	$-.35$	
0		$-.53$	-0	$-.53$		.85	0	.85	
I		.11	$-.07$	.04		.60	.12	.72	
II		0	$-.14$	$-.14$		$-.28$	.23	$-.05$	
III		.05	$-.21$	$-.16$		-1.02	.34	$-.68$	
IV		$-.13$	$-.26$	$-.39$		$-.32$	.43	.11	
V		$-.32$	$-.30$	$-.62$		.41	.49	.08	
VI		.40	$-.32$	.08		$-.69$	.52	$-.17$	
VII		.33	$-.30$	.03		$-.57$	.49	$-.08$	
VIII		.27	$-.26$	.01		$-.46$	.43	$-.03$	
IX		.20	$-.21$	$-.01$		$-.33$	.34	.01	
X		.13	$-.14$	$-.01$		$-.23$	.23	0	
XI		.07	$-.07$	0		$-.12$	.12	0	
XII		0	-0	0		-0	0	0	
'XI'		$-.07$	.07	0		.12	$-.11$	0	
'X'		$-.13$	.13	0		.23	$-.22$	0	

Member D_1' $(4-5)$

Load at	Primary	END 4				END 5			
	Stress	V	H	S	$\%$	V	H	S	$\%$
II′	560.2	3.24	.14	3.38	.60	-3.31	.22	-3.09	1.55
I′	840.2	3.68	.07	3.75	.67	-3.79	.11	-3.68	.65
0		.98	-0	.98		$-.99$	-0	$-.99$	
I		$-.11$	$-.08$	$-.19$		1.95	$-.12$	1.83	
II		.18	$-.16$	.34		1.04	$-.24$	.80	
III		.27	$-.23$	.04		.81	$-.35$	.46	
IV		$-.46$	$-.29$	$-.75$		1.25	$-.44$	.81	
V		$-.20$	$-.33$	$-.53$		1.23	$-.51$	.72	
VI		.10	$-.45$	$-.35$		.69	$-.54$	.15	
VII		.09	$-.33$	$-.24$		.58	$-.51$	.07	
VIII		.07	$-.29$	$-.22$		.46	$-.44$	.02	
IX		.05	$-.23$	$-.18$		.35	$-.35$	0	
X		.04	$-.16$	$-.12$		.23	$-.24$	$-.01$	
XI		.02	$-.08$	$-.06$		.12	$-.12$	0	
XII		0	-0	0		0	0	0	
'XI'		$-.02$	.07	.05		$-.12$	.11	$-.01$	
'X'		$-.04$	.14	.10		$-.23$	.22	$-.01$	

Member D_1 $(5-7)$

Load at	Primary	END 5				END 7			
	Stress	V	H	S	$\%$	V	H	S	$\%$
II′	-56.6	2.39	-1.26	1.13	2.00	-1.87	1.44	$-.43$	.76

I′	−26.3	5.17	−.64	4.53	1.72	−7.28	.73	−6.55	2.48
0	−4.0	−.33	.02	−.31	7.75	.65	−.02	.64	16.00
I	847.3	−4.71	.72	−3.99	.47	4.50	−.79	3.71	.44
II	585.9	−4.23	1.39	−2.84	.48	5.23	−1.58	4.65	.79
III	336.0	−3.45	2.01	−1.44	.41	4.10	−2.30	1.80	.52
IV	106.1	−3.33	2.56	−.77	.73	3.65	−2.92	.73	.69
V	−83.7	−3.33	2.94	−.39	.47	3.73	−3.36	.37	.44
VI	−216.4	−2.60	3.10	.50	.24	3.01	−3.54	−.53	.25
VII	−270.4	−2.15	2.94	.78	.29	2.51	−3.36	−.85	.31
VIII	−267.4	−1.73	2.55	.83	.31	2.01	−2.92	−.91	.34
IX	−224.1	−1.30	2.01	.71	.22	1.51	−2.30	−.79	.35
X	−161.0	−.86	1.39	.53	.33	1.01	−1.58	−.57	.35
XI	−86.3	−.43	.72	.29	.33	.50	−.79	−.29	.33
XII	−4.0	−0	.02	.02	.50	0	−.02	−.02	.50
'XI'	65.8	.43	−.64	−.21	.31	−.50	.73	.23	.34
'X'	129.7	.86	−1.26	−.40	.31	−1.01	1.44	.43	.33

Member $D_2(8-9)$

Load at	Primary	END 8				END 9			
	Stress	V	H	S	%	V	H	S	%
II′	−30.8	3.11	−3.12	−.01	.03	−2.51	2.54	.03	.09
I′	−13.0	−.19	−1.58	−1.77	14.30	.99	1.29	2.28	17.50
0	−4.7	0	.04	.04	.85	0	−.03	−.03	.64
I	−10.1	−5.42	1.77	−3.65	36.10	3.50	−1.45	2.05	20.20
II	879.2	−9.73	3.43	−6.30	.72	7.42	−2.80	4.62	.52

III	564.5	-8.77	4.97	-3.80	.67	8.40	-4.06	4.34	.77
IV	273.1	-16.31	6.32	-9.99	3.64	15.52	-5.13	10.39	3.78
V	30.0	-7.13	7.27	.14	.47	5.92	-5.94	-.02	.06
VI	-145.9	-5.85	7.67	1.82	1.26	5.74	-6.25	-.51	.35
VII	-228.3	-4.87	7.27	2.40	1.06	4.78	-5.94	-1.16	.51
VIII	243.4	-3.90	6.32	2.42	.99	3.83	-5.13	-1.30	.53
IX	210.4	-2.92	4.97	2.03	.97	2.87	-4.06	-1.19	.51
X	-153.9	-1.95	3.43	1.48	.96	1.91	-2.80	-.89	.58
XI	-83.9	-.98	1.77	.79	.94	.95	-1.45	-.50	.59
XII	-4.7	-0	.04	.04	.85	0	-.03	-.03	.64
'XI'	60.8	.98	-1.58	-.60	.99	-.95	1.29	.34	.56
'X'	116.7	1.95	-3.12	-1.17	1.00	-1.91	2.54	.63	.54

Member D_3 (10 – 11)

Load at	Primary Stress	END 10				END 11			
		V	H	S	%	V	H	S	%
II′	14.0	2.69	-3.11	-.42	3.00	-1.54	2.58	2.04	1.44
I′	9.9	1.94	-1.57	.37	3.71	-1.45	1.30	-.15	1.52
0	-5.8	0	.04	.04	.69	0	-.03	-.03	.52
I	-38.4	-.25	1.77	1.52	3.94	.14	-1.47	-1.33	3.70
II	-59.7	-7.41	3.42	-3.99	6.69	4.65	-2.84	1.81	3.02
III	627.6	-8.70	4.96	-3.74	.45	5.40	-4.12	1.28	.15
IV	543.3	-9.40	6.27	-3.13	.58	8.47	-5.23	3.24	.60
V	218.3	-6.28	7.25	.97	.44	5.39	-6.02	-.63	.29
VI	-23.9	-6.41	7.65	1.24	5.15	6.02	-6.35	-.27	1.14

	Primary Stress	V	H	S	%	V	H	S	%
VII	−151.3	−5.34	7.25	1.91	1.31	5.02	−6.02	−1.00	.66
VIII	−195.8	−4.27	6.27	2.00	1.02	4.02	−5.23	−1.01	.52
IX	−181.1	−3.21	4.96	1.75	.97	3.01	−4.12	−1.01	.53
X	−137.5	−2.14	3.42	1.28	.93	2.01	−2.84	−.83	.60
XI	−77.3	−1.07	1.77	.70	.91	1.00	−1.47	−.47	.61
XII	−5.8	−0	.04	.04	.69	0	−.03	−.03	.52
'XI'	48.6	1.07	−1.57	−.50	1.01	−1.00	1.30	.30	.61
'X'	91.8	2.14	−3.11	−.97	1.05	−2.01	2.58	.57	.62

Member D_4 (12 – 13)

Load at	Primary Stress	END 12				END 13			
		V	H	S	%	V	H	S	%
II′	98.8	2.40	−2.65	−.25	.25	−.49	1.15	.66	.67
I′	53.0	1.00	−1.34	−.34	.64	−.03	.58	.55	1.01
0	−7.1	0	.03	.03	.42	0	−.01	−.01	.14
I	−88.1	−1.45	1.51	.06	.07	.38	−.65	−.27	.30
II	−155.1	−1.15	2.92	1.77	1.18	3.70	−1.27	2.43	1.58
III	−201.5	−7.63	4.23	−3.30	1.65	2.79	−1.84	.95	.47
IV	995.6	−8.20	5.38	−2.62	.28	1.91	−2.34	−.43	.04
V	554.3	−8.17	6.18	−1.99	.36	4.96	−2.69	2.27	.41
VI	215.0	−7.28	6.52	−.76	.35	4.36	−2.84	1.52	.71
VII	17.4	−5.86	6.18	.32	1.84	3.24	−2.69	.55	3.15
VIII	−78.3	−4.68	5.38	.70	.90	2.59	−2.34	.25	.32
IX	−100.8	−3.51	4.23	.72	.72	1.95	−1.84	.11	.11
X	−88.0	−2.34	2.92	.58	.66	1.30	−1.27	.03	.03

XI	−54.5	−1.17	1.51	.34	.62	.65	−.65	0	0
XII	−7.1	−0	.03	.03	.42	0	−.01	−.01	.14
'XI'	19.4	1.17	−1.34	−.17	.87	−.65	.58	−.10	.52
'X'	31.7	2.34	−2.65	−.31	.98	−1.30	1.15	−.15	.47

Member $D_5(14-15)$

Load at	Primary	END 14				END 15			
	Stress	V	H	S	%	V	H	S	%
II′	243.5	1.91	−2.32	−.41	.17	1.73	−.83	.90	.27
I′	125.5	.98	−1.18	−.20	.16	.90	−.42	.48	.38
0	−7.5	0	.03	.03	.40	0	.01	.01	.13
I	−162.2	−.88	1.34	.46	.28	−.90	.47	−.33	.20
II	−302.4	−2.11	2.56	.45	.15	−1.61	.91	−.70	.23
III	−421.1	−2.01	3.72	1.71	.41	−2.97	1.32	−1.65	.39
IV	−502.7	−7.47	4.73	−2.74	.53	−1.68	1.68	−0	0
V	1070.5	−7.11	5.43	−1.68	.16	−2.49	1.93	−.56	.05
VI	645.7	−5.16	5.72	.56	.09	−.50	2.03	1.53	.23
VII	368.9	−5.25	5.43	.18	.05	.68	1.93	2.61	.71
VIII	198.8	−4.33	4.73	.40	.20	.82	1.68	2.50	1.26
IX	105.0	−3.25	3.72	.53	.50	.62	1.32	1.94	1.84
X	48.4	−2.17	2.56	.39	.80	.41	.91	1.32	2.74
XI	13.2	−1.09	1.34	.25	1.91	.20	.47	.67	5.10
XII	−7.5	0	.03	.03	.39	0	.01	.01	.13
'XI'	−49.9	1.09	1.18	−.09	.18	−.20	−.42	−.62	1.24
'X'	−107.3	2.17	−2.32	−.05	.05	−.41	−.83	−1.24	1.16

Member D_6 (16 – 17)

| Load at | Primary Stress | END 16 | | | | END 17 | | | |
		V	H	S	%	V	H	S	%
II′	344.0	1.63	-1.26	.37	.11	4.50	-3.66	.84	.24
I′	173.8	.79	-.64	.15	.09	2.25	-1.85	.40	.23
0	-3.6	0	.02	.02	.55	0	.05	.05	1.39
I	-191.7	-.80	.72	-.08	.04	-2.25	2.08	-.17	.09
II	-372.8	-1.53	1.38	-.17	.05	-4.60	4.02	-.48	.13
III	-543.3	-2.44	2.01	-.43	.08	-6.75	5.83	-.92	.17
IV	-695.7	-2.67	2.55	-.12	.02	-8.93	7.36	-1.37	.22
V	-810.9	-7.95	2.94	-5.01	-.50	-7.80	8.51	.71	.09
VI	1080.4	-2.26	3.10	.84	.08	-16.15	8.99	-7.16	.66
VII	822.0	-1.30	2.94	1.64	.20	-10.25	8.51	-1.74	.21
VIII	610.6	-1.64	2.55	.91	.15	-7.23	7.36	.13	.02
IX	436.4	-1.53	2.01	.48	.11	-5.03	5.33	.80	.18
X	280.3	-1.02	1.38	.36	.13	-5.35	4.02	.67	.24
XI	134.9	-.51	.72	.21	.16	-1.67	2.08	.41	.30
XII	-3.6	-0	.02	.02	.53	-0	.05	.05	1.39
'XI'	-152.8	.51	-.64	-.13	.09	1.67	-.85	.82	.53
'X'	-309.1	1.02	-1.26	-.24	.08	3.35	-3.66	-.31	.10

Member V_2' (1 – 2)

| Load at | Primary Stress | END 1 | | | | END 2 | | | |
		V	H	S	%	V	H	S	%
II′	-1000.0	5.80	-.27	5.53	.55	-3.79	.09	-3.70	.37

I′	1.30	− .14	1.16	− .21	.05	− .16
0	.17	0	.17	.41	− 0	.41
I	− .31	.15	− .16	.44	− .06	.38
II	− .29	.30	.01	.29	− .11	.18
III	.20	.43	.23	.11	− .16	− .05
IV	− .63	.55	− .08	.75	− .21	.54
V	.93	.63	− .30	.65	− .23	.42
VI	− .03	.67	.64	− .20	− .24	− .04
VII	− .02	.63	.61	− .17	− .23	− .05
VIII	− .02	.55	.53	− .13	− .21	− .08
IX	− .01	.43	.42	− .10	− .16	− .06
X	− .01	.30	.29	− .07	− .11	− .04
XI	− 0	.15	.15	− .04	− .06	− .02
XII	− 0	0	0	− 0	0	0
'XI'	0	− .14	− .14	.04	.05	.09
'X'	.01	− .27	− .27	.07	.09	.16

Member V_1' (3 − 4)

Load at	Primary	END 3				END 4			
	Stress	V	H	S	%	V	H	S	%
II′	− 727.3	4.95	.41	5.36	.74	− 4.92	− .49	− 5.41	.75
I′	− 1000.0	4.58	.21	4.79	.48	− 5.43	− .25	− 5.68	.57
0		− .22	− .01	− .23		− .94	.01	− .93	
I		1.64	− .24	1.40		− 2.11	.28	− 2.03	
II		1.12	− .46	.66		− 1.59	.54	− 1.05	

III	1.28	− .66	.62	− 2.00	.78	− 1.22
IV	.91	− .84	.07	− .94	.99	.05
V	.20	− .97	− .77	− .68	1.15	.47
VI	1.00	− 1.02	− .02	− 1.05	1.21	.05
VII	.83	− .97	− .14	− .89	1.15	.26
VIII	.66	− .84	− .18	− .71	.99	.28
IX	.50	− .66	− .16	− .53	.78	.25
X	.33	− .46	− .13	− .35	.54	.19
XI	.17	− .24	− .07	− .18	.28	.10
XII	0	− .01	− .01	− 0	.01	.01
'XI'	− .17	.21	.04	.18	− .25	− .07
'X'	− .33	.41	.08	.35	− .49	− .14

Member $V_0(5-6)$

Load at	Primary Stress	END 5				END 6			
		V	H	S	%	V	H	S	%
II′	− 449.1	2.11	− .85	1.26	.28	− .61	.73	.08	.02
I′	− 809.7	1.32	− .43	.89	.11	1.22	.37	1.59	.19
0	3.5	.07	.01	.08	2.28	.48	− .01	.47	1.35
I	− 786.3	− 4.31	.48	− 3.83	.51	4.05	− .41	3.64	.48
II	− 522.9	− 2.63	.93	− 1.70	.32	2.36	− .80	1.56	.30
III	− 299.9	− 2.21	1.35	− .86	.28	1.61	− 1.16	.45	.15
IV	− 94.5	− 2.08	1.72	− .24	.25	.92	− 1.47	− .55	.57
V	74.8	− 1.76	1.98	.22	.29	.83	− 1.70	− .87	1.16
VI	193.2	− 1.41	2.08	.87	.35	.89	− 1.47	− .58	.30

VII	241.4	−1.18	1.98	.80	.32	.74	−1.70	−.96	.39
VIII	238.9	−.94	1.72	.78	.32	.59	−1.47	−.88	.37
IX	200.1	−.71	1.35	.74	.37	.44	−1.16	−.72	.36
X	143.7	−.47	.98	.51	.36	.30	−.80	−.50	.35
XI	77.1	−.24	.48	.24	.31	.15	−.41	−.26	.34
XII	3.5	−0	.01	.01	.29	0	−.01	−.01	.29
'XI'	−59.7	.24	−.43	−.19	.32	−.15	.37	.22	.37
'X'	−115.7	.47	−.85	−.38	.33	−.30	.73	.43	.37

Member $V_1(8-7)$

Load at	Primary Stress	END 8				END 7			
		V	H	S	%	V	H	S	%
II′	25.3	3.69	−3.60	.09	.35	−3.60	3.44	−.16	.63
I′	10.7	5.12	−1.82	3.30	31.00	−9.80	1.74	−8.06	75.00
0	3.9	−.48	.04	−.44	11.30	.97	−.04	.93	23.80
I	−991.8	−7.60	2.04	−5.56	.56	8.03	−1.93	6.08	.61
II	−722.1	−12.52	3.95	−8.57	1.18	11.85	−3.78	8.07	1.11
III	−463.6	−9.75	5.74	−4.01	.87	9.31	−5.48	3.83	.83
IV	−224.4	−8.73	7.29	−1.44	.64	7.92	−6.97	.95	.42
V	−24.6	−8.58	8.38	−.20	.81	8.12	−8.01	.11	.45
VI	119.8	−6.50	8.83	2.33	1.94	6.40	−8.45	−2.05	1.70
VII	187.5	−5.41	8.38	2.83	1.51	5.33	−8.01	−2.68	1.43
VIII	199.9	−4.33	7.29	2.96	1.48	4.27	−6.97	−2.70	1.34
IX	172.8	−3.25	5.74	2.49	1.44	3.20	−5.48	−2.28	1.32
X	126.5	−2.17	3.95	1.78	1.40	2.13	−3.78	−1.65	1.30

XI	68.9	-1.09	2.04	.95	1.38	1.07	-1.95	-.88	1.28
XII	3.9	-0	.04	.04	1.01	0	-.04	-.04	1.01
'XI'	-49.9	1.09	-1.82	-.73	1.46	-1.07	1.74	.67	1.34
'X'	-96.0	2.17	-3.60	-1.57	.54	-2.13	3.44	1.31	1.36

Member $V_2(10-9)$

Load at	Primary	END 10				END 9			
	Stress	V	H	S	%	V	H	S	%
II'	-10.0	4.85	-4.30	.55	5.50	-5.45	4.53	-.92	9.20
I'	-7.1	1.45	-2.18	-.73	10.30	-.14	2.29	2.15	30.30
0	4.1	0	.05	.05	1.21	0	-.06	-.06	1.46
I	27.4	-1.46	2.44	.98	3.67	2.71	-2.58	.13	.47
II	-957.4	-11.56	4.73	-6.83	.71	12.93	-4.98	7.95	.83
III	-662.5	-17.85	6.86	-10.99	1.66	18.00	-7.21	10.79	1.62
IV	-388.0	-13.61	8.73	-4.88	1.26	14.20	-9.17	5.03	1.30
V	-156.0	-9.23	10.02	.79	.51	10.35	-10.55	-.20	.13
VI	17.0	-9.72	10.56	.84	4.95	9.90	-11.15	-1.25	7.30
VII	107.9	-8.10	10.02	1.92	1.04	8.26	-10.55	-2.29	2.28
VIII	139.8	-6.48	8.73	2.25	1.61	6.60	-9.17	-2.57	1.83
IX	129.2	-4.87	6.86	1.99	1.54	4.96	-7.21	-2.25	1.74
X	98.2	-3.24	4.73	1.49	1.52	3.30	-4.98	-1.68	1.71
XI	55.1	-1.62	2.44	.82	1.49	1.65	-2.58	-.93	1.69
XII	4.1	-0	.05	.05	1.22	0	-.06	-.06	1.36
'XI'	-34.8	1.62	-2.18	-.56	1.60	-1.65	2.29	.64	1.83
'X'	-65.5	3.24	-4.30	-1.06	1.62	-3.30	4.53	1.23	1.88

Member $V_3(12-11)$

Load at	Primary Stress	END 12				END 11			
		V	H	S	%	V	H	S	%
II′	−57.7	6.82	−8.24	−1.42	2.47	−7.28	8.72	1.44	2.50
I′	−30.9	3.66	−4.17	−.51	1.65	−4.46	4.41	−.05	.16
0	4.2	0	.10	.10	2.37	0	−.11	−.11	2.62
I	51.4	−3.67	4.69	1.02	1.99	3.62	−4.95	−1.33	2.39
II	90.6	−6.07	9.08	3.01	3.31	8.29	−9.58	−1.29	1.43
III	−882.2	−18.70	13.15	−4.55	.52	21.35	−13.90	7.45	.84
IV	−581.7	−29.30	16.72	−12.58	2.16	29.90	−17.70	12.20	2.09
V	−323.8	−24.10	19.21	−4.89	1.51	24.80	−20.40	4.40	1.35
VI	−125.7	−20.25	20.30	.05	.04	20.90	−21.40	−.60	.40
VII	−10.1	−16.95	19.21	2.26	22.30	17.43	−20.40	−2.97	29.30
VIII	45.8	−13.56	16.72	3.16	.69	13.95	−17.70	−3.75	.82
IX	58.9	−10.17	13.16	2.98	5.03	10.45	−13.90	−3.45	5.83
X	51.4	−6.78	9.08	2.30	4.47	6.98	−9.58	−3.40	6.61
XI	31.8	−3.39	4.69	1.30	4.07	3.49	−4.95	−1.46	4.56
XII	4.2	−0	.10	.10	2.38	0	−.11	−.11	2.62
'XI'	−11.3	3.39	−4.17	−.78	6.90	−3.49	4.41	.92	8.20
'X'	−18.5	6.78	−8.24	−1.46	7.90	−6.98	8.72	1.74	9.40

Member $V_4(14-13)$

Load at	Primary Stress	END 14				END 13			
		V	H	S	%	V	H	S	%
II′	−115.7	7.61	−11.30	−3.69	5.19	−8.47	11.93	3.46	2.98

Load at	Primary Stress	V	H	S	%	V	H	S	%
I'	−59.7	3.71	−5.72	−2.01	3.38	−3.95	6.05	2.10	3.51
0	3.5	0	.14	.14	4.00	0	−.15	−.15	4.23
I	77.1	−3.69	6.43	2.74	3.55	4.21	−6.81	−2.60	3.26
II	143.7	−7.77	12.50	4.73	3.29	8.08	−13.17	−5.09	3.55
III	200.1	−9.89	18.08	8.19	4.08	12.20	−19.10	−6.90	3.45
IV	−761.1	−29.50	23.00	−6.50	.85	32.90	−24.25	8.65	1.14
V	−508.6	−42.10	26.40	−15.70	3.08	44.00	−27.90	16.10	3.35
VI	−306.8	−32.35	27.80	−4.55	1.48	34.60	−29.40	5.20	1.69
VIII	−175.2	−27.60	26.40	−1.20	.69	28.60	−27.90	.70	.40
IX	−94.5	−22.00	23.00	1.00	1.05	22.80	−24.25	−1.45	1.54
X	−49.9	−16.50	18.08	1.58	3.15	17.10	−19.10	−2.00	4.00
XI	−6.3	−3.50	6.43	.93	14.80	5.71	−6.81	−1.10	17.40
XII	3.5	−0	.14	.14	4.00	0	−.15	−.15	4.30
'XI'	23.7	5.50	−5.72	−.22	.93	−5.71	6.05	1.66	7.00
'X'	50.9	11.01	−11.30	−.21	.41	−11.42	11.93	.51	1.00

Member $V_5(16-15)$

Load at	Primary Stress	END 16				END 15			
		V	H	S	%	V	H	S	%
II'	−162.5	6.73	−10.58	−3.85	2.36	−7.91	12.73	4.82	2.96
I'	−82.1	3.30	−5.36	−2.06	2.50	−3.93	6.43	3.50	4.25
0	1.6	0	.13	.13	8.10	0	−.16	−.16	10.00
I	89.8	−3.31	5.78	2.47	2.74	3.93	−7.22	−3.29	3.65
II	174.9	−6.53	11.65	5.12	2.83	7.80	−14.00	−6.26	3.57
III	255.7	−10.15	16.90	6.75	2.61	12.05	20.30	−8.25	3.21

	Primary Stress	V	H	S	%	V	H	S	%
IV	328.4	−11.20	21.50	10.30	3.12	14.88	−25.85	−10.97	3.33
V	−604.8	−33.50	24.70	−8.80	1.46	43.10	−29.70	14.40	2.38
VI	−469.6	−57.25	26.05	−31.20	6.63	58.45	−31.25	27.20	5.77
VII	−355.6	−40.70	24.70	−16.00	4.48	47.20	−29.70	17.50	4.88
VIII	−264.2	−33.33	21.50	−11.83	4.48	38.60	−25.85	12.75	4.78
IX	−188.8	−25.38	16.90	−8.48	4.48	29.20	−20.30	8.90	4.72
X	−121.3	−16.88	11.65	−5.23	4.33	19.46	−14.00	5.46	4.51
XI	−58.3	−8.44	5.78	−2.66	4.60	9.73	−7.22	2.51	4.31
XII	1.6	−0	.13	.13	8.11	0	−.16	−.16	10.00
'XI'	66.0	8.44	−5.36	3.08	4.68	−9.73	6.43	−3.30	5.00
'X'	133.7	16.88	−10.58	6.30	4.74	−19.46	12.73	−6.73	5.06

Member $V_6(18-17)$

Load at	Primary Stress	END 18				END 17			
		V	H	S	%	V	H	S	%
II′		−3.93	0	−3.93		7.83	0	7.83	
I′		−1.97	0	−1.97		3.92	0	3.92	
0		0	0	0		0	0	0	
I		1.97	0	1.97		−3.92	0	−3.92	
II		3.93	0	3.93		−7.83	0	−7.83	
III		5.91	0	5.91		−11.75	0	−11.75	
IV		7.40	0	7.40		−14.33	0	−14.33	
V		11.84	0	11.84		−16.40	0	−16.40	
VI	−1000.0	0	0	0	0	0	0	0	0
VII		−11.85	0	−11.85		16.40	0	16.40	

VIII	−7.40	0	−7.40	14.33	0	14.33
IX	−5.91	0	−5.91	11.75	0	11.75
X	−3.93	0	−3.93	7.83	0	7.83
XI	−1.97	0	−1.97	3.92	0	3.92
XII	−0	0	0	0	0	0
'XI'	1.97	0	1.97	−3.92	0	−3.92
'X'	3.93	0	3.93	−7.83	0	−7.83

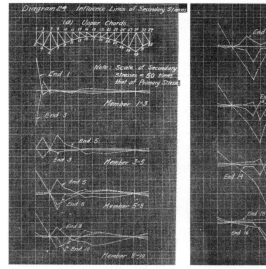
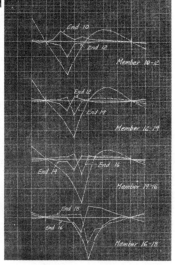

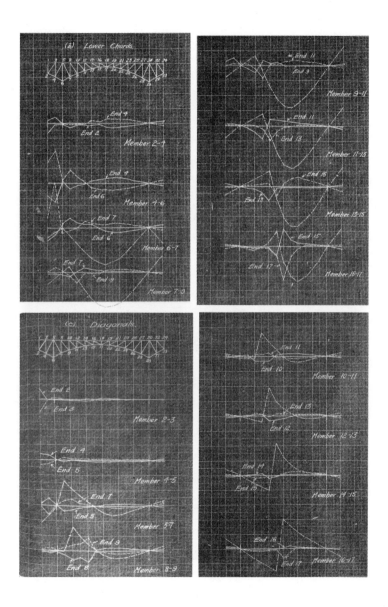

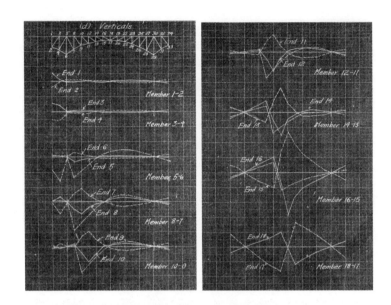

CHAPTER XXXVIII. CONCLUSION

The most important conclusion that can be drawn from the above investigation is that the secondary stresses in two-hinged spandrel-braced arches are very low. This is largely due to the fact that the stresses due to vertical and horizontal forces are about equal to each other and are generally opposite in sign. A glance at Table 64 will show that the percentage of secondary stresses in respect to primary stresses generally runs below 10% , sometimes as low as 1/10% . Only in one case has it been as high as 75% (V_1) but this is due to the exceedingly low primary stresses. These figures might not be absolutely correct because the stresses being small, an error made in the third decimal places of the expression $2\phi_a + \phi_n - 3\psi_{an}$ is sufficient to give a large discrepancy in the final result. However, there is no doubt that two-hinged spandrel-braced arches are very free from secondary stresses because the figures in Table 7 look very uniform and are not likely to be in error to any considerable extent.

Besides that mentioned above, the following conclusions may be drawn:

(1) The secondary stresses in web members are comparatively very high and not negligible like those in simple trusses.

(2) The greatest stress in the end of a member occurs when the load is at that end. This is more plainly seen in chord members but is found irregular in diagonals probably due to the low ratio of $\dfrac{2C}{L}$ and the effect of chord members.

(3) The maximum secondary stresses for most of the members occur under the same positions of live loads as those for the maximum primary stresses.

(4) Secondary stresses in members of cantilever arm are very low for loads on the arch span proper.

(5) Secondary stresses in vertical posts are very high espectially those near the center of the truss due to the shortness of length as compared with the verticals of a simple truss.

CONCLUSION

Before concluding this theses the writer wishes to acknowledge that the work has not been done as completely as it was planned mainly due to the fact that the time at his disposal has not been properly utilized. Nevertheless, the treatment of the subject, brief as thorough, has included all that is of vital importance and enables the writer to arrive at many conclusions which are worth while the painstaking task. To capitulate the following are some of the more important conclusions, for a detailed account of which reference is to be made under the proper headings.

(1) Secondary stresses in two-hinged spandrel braced arches are very low, rarely exceed 10% of primary stresses.

(2) Two-hinged spandrel-braced arches are lighter than simple trusses, both of the same span.

(3) Stresses in arches with cantilever arms are about the

same as those in arches without cantilever arms, but the balanced type has a superior advantage under full live loads.

(4) A slight increase in weight is resulted if the two-hinged arch is erected, as having a crown hinge at the centre of span.

(5) Wind load stresses may be neglected in the design if the unit stresses are allowed to increase 25% above normal.

(6) Effect of web members may be neglected in the preliminary design.

Respectfully submitted, *Thomson Mao.*

博士论文

茅以升全集 ❷

SECONDARY STRESSES IN BRIDGE TRUSSES

Introducing

The Graphic Method of Deformation Contour

and

Its Analytic Solution With Scientific Arrangement

of Computations

An Abstract of a THESIS Presented to the

Division of Science and Engineering

for the degree of

DOCTOR OF ENGINEERING

BY

THOMSON EASON MAO

Carnegie Institute of Technology

Pittsburgh, Pa. , U. S. A.

1919

TO

······························

WITH COMPLIMENTS OF

THE AUTHOR

······························

THIS PAPER IS

RESPECTFULLY DEDICATED TO

HENRY S. JACOBY

UNDER WHOSE INSTRUCTIONS

THE AUTHOR

FIRST ACQUIRED THE KNOWLEDGE OF

SECONDARY STRESSES

PREFACE

This paper is an abstract from a thesis entitled "Secondary Stresses in Framed Structures" presented by the writer to the Division of Science and Engineering, Carnegie Institute of Technology, as a partial requirement for the degree of Doctor of Engineering. In writing the original thesis three objects were kept in view: (1) to treat the subject of secondary stresses in a manner more logical, comprehensive and practical than is found elsewhere, (2) to introduce three new methods, two graphic and one analytic, for analyzing the secondary stresses, and (3) to investigate the effects of secondary stresses upon the design. In this paper two of the new methods are reproduced, with the materials entirely re-arranged and re-written. This is followed by a set of principles to be observed in reducing secondary stresses in bridge design.

The methods for computing secondary stresses have been greatly improved in recent years. Two objections, however, still remain: First, the amount of time involved is often excessive, and second, the lack of a systematic checking device by which the correctness of the various steps of procedure may be ensured. While there are numerous other defects, these two alone are generally sufficient to reduce their practical utility. Ever since the beginning of 1917 when the writer undertook the analysis of secondary stresses in a two-hinged arch, the results of which have been published in the Transactions of the American Society of Civil Engineers, vol. 82, p. 1104, it always occurred to him that there must be some method which is not only shorter and less cumbersome than the current ones but also admits of a check. For two and a half years he had worked on the subject almost incessantly, striving to find some new method that will accomplish both the above mentioned results. At last, much to his satisfaction, the graphic method of deformation contour was obtained, which not only takes less time and checks itself, but also gives remarkably accurate results. Along with this method, almost contemporaneously, the analytic method was evolved. As these methods are new in their field the writer has no hesitation to present them to the engineering world and it is his sincere hope that their usefulness be actually tested by further investigators.

In conclusion, the writer wishes to express his indebtedness to Professors H. R. Thayer and F. M. McCullough for suggestions and improvements which greatly enhance the value of the paper.

T. E. M.

Pittsburgh, Pa. ,

December, 1919.

The above was written at a time when the abstract was to appear in the papers of an engineering society. On account of the author's returning home in China the arrangement was not carried out and the printing of the abstract was delayed. After corresponding with the Carnegie Institute of Technology for nearly a year the decision was finally reached to issue this paper in pamphlet form. This explanation is necessary because of some slight changes made in the original preface.

T. E. M.

CHINESE GOVERNMENT ENGINEERING COLLEGE,

Tangshan, China,

December, 1920.

CONTENTS

PREFACE

CONTENTS

---

①　此为原论文页码。论文中提到的页码均为原打印稿页码。

INTRODUCTION

I . THE NATURE OF SECONDARY STRESS

In an articulated truss the members meeting at panel points may be connected either by pins or by stiff gusset plates. In the former arrangement the members are more or less free to turn about their joints and the angles between them may vary with the deformation of the truss. In the latter case, however, this is impossible, as the relative angles between the members must remain constant. This "Rigidity of Joints" is the principal cause of the so-called secondary stresses.

Let a bridge truss with rigid joints be loaded in any manner. The various members will then change in length and the panel points will deflect. As the truss is made of triangles and a triangle cannot retain its shape after the sides have changed in length, the angles made by the members at the joints must also vary to accommodate the change. But the members cannot conform to this angle because the joints are rigid. The triangles must, therefore, lose their original forms and the sides must bend, as a result of the rotation of joints. Bending moments are thus produced in the members, reaching a maximum near the joints. [1] The fibre stresses resulting therefrom constitute the secondary stresses due to rigidity of joints.

II. THE TWO FUNDAMENTAL EQUATIONS

To analyze secondary stresses, certain conditions must be known that will give equations sufficient for the total unknowns. As a bridge is a static structure, one condition must be derived from the equilibrium of forces. It being also an elastic structure, another condition may be obtained from the elastic properties of the material of which the bridge is built. These two conditions give rise to two fundamental equations which, as will be seen

---

[1] Taking the bending of the members into account, the maximum moments in compression members do not occur at the ends but somewhere between the ends.

later, are sufficient for solving the total unknowns.

1. The Equilibrium Equation

Since the truss is always in static equilibrium the resulting moment about any panel joint must be zero. [1] Hence, if the members 12, 13, 14, etc. of a truss meet at joint 1, and if M_{12}, M_{13}, M_{14}, etc. are the moments of the respective members about joint 1, then,

$$M_{12} + M_{13} + M_{14} + \ldots = 0,$$

Or, in general, $\quad \sum M$ about any joint $= 0$.

Let f = secondary stress produced by M,

$\quad I$ = moment of inertia of the member,

$\quad y$ = extreme fiber distance from the neutral axis for the same member,

$\quad S$ = section modulus of the member $= \dfrac{I}{y}$,

Then, $\quad M = f \dfrac{I}{y} = fS$, and the above equation reduces to

$$\sum fS \text{ around any joint} = 0, \tag{1}$$

which is the first fundamental equation known as the

---

[1] Eccentric moment considered as external.

"equilibrium equation."

2. The Deformation Equation

As noted before, in a truss with rigid joints the deformations in members can not be taken care of by the change of angles, which the members make at the joints. On the other hand, these angles remain constant, and provision for deformations is made in some other way. Since the members are of elastic material, the simplest and easiest way would be to bend them; but they cannot bend without being subject to some outside influence and this outside influence is what produces the internal stress. In other words, some work must be performed on the member to produce this bending and as a consequence some corresponding internal work must be set up. Take, for illustration (Fig. 1) a beam, 12, subject to the influence of moment M which increases from 0 at 1 to M_2 at 2. (This is equivalent to a force equal to $\dfrac{M_2}{L}$ placed at end 1, with end 2 fixed). The effect of M on the beam is to rotate end 2 through an angle T_{21}, and end 1, through an angle T_{12}. These rotations are brought about at the expense of an external work equal in amount to $\dfrac{1}{2}M_2T_{21}$. This induces an internal work in the beam equal to

$$\frac{1}{2}\int_0^L \frac{M^2 \delta x}{EI} = \frac{1}{2}\frac{1}{EI}\int_0^L \left(M_2 \frac{x}{L}\right)^2 \delta x = \frac{1}{2}\frac{M_2^2 L}{3EI} = \frac{1}{2}M_2 T_{21} ,$$

or
$$T_{21} = \frac{M_2 L}{3EI}. \tag{a}$$

That is, the end 2 of beam 12 rotates through an angle T_{21} at the expense of an external moment M_2.

Similarly, the bending in truss members due to rigidity of joints is also brought about at the expense of external moments which become internal moments when the joint is considered as a whole. This bending may be considered as the result of two different rotations: the first due to the deformation of truss members and the other due to the rigidity of joints. Let 12, 13, 14, 15 and 16 be five members meeting at joint 1, which is rigid (Fig. 2). After loading, on account of the deformations in members, the axis lines 12, 13, 14, etc., would have been displaced to 12′, 13′, etc. had the joint been a frictionless pin. The angles, 212′ = H_{12}, 313′ = H_{13}, etc., thus formed, are functions of the deformations and are different for different members. They could not, therefore, be actually realized in a truss with rigid joints, where a member is prevented from rotating relatively to the others. On the other hand, all the members must rotate through the same angle and the total rotations of members 12, 13, 14, etc., are not H_{12}, H_{13}, etc., but $B_1 — H_{12}$, $B_1 — H_{13}$, etc., where B_1 is the rotation of joint 1—the same for all members. This same result may also be obtained by considering: first, the axes $12'$, $13'$, etc., brought back to their respective original positions 12, 13, etc., and then, simultaneously, all the axes turned through a common angle, B_1.

These rotations, as mentioned before, must be produced at the expense of moments, and the work thus performed on members 12, 13, etc., is

$$W_{12} = \frac{1}{2}M_{12}(B_1 - H_{12}),$$

$$W_{13} = \frac{1}{2}M_{13}(B_1 - H_{13}), \text{etc.}$$

Since B_1 is the same for all the members at joint 1,

$$\frac{W_{12}}{M_{12}} - \frac{W_{13}}{M_{13}} = \frac{1}{2}(H_{13} - H_{12}),$$ (b)

$$\frac{W_{13}}{M_{13}} - \frac{W_{14}}{M_{14}} = \frac{1}{2}(H_{14} - H_{13}), \text{etc.}$$

Now, $H_{12} - H_{13}$ = angles $213 - 2'13'$ = change of angle between members 12 and 13 as a result of the loading. This change of angle is a function of the primary stresses in the members and also the proportions of the truss. If P_{12}, P_{13}, and P_{23} are the *unit* primary stresses in sides 12, 13, and 23 of triangle 123 and δA_{213}, δA_{132}, and δA_{321} are the changes in angles 213, 132, and 321 of the triangle it can be found by calculus that

$$E\delta A_{213} = (P_{23} - P_{13})\cot 132 + (P_{23} - P_{12})\cot 123,$$

$$E\delta A_{132} = (P_{12} - P_{23})\cot 132 + (P_{12} - P_{13})\cot 213,$$

$$E\delta A_{321} = (P_{13} - P_{12})\cot 213 + (P_{13} - P_{23})\cot 132. \quad (2)$$

As the terms in equation (2) are all known, the change of angles could be easily computed for the truss and $H_{13} - H_{12} = \delta A_{213}$, $H_{14} - H_{13} = \delta A_{314}$, etc. , are therefore known. Substitute these values of δA for H in equation (b):

$$\frac{W_{12}}{M_{12}} - \frac{W_{13}}{M_{13}} = \frac{1}{2}\delta A_{213},$$

$$\frac{W_{13}}{M_{13}} - \frac{W_{14}}{M_{14}} = \frac{1}{2}\delta A_{314}, \text{etc.}$$
(c)

Since the expression for internal work includes the total work done on the member, not only by M_{12} at end 1 of member 12 but also by M_{21} at end 2, the total external work on the member is

$$W = W_{12} + W_{21} = \frac{1}{2}(M_{12}T_{12} + M_{21}T_{21}),$$

and the internal work is

$$W = \frac{1}{2}\int_0^L \frac{M_x^2 \delta x}{EI},$$

where (Fig. 3) $M_x = M_a - M_b = M_{12} - \frac{x}{L}(M_{12} + M_{21})$.

Substitute this value of M_x in the above equation and integrate:

$$W = W_{12} + W_{21} = \frac{1}{2}\left\{\frac{1}{EI}\left(\frac{M_{12}^2 L}{3} + \frac{M_{21}^2 L}{3} - \frac{M_{12} M_{21} L}{3}\right)\right\}. \quad (d)$$

It will be seen in equation (d) that the first two terms of the right hand member represent the work done on ends 1 and 2 of beam 12 by M_{12} and M_{21} respectively due to angular rotations caused by the respective moments—by equation (a)—and the last term is the sum of the work done by M_{12} at 1 and M_{21} at 2 due to the angular rotation caused by the moment at the other end. If δ_{12} is the angular rotation at 1 caused by a unit moment at 2 and δ_{21} is the angular rotation at 2 caused by a unit moment at 1, the total angular rotation at 1 due to M_{21} at 2 will be $M_{21}\delta_{12}$ and that at 2 due to M_{12} at 1 will be $M_{12}\delta_{21}$. The work done by M_{12} at 1 due to the action of M_{21} at 2 is therefore $\frac{1}{2}M_{12}M_{21}\delta_{12}$ and that by M_{21} at 2 due to the action of M_{12} at 1 is $\frac{1}{2}M_{21}M_{12}\delta_{21}$. But by

Maxwell's theorem of reciprocal deflections, $\delta_{12} = \delta_{21}$, each being

equal to $\dfrac{L}{6EI}$, and hence these two kinds of work done are equal.

The difference of W_{12} and W_{21} is therefore simply the difference of the first two terms in the right hand side of the above equation and

$$W_{12} - W_{21} = \frac{1}{2}\left\{\frac{1}{EI}\left(\frac{M_{12}^2 L}{3} - \frac{M_{21}^2 L}{3}\right)\right\}. \tag{e}$$

Solving simultaneously equation (d) and (e),

$$W_{12} = \frac{1}{2}\frac{M_{12}L}{6EI}(2M_{12} - M_{21}).$$

Similarly, $\qquad W_{13} = \dfrac{1}{2}\dfrac{M_{13}L}{6EI}(2M_{13} - M_{31}).$

Substituting these values of W in equation (c), there is obtained

$$\frac{L_{12}}{6EI_{12}}(2M_{12} - M_{21}) - \frac{L_{13}}{6EI_{13}}(2M_{13} - M_{31}) = \delta A_{213}.$$

Replacing M by $f S$ where $S = \dfrac{I}{y}$,

$$\frac{L_{12}}{y_{12}}(2f_{12} - f_{21}) - \frac{L_{13}}{y_{13}}(2f_{13} - f_{31}) = 6E\delta A_{213},$$

or, in general, for members lm and lp,

$$\frac{L_{lm}}{y_{lm}}(2f_{lm} - f_{ml}) - \frac{L_{lp}}{y_{lp}}(2f_{lp} - f_{pl}) = 6E\delta A_{plm}, \tag{3}$$

where f and δA carry their own signs.

This is the second fundamental equation known as the "deformation equation."

III. FORMULAS AND EQUATIONS

Before proceeding further, a few words are necessary regarding the signs of the quantities in the deformation equation, (3). To secure a uniform system a positive f will be understood to be due to positive moment, which is taken as counter clockwise in this paper. This means that the value of f, together with its signs, applies to that fiber of the member which is first met with in passing around the joint in a clockwise direction (Fig. 4). For example, in Fig. 5, if M_{12} is positive and M_{13} negative, the top fiber of 12 at 1 will be in tension and that of 13 at 1 will be in compression. Next, consider the signs of δA. As is customary, it will be positive if the angle A is changed to a larger angle, as in Fig. 6, if H_{12} is greater than H_{13} or T_{13} is greater than T_{12}. Since both T_{12} and T_{13} must have the same signs as M_{12} and M_{13}, respectively, they will be positive if they are made in a positive direction; that is, if the dotted lines 12′ and 13′ are moved to 12″

and 13″ in a counter clockwise direction. Hence, if all the quantities carry their own signs, $T_{13} - T_{12} = \delta A$ and $H_{12} - H_{13} = \delta A$. Comparing this with equation (b) from which the deformation equation is derived it will be seen that the sign of $E\delta A$ in equation (3) should be negative for the present convention of signs, and that

$$\frac{L_{lm}}{y_{lm}}(2f_{lm} - f_{ml}) - \frac{L_{lp}}{y_{lp}}(2f_{lp} - f_{pl}) = -6E\delta A_{plm},$$

if the member lm is met with before member lp when passing around joint 1 in a clockwise direction.

Therefore, $\dfrac{L_{lp}}{y_{lp}}(2f_{lp} - f_{pl}) = \dfrac{L_{lm}}{y_{lm}}(2f_{lm} - f_{ml}) + 6E\delta A_{plm}.$

Let $\qquad \dfrac{L_{lp}}{y_{lp}} = U_{lp}, \dfrac{L_{lm}}{y_{lm}} = U_{lm}$ and $6E\delta A_{plm} = K_{plm}.$ $\qquad$ (4)

The above equation then reduces to

$$U_{lp}(2f_{lp} - f_{pl}) = U_{lm}(2f_{lm} - f_{ml}) + K_{plm}. \qquad (5)$$

This modified form of the deformation equation will be used exclusively in the rest of the discussion. To aid the memory it is necessary only to note that the quantity $(2f_{lm} - f_{ml}) U_{lm}$ of a member lm, when added to K_{plm} in a clockwise direction, will give the corresponding quantity $(2f_{lp} - f_{pl}) U_{lp}$ of the next member lp

met with in the same direction.

From equation (5),

$$2f_{lp} - f_{pl} = \frac{U_{lm}}{U_{lp}}(2f_{lm} - f_{ml}) + \frac{K_{plm}}{U_{lp}}, \tag{6}$$

$$f_{pl} = 2f_{lp} - \frac{U_{lm}}{U_{lp}}(2f_{lm} - f_{ml}) - \frac{K_{plm}}{U_{lp}}. \tag{7}$$

Apply equation (6) to member pm, of triangle plm, Fig. 7:

$$2f_{mp} - f_{pm} = \frac{U_{lm}}{U_{pm}}(2f_{ml} - f_{lm}) + \frac{K_{lmp}}{U_{pm}}.$$

$$2f_{pm} - f_{mp} = \frac{U_{lp}}{U_{pm}}(2f_{pl} - f_{lp}) - \frac{K_{lpm}}{U_{pm}}.$$

Solving simultaneously for f_{mp} and f_{pm},

$$f_{mp} = \frac{2}{3}\left\{\frac{U_{lm}}{U_{mp}}(2f_{ml} - f_{lm}) + \frac{K_{lmp}}{U_{pm}}\right\} + \frac{1}{3}\left\{\frac{U_{lp}}{U_{pm}}(2f_{pl} - f_{lp}) - \frac{K_{lpm}}{U_{pm}}\right\},$$

$$f_{pm} = \frac{1}{3}\left\{\frac{U_{lm}}{U_{mp}}(2f_{ml} - f_{lm}) + \frac{K_{lmp}}{U_{pm}}\right\} + \frac{2}{3}\left\{\frac{U_{lp}}{U_{pm}}(2f_{pl} - f_{lp}) - \frac{K_{lpm}}{U_{pm}}\right\}.$$

$$\tag{8}$$

As a means of abbreviation, a new notation, r, will now be introduced to replace the difference of f, as defined by the following equation:

$$r_{lm} = 2f_{lm} - f_{ml},$$

$$r_{ml} = 2f_{ml} - f_{lm}. \tag{9}$$

On account of its very frequent occurrence in later discussions, the value of r_{lm} for member lm at l will be known as the "deformation modulus" of member lm at l.

Substitute r in the above equations:

$$r_{lp} = \frac{U_{lm}}{U_{lp}} r_{lm} + \frac{K_{plm}}{U_{lp}}, \tag{10}$$

$$f_{pl} = 2f_{lp} - \frac{U_{lm}}{U_{lp}} r_{lm} - \frac{K_{plm}}{U_{lp}}, \tag{11}$$

$$\left. \begin{aligned} f_{mp} &= \frac{2}{3}\left\{ \frac{U_{lm}}{U_{pm}} r_{ml} + \frac{K_{lmp}}{U_{pm}} \right\} + \frac{1}{3}\left\{ \frac{U_{lp}}{U_{pm}} r_{pl} - \frac{K_{lpm}}{U_{pm}} \right\} \\ f_{pm} &= \frac{1}{3}\left\{ \frac{U_{lm}}{U_{pm}} r_{ml} + \frac{K_{lmp}}{U_{pm}} \right\} + \frac{2}{3}\left\{ \frac{U_{lp}}{U_{pm}} r_{pl} - \frac{K_{lpm}}{U_{pm}} \right\} \end{aligned} \right\} \tag{12}$$

Solving equation (9) for f,

$$f_{lm} = \frac{1}{3}(2r_{lm} + r_{ml}),$$

$$f_{ml} = \frac{1}{3}(2r_{ml} + r_{lm}). \tag{13}$$

From equation (13),

$$r_{ml} = 3f_{lm} - 2r_{lm}. \tag{14}$$

The above equations, which form the basis of the methods developed in this paper, are applied graphically to the "Method of Deformation Contour" in the following sections.

GRAPHIC ANALYSIS OF SECONDARY STRESSES BY THE METHOD OF DEFORMATION CONTOUR

Ⅳ. FUNDAMENTAL CONCEPTIONS

It has been shown that when a riveted truss is deformed, the members constituting the truss will be bent and bending moments developed along their axes. The moment in any member varies from end to end in a linear equation[①] and may be represented as such by a straight line. Since $M = fS$, and S is a constant for the member, the variation of f can also be represented by a straight line. More directly—*if S be considered as a stress acting in the member, and M the moment it produces about any point, then the*

[①] Neglecting the moment of primary stress due to deflection of the member from its straight axis.

offset of the stress from the point gives the value of f.

Let 12 (Fig. 8) be a member connecting joints 1 and 2, and S its section modulus. If S be assumed as a compressive stress acting in this member, like the primary stress, it will produce moments at 1 and 2 if its line of action is displaced from axis 12, and the stresses produced at 1 and 2 will then be represented by the offsets f_{12} and f_{21}. Particularly, if the moments about 1 and 2 are those due to rigidity of joints, considering member 12 as a part of a riveted truss, the offsets f_{12} and f_{21} will be the secondary stresses.

The line of action of S will be known as the "secondary stress line", meaning that the secondary stress is given by the offset to this line from the axis of the member. [1]

Let the above conception be extended to every member of the truss. There will then be as many secondary stress lines as there are members. Each line will be displaced to an extent proportional to the bending moments produced. If, therefore, the lines are so located as to satisfy the various imposed conditions as

---

[1] For the fiber where the secondary stress has the same sign as the moment.

a result of the rigidity of joints the offsets of the secondary stress lines will give directly the secondary stresses. This conception is fundamental.

The first condition that the secondary stress lines are subject to is that the total resulting moment around any joint must be zero. Graphically, this means that if a force diagram be drawn of the secondary stress lines around any joint and an equilibrium polygon be likewise constructed for the joint, these two figures must respectively close. This, however, is impossible for the convention of secondary stress lines adopted above because the values of S, being constant, cannot be made to balance each other. Thus, in Fig. 9, stresses S_{12} and S_{13} cannot balance each other unless they are on the same straight line—a condition which is impossible. To overcome this difficulty, there will be introduced an external ideal force, R, with such a magnitude and direction that equilibrium is maintained around the joint. In other words, this force must be equal to the resultant of the secondary stress lines acting on the joint but with opposite sense in direction. Since there can be no external moment around the joint, its line of action must pass through the joint in consideration. Thus in Fig. 9 a force R must be introduced to balance the stresses S_{12} and S_{13}.

With the conceptions of S and R thus established it is now possible to draw a force diagram around every joint of the truss and to construct from this an equilibrium polygon. If these two polygons are made to close, this everywhere satisfies the first condition that the sum of internal moments around any joint must be zero. Further, from the condition that the total external moment on the truss is also zero, it may be inferred that both the force and equilibrium polygons of the external ideal forces, R, must respectively close.

The second condition which must be satisfied by the secondary stress lines is evidently derived from the rigidity of joints. This is accomplished by the use of the "deformation contour," a detailed description of which is given later.

V. THE EQUILIBRIUM CONTOUR

In a graphic analysis of secondary stresses the two conditions for locating secondary stress lines must be satisfied simultaneously at every step of the procedure. This necessitates the drawing of a great number of force diagrams required for each joint of the truss, on account of the continuously changing positions of the secondary stress lines. From a practical point of view, therefore, the conception of secondary stress lines as established in the last section is not suitable during the process of construction, for a speedy method demands that the direction of these lines be kept constant and that only one force diagram be required for each joint. This, however, does not mean the abandonment of the secondary stress lines, for at the end of the procedure these lines must be drawn in for every member of the truss, to show the distribution of secondary stresses.

The lines representing forces S and having a fixed direction for each member will be called hereafter the "section lines" to distinguish them from the secondary stress lines. The fixed direction of these section lines is best chosen perpendicular to the axis of the member, for then the offset (equal to f) would be found along the axis. For example, let S_{12} cut member 12 at o, Fig. 10. Stress f_{12} is then equal to 1 c, positive if c is between 1 and 2 and negative if on the other side of 1. For end 2, the same force S_{12} may be shifted to a point the distance of which from the end gives stress f_{21}. The direction of the section lines should be such that they give a counter clockwise moment for the joint whenever the value of f is positive; that is, when the lines cut the members somewhere between the joints. Thus, in the above example, the direction of the section line for end 1 should be upward and for end 2 downward. Note that these directions are constant for both positive and negative values of f, being always upward for end 1, whether on the right or left side of end 1 — similarly for the line for end 2.

Consider a joint as shown in Fig. 11. As the section moduli of the members are given, a force diagram may be drawn of the section lines, with their directions perpendicular to the respective members. Since a counter clockwise moment gives a positive f, the lines for the successive members should form a contour around joint 1 in a counter clockwise direction, as shown. The closing line of the force diagram, fe, corresponds to the external ideal force, R, for the secondary stress lines considered in the preceding section and must therefore be applied through joint 1, for the external moment about that joint is zero. Since the direction of R is constant, it may be drawn on the truss diagram as a part of the truss.

Since the condition of equilibrium of joints requires that the equilibrium polygon for section lines, S, and external ideal force, R, be closed for the joint, the stress in any one member may be found if it is known for all the other members meeting at the same joint. For example, suppose in Fig. 11 the stresses in all members are known except that in member 16. The section lines for members 12, 13, 14, and 15 may therefore be located from the values of f. The direction of the section line for member 16 is known from the force diagram and is seen to be upward. Its exact location on the truss may be found from an equilibrium polygon by taking some point as the pole and drawing the rays. To facilitate the construction, take f as the pole and draw the rays fb, fc, and fd. At the intersection, r, of forces FA and AB, draw a line parallel to fb, meeting force BC at s. Through s draw a line parallel to fc, meeting force CD at t. Through t draw a line parallel to fd, meeting force $R = FE$ at v. Then the section line for member 16 must pass through v. Drop a perpendicular from v on side 12, the prolongation of side 16. The value of f_{16} is then given by the distance $1w$. It is seen to be negative as it is on the

668 |

opposite side of 1 from 6.

By a similar process the stress in any member at any joint may be found if the stresses in all the other members meeting at the same joint are known. The contour formed by the lines parallel to the rays in the force diagram will be known as the "equilibrium contour," as *prstv* in Fig. 11.

If there are eccentric connections at the joint the moment thus produced may be taken care of by giving force R an offset equal to the moment divided by the value of R.

VI. THE DEFORMATION CONTOUR

1. Deformation and Property lines

Consider equation (10) applied to vertex 1 of triangle *plm*, in Fig. 12.

$$r_{lp} = \frac{U_{lm}}{U_{lp}} r_{lm} + \frac{K_{plm}}{U_{lp}}.$$

As this is a linear equation connecting r_{lp} and r_{lm}, it may be represented by a straight line, *ab*, if the sides, *lp* and *lm*, are considered as the co-ordinate axes of r_{lp} and r_{lm}. Positive values of r_{lm} will be laid off toward *m* from *l* with negative values in the opposite direction. Similarly for r_{lp}. To locate the line *ab* the simplest method would be to find the intercepts on the two axes.

$$\text{If } r_{lm} = 0, \; r_{lp} = \frac{K_{plm}}{U_{lp}}.$$

$$\text{If } r_{lp} = 0, \; r_{lm} = -\frac{K_{plm}}{U_{lm}}.$$

That is, the intercepts on the two adjacent sides are respectively equal to the value of K_{plm} divided by the values of U for the corresponding sides. The signs of the intercepts are determined by the fact that if K is positive the positive value of $\frac{K}{U}$ is found on the side which is first met with by passing around joint l in a counter clockwise direction; and the negative value of $\frac{K}{U}$ on the side next met with in the same direction. Accordingly the value of $\frac{K_{plm}}{U_{lp}}$ is laid off toward p while that of $\frac{K_{plm}}{U_{lm}}$ is laid off away from m, as shown. This is for positive K; if it be negative the process should be reversed. From the intercepts thus obtained, the straight line, ab, may be located by passing through their ends. This gives the relations between r_{lp} and r_{lm} as represented by equation (10). For any value of r_{lm}, to find the corresponding value of r_{lp} it is necessary only to draw a line from the end of *segment* r_{lm} parallel to side lp, and from the intersection of this line with ab draw a line parallel to side lm. The intercept thus obtained on side lp gives the value of r_{lp} required.

By a similar method, a line cd may be drawn to represent the relations between r_{pl} and r_{pm}, positive values of r_{pl} being measured from p toward l and negative values away from l. A third line, ef, may likewise be drawn for the vertex, m, of triangle plm. It is therefore evident that each triangle of a truss has three lines to represent equation (10) and they completely determine the deformation equation of Section Ⅱ. These lines will hence be known as the "deformation lines" for the vertices of the triangle and will be designated in the figure by a letter denoting the triangle and a numeral denoting the joint. Thus, K_{A1} would designate the line which belongs to triangle A and joint 1.

It will be seen from equation (10) that if $K_{plm} = 0$, these deformation lines will pass through the origin of the axes, that is, the vertex of the triangle—joint l in this case. These lines will be known as the "property lines" of the truss since they represent the ratios of U which are the property of the truss. In other words, they are independent of the loading.

One feature of the property lines, which deserves attention, is that in any triangle the three property lines must meet in one point. This furnishes a good check on the deformation lines as they are parallel to the property lines.

2. The Deformation Contour

Suppose that, by the method just described, the deformation lines are constructed for all those triangles of a truss which have a common vertex; as, for example, those at joint 5, Fig. 13. Then, if the value of r_{53} be known, that of r_{52} may be obtained from line K_{B5} by drawing lines parallel to sides 52 and 53. Similarly, from the value of r_{52} thus obtained, that of r_{54} may be found from line K_{C5}. By continuing the same process r_{57} is eventually obtained with the aid of lines K_{D5} and K_{E5}. Hence—and this is important—if the value of the deformation modulus, r, of any member at any joint be known, it is known for all the other members at the same joint by the use of the deformation lines.

In order to check the result and also close the polygon, a deformation line, K_5, will be drawn for joint 5, giving the relationship between r_{53} and r_{57} considering the change in the angle that is outside of the truss, or the negative of the total changes in angles 352, 254, 456, and 657. Since both r_{53} and r_{57} are on the same straight line this deformation line K_5 could not be obtained by the usual Cartesian co-ordinates, but must be found as follows: Find a pair of corresponding values of r_{53} and r_{57} and at the end of each segment draw a 45-degree line directed toward the other. These two lines will intersect in a point. This, with another point obtained in a similar manner, will then determine the deformation line K_5. For a joint where the outside members do not lie on the same straight line, the deformation lines may be obtained in the usual way. (See joint 2, Plate Ⅲ).

The continuous broken lines (including the two 45-degree lines), drawn for each joint of the truss and parallel to the members, with their intersection points on the deformation lines, form a closed figure known as the "deformation contour." In Fig. 13 the deformation contour for joint 5 and r_{53} is *abcdefghijk*. For any other value of r_{53} the deformation contour for joint 5 will be parallel to the one as shown. That deformation contour which gives correct values of r, and consequently the actual secondary stresses, is known as the "correct deformation contour."

The deformation contour entirely expresses the deformation

equation of Section Ⅱ, and the fact that it must close for any joint forms the second condition imposed on the location of the secondary stress lines. In the graphic analysis, therefore, *the two fundamental equations are replaced by the closing requirements of the equilibrium and deformation contours.*

To fix the idea of the deformation contour it is necessary to give it a value equal to the deformation modulus, r, of one of the members meeting at the joint. For reasons stated later this member will be chosen as the chord at the left side of the joint in consideration. This will be known as the "reference member for the deformation contour." Thus the value of the contour for joint 5 just considered will be designated by r_{53}, 53 being the reference member for joint 5.

In order to utilize the equilibrium and deformation contours for the graphical analysis of secondary stresses, a relation will now be established which forms the basis of the method of deformation contour. Let a truss shown in Fig. 14 be analyzed for the secondary stresses. Beginning with triangle A suppose the deformation contours for joints 1 and 2 are known, and let it be required to located the deformation contour for joint 3.

From equation (14), $r_{31} = 3f_{13} - 2r_{13}$,

where r_{13} is known from the given contour.

Since $$f_{13} = -\frac{S_{12}}{S_{13}}f_{12},$$ by equation(1)

$$f_{13} = -\frac{S_{12}}{S_{13}} \cdot \frac{2r_{12} + r_{21}}{3}.$$ by equation(13)

Also, by equation (10), $r_{13} = ar_{12} + b$,

where a and b are constants.

Therefore, $r_{31} = -\dfrac{S_{12}}{S_{13}}(2r_{12} + r_{21}) - 2(ar_{12} + b)$.

That is, the value of r_{31} may be found from r_{12} and r_{21} by a linear equation, or

$$r_{31} = mr_{12} + nr_{21} + p,$$

where m, n and p are constants.

Next, consider the relations between r_{53} and r_{35}. By equation (14),

$$r_{53} = 3f_{35} - 2r_{35}.$$

Since $$f_{35} = -\frac{S_{23}}{S_{35}}f_{32} - \frac{S_{13}}{S_{35}}f_{31}$$

$$= -\frac{S_{23}}{S_{35}} \cdot \frac{2r_{32} + r_{23}}{3} - \frac{S_{13}}{S_{35}} \cdot \frac{2r_{31} + r_{13}}{3},$$

it is seen that r_{53} is a linear function of r_{35} and r_{21} because the relations between r_{32} and r_{31}; r_{13} and r_{31}; r_{31} and r_{35}; and r_{23} and r_{21} are all linear. Therefore,

$$r_{53} = m'r_{35} + n'r_{21} + p',$$

where m', n' and p' are constants.

Similarly, consider the relations between r_{42} and r_{24}.

$$r_{42} = 3f_{24} - 2r_{24},$$

$$f_{24} = -\frac{S_{25}f_{25} + S_{23}f_{23} + S_{21}f_{21}}{S_{24}}$$

$$= -\frac{S_{25}}{S_{24}} \cdot \frac{2r_{25} + r_{52}}{3} - \frac{S_{23}}{S_{24}} \cdot \frac{2r_{23} + r_{32}}{3} - \frac{S_{21}}{S_{24}} \cdot \frac{2r_{21} + r_{12}}{3}.$$

Now the following relations are linear: r_{25} and r_{24}; r_{52} and r_{53}; r_{23} and r_{24}; r_{32} and r_{35} (and hence r_{32} and r_{53}); r_{21} and r_{24}; r_{12} and r_{13} (and hence r_{12} and r_{31}, r_{12} and r_{35}, and finally r_{12} and r_{53}). Hence the relations between r_{42}, r_{24} and r_{53} are also linear and it follows that

$$r_{42} = m''r_{24} + n''r_{53} + p'',$$

where m'', n'' and p'' are constants.

By a similar process it can be shown that for any triangle of the truss a linear relation can always be expressed between the values of r at the three vertices. In general, for any triangle plm (Fig. 7) of a truss,

$$r_{ml} = gr_{lm} + hr_{pn} + k \qquad (15)$$

where g, h and k are constants and r_{pn} is the deformation modulus for the reference member of joint p. By a successive application of equation (15) the value of the deformation contour for any joint of the truss may be found from two other given contours, say r_{12} and r_{21}.

By the aid of equation (15) the following principle may now be formulated:

"*In any framed structure composed of triangular elements the values of deformation contours for the joints of the structure are related to each other by a linear equation.*"[1] If the least number of members entering into a joint of the structure is n, the value of each of the deformation contours may be expressed linearly in terms of n others. In a bridge truss, n is generally equal to 2 so that *the deformation contours of any three joints of the truss are connected by a linear equation and, by the process of successive substitution, all the deformation contours of the truss can be made to depend, by linear expressions, on only two other deformation*

---

① The effect of the primary stress on the deflection of the member is neglected.

contours.

The use of the above principle overcomes the necessity of solving a large set of simultaneous equations, which is always considered the most laborious part in the solution of secondary stresses. It is due to this principle that the methods presented in this paper take much less time than the four current solutions in general use.

Having thus completed the necessary discussions preparatory to the method of deformation contour, the graphic solution may now be applied to the analysis of secondary stresses.

Ⅶ. GRAPHIC ANALYSIS BY THE METHOD OF DEFORMATION CONTOUR

1. Graphic Representation of a Linear Equation involving Three Variables

In the previous section the values of deformation contours for any three joints in a truss are shown to have been connected by a linear equation. If r_{53}, r_{35} and r_{21} are the contours at the vertices of a triangle 235, Fig. 15, then, by equation (15),

$$r_{53} = ar_{35} + br_{21} + c.$$

If the value of one of the variables, say r_{21}, be assumed constant or arbitrarily assigned, the other two, r_{35} and r_{53}, can be graphically represented by a straight line. This, however, implies that the two variables are referred to two axes which are not in the same straight line as in the present case. Here, both r_{35} and r_{53} are referred to line 35. To overcome this difficulty and facilitate the construction the following device has been evolved.

Find a pair of values of r_{35} and r_{53} connected by the above equation, with r_{21} arbitrarily assigned, and locate the points a and a' by making $3a = r_{35}$ and $5a' = r_{53}$. From a draw a line at 45 degrees with 35 and directed toward the right of joint 3 and below the truss. Similarly from a' draw a line at 45 degrees with 53 and directed toward the left of joint 5, also below the truss. These two lines will intersect at a point A. Next find another pair of values of r_{35} and r_{53}, with the same r_{21} and locate the point b and b'. From b and b' draw lines at 45 degrees with 35 and directed toward each other, intersecting at B. Draw a straight line through A and B. This line will then give the relations between r_{35} and r_{53} for the assigned value of r_{21}. For any value of r_{35}, say $3c$, to find the corresponding value of r_{53} draw a line from c inclined at 45 degrees with 35 and directed toward end 5. Let this line intercept the line AB at C. From C draw a line at 45 degrees with 35 and directed also toward 5. Let this intercept 35 at c'. Then r_{53} is given by the segment $5c'$. The above constructions can be easily proven by analytic geometry, as follows:

Let the equation between r_{35} and r_{53} with arbitrarily assigned value of r_{21} be

$$r_{53} = ar_{35} + f, \tag{a}$$

where $f = br_{21} + c =$ constant. Let P, Fig. 16—the point obtained by the above construction with any two corresponding values of r_{35} and r_{53}—be referred to 35 and $3Y$ as X and Y axes, so that its co-ordinates are x and y. From the figure,

$$y = \frac{1}{2}(d - r_{35} - r_{35}), \tag{b}$$

where r_{35} and r_{53} carry their own signs. Also,

$$x = r_{35} + y. \tag{c}$$

Substituting the value of r_{53} from (a) in (b) and eliminating r_{35} between the resulting equation and (c) there is obtained a relation between x and y:

$$y = \frac{a+1}{a-1}x - \frac{d}{a-1} + \frac{f}{a-1},$$

$$y = \frac{a+1}{a-1}x - \frac{d}{a-1} + \frac{br_{21}+c}{a-1},$$

$$y = \frac{a+1}{a-1}x + \frac{br_{21}-(d-c)}{a-1}. \tag{d}$$

As this is a linear equation, the locus of P is a straight line.

It is evident from equation (d) that the slope of the line is a function of a alone, and independent of r_{21}. That is, the slope of the line is constant for all the assigned values of r_{21} and may be most readily obtained by making $r_{21} = 0$. This line will be known as the "base line" for triangle 235, for the reason that it expresses the relation between the r's of base 35 with r at the vertex of the triangle $= 0$.

When the value of r_{21} is not zero the straight line given by equation (d) will occupy a new position parallel to itself with the y-intercept given by the following equation:

$$p = \frac{b}{a-1}r_{21} - \frac{d-c}{a-1}, \tag{e}$$

which is also linear and may be represented by a straight line. To construct this line proceed as follows: Prolong member 23 until it intersects the base line at O, Fig. 15. Assume a value for $r_{21} = 2e$, preferably as large as possible, and by the above method find line EF representing the relations of r_{35} and r_{53} for $r_{21} = 2e$. This line, as proven above, must be parallel to the base line, AB. From e draw a line parallel to member 23 and intersect EF at E. Join OE. This line OE then gives the y-intercept p for any value of r_{21}. If $2g$ is the value of r_{21} the intercept is found by drawing a line through g parallel to member 23 until it meets OE at G, and through G drawing a line parallel to the base line, AB. This line GH then gives the relation between r_{35} and r_{53} with $r_{21} = 2g$. The line OE giving relations between p and r_{21} at the vertex of the triangle, will be known as the "vertex line."

To utilize the base line and vertex line in finding r_{53} from r_{35} and r_{21} proceed as follows: Let $r_{21} = 2g$ (negative) and $r_{35} = 3h$. Through g draw a vertical line cutting the vertex line at G. Through G draw a line GH parallel to the base line, AB. Through h draw a downward, 45-degree line cutting GH at H. From H draw an upward, 45-degree line directed toward joint 5, cutting 35 at h'. Then r_{53} for the above values of r_{35} and r_{21} is given by $5h'$ (negative). Similarly, the values of r_{53} for any other combinations of r_{35} and r_{21} may be found.

By reversing the above process the value of r_{21} may be found from any given values of r_{35} and r_{53}. For these two values, say $3h$ and $5h'$, respectively, determine a point, H, by 45-degree lines and through H draw a line parallel to the base line, cutting the vertex line at G. Through G draw a vertical line cutting member 21 at g. Then $2g$ is the value of r_{21} required.

By the same method the base line and vertex line may be constructed for any other triangle of the truss. In case the triangle has a base on the top chord of the truss, it is advisable to draw the 45-degree lines directed upward instead of downward as for member 35.

As a means of standardization all the base lines will be shown by heavy full lines and the vertex lines by heavy dashed lines.

2. Construction of Base Lines and Vertex Lines

Since the base line is a straight line it can be determined by two points. A point for the vertex line is given by the intersection of the base line with the vertical member produced, so that only one more point is necessary for the vertex line. Theoretically, therefore, three points are sufficient for constructing both the base line and the vertex line. For practical purposes, however, it is advisable to find one more point for each of the two lines—both as a check and as a means of improving the accuracy of the results. These points may be so chosen as to require the least amount of work.

Let the base and vertex lines be constructed for a truss as shown in Fig. 17. This is a Warren truss with verticals but the method is applicable to all kinds of trusses, the only difference being found in the order of procedure. This is governed by the number of members meeting at the joints. It is well to remark here that the base of a triangle is always chosen as the member which forms the outline of the truss.

(1) Base Line and Vertex Line for Triangle A.

Here, either member 12 or 13 may be chosen as the base. While only one is necessary in the solution of the truss, it is advisable to have another in order to check the final results.

(a) Member 13 as Base.

Since the vertex for triangle A is joint 2, the base line is obtained by taking $r_{21} = 0$, as 21 is the reference member for joint 2. To construct this base line three points are necessary and these are determined by the corresponding values of r_{13} and r_{31}

connected by equation(14). The detailed process is as follows. Assume a value for r_{13} and complete the deformation contour for joint 1, giving r_{12}. Also, for $r_{21} = 0$, complete the deformation contour for joint 2, giving r_{21}, r_{23}, r_{25}, and r_{24}. From the values of r_{12} and r_{21} the stress f_{12} may be found by equation (13). Since equation (14) calls for $3f_{13}$ it must be found from f_{12}. This is best done by laying off $2r_{12} + r_{21}$ from joint 1—careful attention being given to the signs—and completing the equilibrium contour for joint 1, obtaining $3f_{13}$. From $3f_{13}$ subtract $2r_{13}$ and the difference will be r_{31}.

The value of r_{31} thus obtained, together with the assumed value of r_{13}, determines a point for the base line, use being made of the 45-degree lines as described in the previous article. By a similar process, still keeping $r_{21} = 0$, two more points may be obtained which, together with the one previously found, must lie on the same straight line. This gives the base line, ab, required.

To find the vertex line corresponding to the base line just found, keep one of the assumed values of r_{13} as a constant and find a pair of values of r_{21} and r_{31} which correspond. To do this, first give to r_{21} a value which is fairly large and complete the deformation contour for joint 2 with this value of r_{21}. The contour for r_{13} has already been drawn for constructing the base line so that r_{12} and r_{13} are known. From these contours obtain $2r_{12} + r_{21}$ $= 3f_{12}$ and from $3f_{12}$ obtain $3f_{13}$. Subtracting $2r_{13}$ from $3f_{13}$, r_{31} is obtained. This r_{31} and the assumed r_{13} determine a point. Keeping the value of r_{21} constant, find another pair of values of r_{13} and r_{31}, determining another point. These two points must lie on a straight line, cd, that is parallel to the base line. Through the end of segment r_{21} thus assumed, draw a vertical line cutting cd at c. Let the base line, ab, cut member 23 produced at a. Join ac. This, then, is the vertex line required.

These base line and vertex line are indicated in the figure, as "Triangle A, Base 13. "

(b) Member 12 as Base.

Here, joint 3 is the vertex of the triangle and the base line is obtained by taking $r_{31} = 0$, as 31 is the reference member for joint 3. To find a pair of values of r_{12} and r_{21}, first assume a value for r_{12}. For this r_{12} complete the deformation contour for joint 1, obtaining r_{13}. Find $3f_{13} = 2r_{13} + r_{31} = 2r_{13}$. Construct an equilibrium contour for joint 1, and $3f_{12}$ is found from $3f_{13}$. From

$3f_{12}$ and $2r_{12}$, the value of r_{21} is obtained. This r_{21} and the assumed r_{12} determine a point on the base line, use being made of the lines inclined at 45 degrees with member 12. Similarly, two more points may be obtained and the base line, ef, is completely known. Find the intersection of this line with a line drawn through joint 3 and perpendicular to member 12, at e.

To locate the vertex line, assign a fairly large value to r_{31} and, keeping it constant, obtain two pairs of corresponding values of r_{21} and r_{12}. These two points must lie on a line, gh, parallel to ef. Find the intersection, g, of this line, gh, with a perpendicular to member 12 through the end of segment r_{31}, assumed. The vertex line is then given by a line drawn through points e and g.

These base line and vertex line are indicated in the figure, as "Triangle A, Base 12."

To find the corresponding values of r_{12} and r_{21} for any assumed value of r_{31} drop a perpendicular to 12 from the end of segment r_{31} and find the intersection, g, of this perpendicular with the vertex line, eg. Draw a line through g parallel to the base line, ef. This line then gives the relations between r_{12} and r_{21} for the assumed r_{31}, the corresponding values being given by 45-

degree lines.

(2) Base Line and Vertex Line for Triangle B.

The base for triangle B is member 35, and the vertex, is joint 2, so that the base line is obtained by making $r_{21} = 0$, the deformation contour for which has already been drawn. As before, assume a value for r_{35} and find the corresponding value for r_{53}. These two values will determine a point on the base line. On account of the greater number of members entering into joint 3 the process is a little more complicated than for triangle A but the principle will be the same.

For the assumed value of r_{35} construct a deformation contour for joint 3, giving r_{32} and r_{31}. Since the value of r_{53} depends on that of f_{35}, which, in turn, depends on f_{32} and f_{31}, or indirectly, on r_{32}, r_{23}, r_{31} and r_{13}, it is necessary to find the value of r_{13} for the corresponding values of r_{31} and r_{21}, as r_{23}, r_{32} and r_{31} are all known. To obtain this r_{13}, use is to be made of the base line and vertex line for triangle A, base 13, precisely as previously explained. For this reason it is advisable to choose $r_{21} = 0$ instead of $r_{23} = 0$ for this triangle, for then the base line for triangle A can be used. Now that the values of r for members 13 and 23 are known, the sums $2r_{32} + r_{23} = 3f_{32}$, and $2r_{31} + r_{13} = 3f_{31}$, may be obtained and the value of $3f_{35}$ obtained from an equilibrium contour for joint 3. Then, r_{53} is found from the difference of $3f_{35}$ and $2r_{35}$. These corresponding values of r_{35} and r_{53} determine a point on the base line for triangle B. Similarly, two more points may be obtained and the base line is completely known. For the vertex line, give r_{21} a value which is fairly large—preferably the same value used in constructing the vertex line for triangle A, base 13. This changed value of r_{21} affects r_{23} directly and r_{13} indirectly, for then the base line for triangle A, cannot be used in finding r_{13} from r_{31} and r_{21}. Instead, the vertex line for triangle A should be used for the assumed value of r_{21}. With this exception, the remainder of the method is the same as already described.

(3) Base Line and Vertex Line for Triangle C.

The base for triangle C is member 24, and the vertex is joint 5. The base line is therefore obtained by making $r_{53} = 0$. Since the value of r_{42} depends on $3f_{24}$ and $2r_{24}$, the latter being arbitrarily assigned, it is necessary to find f_{24}, or indirectly, f_{25}, f_{23}, and f_{21}. These values of f are found from r_{25} and r_{52}; r_{23} and r_{32}; and r_{21} and r_{12}. The deformation contour for joint 2 for the assumed r_{24} gives r_{25}, r_{23}, and r_{21}, while that for joint 5 with $r_{53} = 0$ gives r_{52}. The only r's unknown, therefore, are r_{32} and r_{12}. To find r_{32}, first obtain r_{35} from the base line and vertex line for triangle B, from r_{21} and $r_{53} = 0$, then complete the deformation contour for joint 3 for this value of r_{35}. The contour will then give r_{32}. Lastly, the value of r_{12} is found from r_{13}, which, in turn, is found from the base line and vertex line for triangle A, where r_{21} and r_{31} are known. With these values of r known, $3f_{25}$, $3f_{23}$, and $3f_{21}$ may be obtained by addition, and $3f_{24}$ by an equilibrium contour for joint 2. The resultant, $3f_{24}$, then furnishes r_{42} for the assumed $2r_{24}$. These two values of r determine a point on the base line for triangle C. By a similar method two more points may be obtained, still keeping $r_{53} = 0$, and the base line is then completely located. As a means of facilitating the construction it is well to have one value of r_{24} which will give $r_{21} = 0$.

The vertex line for triangle C is obtained in much the same way as for triangles A and B.

By a similar process the base lines and vertex lines could be located for all the triangles of the truss. For a triangle like E, where five members enter into joint 5, the process is necessarily complicated but that is about the extreme case that is likely to occur in ordinary trusses. ①

It will be found in practice that, with the exception of the triangles at the two ends of the truss, the base line and the vertex line are always very close together and in many cases the vertex lines could be omitted entirely. In this case the base line is to be used for all values of r at the vertex.

For a symmetrical truss with symmetrical loading the base line and the vertex line need be constructed for only one half of the truss, as for the other half they are identical in form. If the truss or the loading, or both, are not symmetrical the base line and the vertex line must be obtained for every triangle of the truss.

A few words are now necessary regarding the selection of the reference members. They are seen to be the chord members, end posts included, that are at the left side of the joints under consideration. This is due to the fact that the above process of constructing base lines and vertex lines advances from left to right

---

① The amount of work required is proportional only to the number of members entering the joint, and not to the size of the truss.

and only a part of the deformation contour that is on the left side of the joint need be constructed.

3. Solution of the Problem

Let the base lines and vertex lines for all the triangles of the truss be constructed by the methods of the previous article. At the right end of the truss there are then drawn the base lines and vertex lines for triangles B', C', etc. For triangle A' three sets of base and vertex lines are drawn; first, for side $1'3'$ as base; second, for side $1'2'$ as base, considering equilibrium about joint $1'$; and, third, also $1'2'$ as base but considering equilibrium about joint $2'$. These three sets of lines are indicated as $1'3'$, $1'2'_{1'}$, and $1'2'_{2'}$ in the figure. Since they all represent the same relations among $r_{3'1'}$, $r_{1'2'}$, and $r_{2'1'}$ they may be used for solving the three unknowns as, analytically, they represent three equations. To perform this solution, first assume a value for $r_{3'1'}$ and apply it to sets $1'2'_{1'}$, and $1'2'_{2'}$, giving two lines parallel to their respective base lines. These two lines intersect at a point which will give a pair of values of $r_{1'2'}$ and $r_{2'1'}$ for the assumed $r_{3'1'}$. Next apply to set $1'3'$ the values of $r_{1'2'}$ and $r_{2'1'}$ thus found, and find the corresponding value of $r_{3'1'}$. This value should check with that previously assumed; if not, another trial may be made. Repeat the process until the values of $r_{3'1'}$, $r_{1'2'}$ and $r_{2'1'}$ all satisfy the three sets of base and vertex lines. These are the values that will give the correct secondary stresses. From the value of $r_{2'1'}$, that of $r_{2'4'}$ may be found from a deformation contour for joint $2'$. Similarly, the value of $r_{3'5'}$ may be found from $r_{3'1'}$ by a deformation contour for joint $3'$. These two contours are the correct deformation contours for the problem, since from them the correct secondary stresses may be obtained. From the values of $r_{2'4'}$ and $r_{3'5'}$, that of $r_{5'3'}$ may be obtained from the base line and vertex line for triangle B', and, consequently, the correct deformation contour for joint $5'$. Repeating the process the correct contours are obtained for all the joints of the truss. To check the accuracy of the results the values of r_{31}, r_{12}, and r_{21} must satisfy the base line and vertex line for triangle A with member 12 as base, as this set of lines has not been used in constructing the correct deformation contours. This insures the correctness of every step in the procedure and is one of the important features of the method of deformation contour.

From the correct deformation contours the actual secondary

stresses may be found by equation (13). This may be accomplished graphically by adding to segment r_{ln} of member ln a distance equal to r_{ln} plus r_{nl}, with proper signs, and measuring off the resulting segment with a scale three times as large as that used for r. It will be seen that on account of the large scale used for measuring f, the result could be read in three significant figures—a result accurate enough for all practical purposes.

When the loading and truss are symmetrical, the stresses at the two ends of the member of symmetry will be zero, hence the values of r at those two ends are also zero. This locates two correct deformation contours for the two joints of symmetry, from which the problem may be solved as before.

To obtain a clearer view of the distribution of secondary stresses in the truss, and also the way they affect each other, it is advisable to draw in the secondary stress lines for the different members and their force diagram, as mentioned in Section Ⅳ.

4. Checks

The following checks are found in this method of deformation contour:

(1) In constructing the values of K, the sum of the K's in each triangle must be zero.

(2) All the property lines in a triangle must meet in one point.

(3) All the deformation lines must be parallel to the property lines in the same triangle.

(4) All deformation contours must close for any value of r.

(5) All equilibrium contours must close.

(6) The base line and vertex line for any triangle are checked by constructing one extra point for each line. These extra points must lie on the respective lines.

(7) The correct deformation contours are checked by the base line and vertex line for the end triangle of the truss, with the end post as base.

(8) For any joint of the truss the equilibrium polygon for the secondary stress lines, S, and external ideal force, R, must close.

VIII. DETAILS OF GRAPHICAL CONSTRUCTION

The details of graphical construction for figures discussed in previous sections will now be given. To obtain a clearer view of the methods, constant references should be made to the drawings shown in Plate Ⅲ.

1. Preliminary Considerations

(a) Scale.

From equations (6) and (9) it is evident that both f and r are of the same dimensional degree as that of $\dfrac{K}{U}$, or $\dfrac{6E\delta A}{\dfrac{L}{y}}$, and the values of f and r should be measured off by a scale the same as the one used in constructing the deformation lines. Since L is always expressed in feet and y in inches, L and y may be represented by the same unit if $6E\delta A$ is changed into $\dfrac{1}{2}E\delta A$. Hence, with the understanding that y is expressed in inches but represented as if it were given in feet, the value of K should be understood as $\dfrac{E\delta A}{2}$, instead of $6E\delta A$. This modified value of K will be used throughout the example worked out in next section. As the value of $E\delta A$ is found from the primary stresses by equation (2) the value of K will be given directly if only one half of the primary stress is considered.

The scale of S, the section modulus, used in constructing force diagrams may be anything independent of f, as it is only the direction that is required.

(b) Sign.

In the present paper there are two instances where a conception of sign is necessary. One is rotational and the other is linear. The first applies to moments, angles, circular sections, and directions of contours while the second applies to the secondary stress lines, their offsets, f, deformation moduli, r, and the co-ordinates of variables. The rotational direction will always be considered positive if it is *counter clockwise*. For linear directions, with the exception of secondary stress lines and their offsets, a segment ax (Fig. 18) measured on the line ab from a to x, is always positive if x is laid off from a toward b, whether x is inside of ab or outside of b, in the latter case ax is greater than ab. Should ax be laid off from a away from b the segment will be negative. Similarly, if x is measured from b.

For the sake of uniformity all secondary stress lines will be considered as compressive stresses in the members. Under this convention if the lines in members 12 and 13 are as shown in Fig. 9, f_{12} and f_{21} will be negative (the moment is clockwise about 1 and also about 2) and f_{13} and f_{31} will be positive(the moments about 1 and 3 are counter clockwise). Hence, if the positions of secondary stress lines are known, both the sign and magnitude of f are at once determined. The value of f is for that fiber of the member where the stress has the same sign as the moment.

(c) Truss Diagrams.

Two truss diagrams are necessary in this method. One of them is used to record the given information and to construct the values of K while the other is for the deformation lines and the solution of the truss. On each of these diagrams there should be drawn certain figures that are preparatory to the analysis: namely, the "y-circles" and "U-lines" for the first diagram and the property lines for the second diagram.

The y-circle is drawn at the end of a member, with a radius equal to the value of y of the member and measured off with a scale equal to that used in constructing the truss diagram in feet, although the value of y is expressed in inches. From the other end of the member draw a line tangent to this circle. This line will give the ratio of $\dfrac{L}{y} = U$ and will be known as the U-line. To find the ratio of $\dfrac{K}{U}$ if K is known, lay off a segment equal to K

from the end of the member from which the U-line is drawn. At the end of segment K draw a circle tangent to the U-line. The radius of this circle then gives the ratio required.

To construct the property lines in the second truss diagram, use is made of the U-lines in the first diagram. Suppose the property line in triangle A and joint 1 is required (Plate Ⅲ). On the first diagram where the U-lines are drawn describe any arc with joint 1 as center, cutting member 12 at a and member 13 at b. It is preferable to make the radius of the arc as large as possible in order to secure accuracy. With a and b as centers draw two arcs tangent to the respective U-lines. Let the radius having a as center be called u_{12} and that having b as center be called u_{13}. On the second truss diagram, in triangle A, lay off the segment u_{12} on side 13 (not side 12), letting the end point be c. With c as center draw an arc with radius equal to u_{13}. Through c draw a line parallel to side 13 cutting the arc at d. Join d to joint 1. This, then, is the required property line for triangle A, joint 1. Similary, all the other property lines in triangle A may be obtained. As a check, these three property lines should meet in one point. This point is enclosed in a small circle in Plate Ⅲ.

Exactly as described before, the y-circles and U-lines may be constructed for every member of the first diagram, and the property lines for every triangle of the second diagram.

2. Force Diagram for Equilibrium Contour

As mentioned before, the force diagram for equilibrium contours need be drawn only once for the whole process as the directions of the section lines are constant, being perpendicular to the respective members. For this reason it is advisable to construct the force diagram before the solution of the truss, in order to record the values of S in a graphical form. In solving a truss as shown in Plate Ⅲ first draw a line normal to member 12 and measure off $ga = S_{12}$. Then from a draw a line normal to member 23 and measure off $ab = S_{23}$. From b draw a line normal to member 25 and measure off $bc = S_{25}$. Continue for all the web members of the truss until points e and f are obtained. The directions of the forces are determined by the fact that the section lines around any joint should form a contour in a counter clockwise direction, if all the values of f are positive. Only one half of the truss is considered here as the other half will be symmetrical. Now from a draw a line normal to member 13 and make $ah = S_{13}$. The point h should be on the under side of a so that force ah will produce a counter clockwise moment about joint 1 for positive f. Similarly, from points b, c, d, and e draw lines perpendicular to the respective chord members and measure off the segments bk, cj, etc., equal to S_{35}, S_{24}, etc. Next join the end points l, j, g, h, etc., with dotted lines as shown. These will then give the directions of the external ideal forces, R, which are to be applied through the corresponding joints in the truss. Lastly, the rays gb, gc, etc., are completed. They are the lines which join the left end of the external force, R, to the inner points of the diagram.

3. Construction for values of K

From equation (2),

$$E\delta A_{213} = P_{23}(\cot 132 + \cot 123) - P_{13}\cot 132 - P_{12}\cot 123,$$

$$E\delta A_{132} = P_{12}(\cot 123 + \cot 213) - P_{23}\cot 123 - P_{13}\cot 213,$$

$$E\delta A_{321} = P_{13}(\cot 213 + \cot 132) - P_{12}\cot 213 - P_{23}\cot 132.$$

In triangle 123, Fig. 19, through 1 draw a line perpendicular to side 23 with a length equal to P_{23} and through its end draw a line parallel to 23, giving segments a and b. Similarly find the segments c, d, e, and f. Then,

$$E\ \delta A_{213} = (a + b) - (f + d) ,$$

$$E\delta A_{132} = (c + d) - (a + e) ,$$

$$E\delta A_{321} = (e + f) - (c + b) ,$$

or, the change of angle 213 is equal to the sum of segments a and b belonging to this angle, minus the sum of the segments of the other two angles, f and d, which are not adjacent to those of the angle under consideration—that is, a and b.

To secure a systematic arrangement both in sign and magnitude the following scheme is advisable. The example is for a triangle with a right angle but the method could easily be extended to any other kind of triangle. Lay off the primary stresses per square inch as shown in Fig. 20. Parallel to the sides of the triangle draw lines through the ends of the segments, obtaining a, b, c, and d, which are of the same signs as the corresponding primary stresses. There are thus obtained three small triangles $13_1 2_1, 21_2 3_2, 32_3 1_3$, similar to the large triangle 123. The consecutive sides of the small triangles may now be followed as a contour in a certain adopted direction, say counter clockwise, as shown. The change of angles may then be found as follows, taking as an example, 213, Fig. 20: At end 2_1 of segment

a, met with in a counter clockwise direction after 3_1 when following the contour 13_12_1, measure off 2_1A1 equal to the segment c of triangle 32_31_3 which is adjacent to side 23—the side parallel to 2_13_1. If the sign of c is the same as that of a it is laid off from 2_1 toward 3_1, but if otherwise, it is laid off in the direction away from 3_1, as shown in the figure. Then the distance 3_1A1, the segment from the initial point 3_1 to the end point $A1$, gives the value of $E\delta A_{213}$. The sign of 3_1A1 is the same as that of a because $A1$ is reached from 3_1 in a counter clockwise direction. In case c and a have the same sign and c is greater than a, then segment c is laid off from 2_1 toward 3_1 and point $A1$ is found below 3_1. The value of $E\delta A_{213}$ is then opposite in sign to that of a, because the point $A1$ is reached from 3_1 in a negative direction, clockwise, when following the contour 13_12_1. Similarly the change in angle 123 may be found from the segments b and d, the latter being that part of side 1_32_3 of triangle 31_32_3 which is adjacent to the side to the large triangle parallel to 1_23_2.

To find $E\delta A_{132}$ lay off segments a and b from end 1_3 of side 1_32_3, care being taken of the signs. If a or b, or both, are of the same signs as c and d the said segments should be laid off toward

2_3, instead of away from 2_3, as shown, which is for a and b having different signs from c and d. The rest of the procedure is the same as for the other two angles.

Since the sum of the changes of angles in a triangle must be zero, the values of $E\delta A$ thus found are subject to a unique check.

As mentioned before the value of K is taken as $\dfrac{E\delta A}{2}$ on account of the y-circles being magnified 12 times. To give this directly from the change of angle, the primary stresses may be laid off on a scale that is one half as great as that used in measuring the segments 3_1A1, etc, for then $3_1A1 = K_{A1}$; $1_2A2 = K_{A2}$, and $2_3A3 = K_{A3}$.

4. Deformation Lines

From the values of K thus obtained, the ratios of $\dfrac{K}{U}$ could be found from the U-lines as explained in Article 1, (c). On the second truss diagram where the property lines have already been drawn, lay off the values of $\dfrac{K}{U}$ on the corresponding sides, with proper regard to the signs. The deformation lines may then be drawn through the ends of these segments. They should be parallel to the property lines drawn in the same triangle. In Plate III the deformation lines are indicated by a shaded triangle which the deformation lines formed for the triangle of the truss.

5. Base Lines and Vertex Lines

The detailed descriptions for base lines and vertex lines have been given in Section VII, Article 2. As a preliminary it is necessary to construct the deformation contour, the graphical addition of $3f_{lm} = 2r_{lm} + r_{ml}$, the equilibrium contour and the graphical subtraction of $r_{ml} = 3f_{lm} - 2r_{lm}$. The two contours have been considered in detail in previous articles so that the only explanations necessary are for the graphical addition and subtraction. To perform $2r_{lm} + r_{ml}$, first find the value of $2r_{lm}$ by adding to r_{lm} a segment equal to itself, and then adding or subtracting r_{ml}, depending on its sign. Attention is here called to the discussion on linear directions in Article 1, (b). To obtain $3f_{lm} - 2r_{lm}$ first find $2r_{lm}$ as just described. Since both r_{lm} and f_{lm} are referred to joint l as origin, the difference is then given by the distance from the end of segment $2r_{lm}$ to the end of segment $3f_{lm}$ (this is given by the intersection of section line lm and member lm), both in sign and in magnitude. It is positive if the segment is in the direction from l to m, and negative if from m to l.

6. Secondary stress Lines

From the values of correct deformation moduli secondary stresses, f, may be obtained from equation (13) by graphical addition. More directly, the segments $2r_{lm}$ and r_{ml} may be added to each other and the resulting segment measured with a scale three times as large as for r, or for $\dfrac{K}{U}$. From the values of f thus obtained, the offsets of secondary stress lines are therefore known and the latter may be drawn in by tangent to arcs drawn with the values of f as radii. From the directions of secondary stress lines, a force diagram of S may be drawn to show the effect of section moduli upon the secondary stresses.

IX. EXAMPLE

To illustrate the method of deformation contour a truss will now be analyzed for secondary stresses with the given values of P and S as shown in Plate Ⅲ. This is the same problem as that solved by Kunz in illustrating the method of Mohr in Engineering News. vol. 66, p. 397, and by Turneaure in illustrating the method of Manderla in Engineering News, vol. 68, p. 438. A fair

706 |

comparison can therefore be made of the three methods, and their relative merits determined therefrom.

X. CONSTRUCTION OF INFLUENCE LINES

In constructing influence lines the only change that is necessary in the above method is found in the value of K. Since in equation (10) the coefficient of r_{lm} is a constant for the truss, the slope of the deformation lines which represent the equation is a constant and the deformation lines for all kinds of loading in the same triangle and for the same joint are parallel to each other. Similarly in equation (15) the values of g and h are also constants for the truss and the base line and vertex line for any triangle are respectively parallel to each other for different kinds of loading. If the said lines are drawn for any one kind of loading, the slopes of these lines will be known and for each additional loading only one extra point is necessary for determining the lines. The present method is therefore very well suited to the construction of influence lines. If Muller – Breslau's method of keeping a constant left hand reaction be used, the method can be made still simpler.

ANALYTIC SOLUTION OF SECONDARY STRESSES WITH SCIENTIFIC ARRANGEMENT OF COMPUTATIONS

XI. GENERAL PROCEDURES

From the equations derived in Section III and the principle of deformation contour discussed in Section VI an analytic solution may be made of the secondary stresses without the use of a large set of simultaneous equations, as required in the ordinary methods.

First of all let the graphical construction of deformation contours be reduced to an analytic form. Apply equation (10) to joint 5 of Fig. 13:

$$r_{52} = \frac{U_{35}}{U_{25}} r_{53} + \frac{K_{B5}}{U_{25}},$$

$$r_{54} = \frac{U_{25}}{U_{45}}r_{52} + \frac{K_{C5}}{U_{45}},$$

$$r_{56} = \frac{U_{45}}{U_{56}}r_{54} + \frac{K_{D5}}{U_{56}},$$

$$r_{57} = \frac{U_{56}}{U_{57}}r_{56} + \frac{K_{E5}}{U_{57}}.$$

By successive substitution,

$$r_{57} = \frac{U_{35}}{U_{57}}r_{53} + \frac{K_{B5} + K_{C5} + K_{D5} + K_{E5}}{U_{57}},$$

or, in general,

$$r_{lp} = \frac{U_{lm}}{U_{lp}}r_{lm} + \frac{\sum K}{U_{lp}}, \qquad (16)$$

where $\sum K$ is the sum of K's for the angles successively met with in passing around joint l in a clockwise direction, from member lm to member lp. It should be negative if member lp is met with after member lm in a counter clockwise direction. Equation (16) takes the place of the deformation contour in the analytic solution and is to be used in finding the deformation modulus of a member at any joint when it is known for another member at the same joint. For convenience, this process is oftentimes spoken of as passing a deformation contour around the joint.

The general procedure for the analytic method is best

understood by applying it to a truss as shown in Fig. 14. Since the principle of deformation contour states that the deformation modulus for any member in a bridge truss can be expressed in terms of two others, let these moduli be for member 12 and taken as two unknowns; or, more directly, take $f_{12} = x$, and $f_{21} = y$. Then by equation (9), $r_{12} = 2x - y$, and $r_{21} = 2y - x$. From these two values of r find those for the other members meeting at joints 1 and 2, by equation (16), as the values of U and K are easily computed from the given information. Applying equilibrium equation (1) to joint 1, f_{13} may be obtained from x, giving the stress at one end of member 13. To find the stress at the other end, f_{31}, first subtract f_{13} from r_{13}, obtaining $f_{13} - f_{31}$ ($r_{13} = 2f_{13} - f_{31}$), and then subtract $f_{13} - f_{31}$ from f_{13} giving f_{31}. Subtracting again $f_{13} - f_{31}$ from f_{31}, $r_{31} = 2f_{31} - f_{13}$ is obtained. From r_{31} thus found obtain r_{32} and r_{35} by equation (16). Now, for member 23, the values r_{23} and r_{32} are known and therefore the stresses f_{23} and f_{32} can be found from equation (13). At joint 3 the stresses in all members are known except that in 35, which may be found by an equilibrium equation around joint 3:

$$S_{31}f_{31} + S_{32}f_{32} + S_{35}f_{35} = 0.$$

By a process similar to that for member 13, there is obtained r_{53}, from which, by equation (16), are determined the values of r for all the members meeting at joint 5.

The stresses in member 25 may now be found from r_{25} and r_{52}. This makes known all the stresses around joint 2 except that of member 24 which can be found from the equilibrium equation. From the value of r_{24}, that of r_{42} may be found from f_{24} by the same method as for member 13. A deformation contour can therefore be passed for joint 4 with this value of r_{42}. The above process may be repeated for all the members of the truss until, eventually, the stresses in members around joints 2′ and 1′ are obtained from the equilibrium equations for joints 2′ and 1′. Up to this point the quantities $K_{A'1'}$ and $K_{A'2'}$ have not been used and it is necessary to test the deformation equations involving these two quantities. Applying equation (5) to joints 2′ and 1′ of triangle A', these are obtained two equations which furnish the means for solving x and y. After they are obtained, all the other stresses being expressed in terms of x and y may be found by substitution.

If the loading is symmetrical about a member at the center of the truss, the stresses at the two ends of that member will be zero. Since these stresses are expressed by linear equations in terms of x and y, the latter can be found by making the stresses equal to zero.

In the analytic method of solution the following points should be noted:

(1) In applying equation (16) to members which are reached from the reference member lm in a counter clockwise direction within the truss the sign of $\sum K$ should be reversed.

(2) The equilibrium equation cannot be applied to a joint until the stresses at the joint are known in all members except one.

(3) Apply equation (5) to any triangle, plm, (Fig. 7);

$$U_{mp}(2f_{mp} - f_{pm}) = U_{lm}(2f_{ml} - f_{lm}) + K_{lmp},$$

$$U_{lp}(2f_{pl} - f_{lp}) = U_{mp}(2f_{pm} - f_{mp}) + K_{mpl},$$

$$U_{lm}(2f_{lm} - f_{ml}) = U_{lp}(2f_{lp} - f_{pl}) + K_{plm}.$$

Adding together, noting that $\sum K$ should be zero in a triangle,

$$U_{lm}(f_{lm} - f_{ml}) + U_{mp}(f_{mp} - f_{pm}) + U_{pl}(f_{pl} - f_{lp}) = 0. \quad (17)$$

This equation must be satisfied as soon as the stresses in a triangle are completely known. This is one of the important features of the analytic method.

(4) Particular attention should be paid to the solution of equilibrium equations and also to the correct substitution of the values of U. Mistakes made here could not be detected by equation(17).

(5) The values of f found from equation (13), $f_{lm} = \dfrac{2r_{lm} + r_{ml}}{3}$, can be obtained in a simpler way as follows: First find

$\dfrac{r_{lm}}{3}$ from r_{lp} of reference member lp, using $3U_{lm}$ instead of U_{lm}. This gives

$$\frac{r_{lm}}{3} = \frac{U_{lp}}{3U_{lm}} r_{lp} + \frac{\Sigma K}{3U_{lm}},$$

from equation (16). Similarly, find $\dfrac{r_{ml}}{3}$. Then obtain the sum of

$\dfrac{r_{lm}}{3}$ and $\dfrac{r_{ml}}{3}$, which when added to $\dfrac{r_{lm}}{3}$ gives f_{lm}, and when added to

$\dfrac{r_{ml}}{3}$ gives f_{ml}. Since only the stresses in web members are found from the values of r alone, only the web members need be computed for $3U$.

XII. SCIENTIFIC ARRANGEMENT OF COMPUTATIONS

As is well known, an analytic method while more accurate takes longer time. It is therefore necessary to devise some method by which the amount of work will be reduced to a minimum. The computations for secondary stresses by the ordinary methods have been systematized to a great extent in recent years, but there is still room for improvement, mainly in the arrangement of computations. Among the many items which are inefficient and are found in all of the ordinary methods may be mentioned the following:

(1) All these analytic methods need some graphic construction, one way or the other, to show the proper location of certain computed quantities.

(2) In all ordinary methods the computations are arranged in tabular form in which at least one or two columns of figures are copied from previous tables. This not only wastes time and energy but also invites error.

(3) Only the final results are shown in the tables, the rough computations and figuring, like addition and subtraction, are

made on a separate paper, which is not filed, and are likely to be performed in an unsystematic way.

(4) Time is wasted in copying the results of rough figuring, thus increasing the liability of mistakes.

(5) There is no method of checking the computations except by repeating the figuring, which not only takes time but also requires a different computer.

(6) The figures arranged in a table do not convey as much idea as when marked on a truss diagram, as a semi-graphical method is always superior to analytic tabular forms.

To get rid of the above objections, which are very common, a simple and self evident scheme would be to dispense entirely with the idea of tables, and to record all the results on the truss diagram direct. A second thought makes it plain that to save time it is also advisable to include all the rough computation on the diagram. By this arrangement it is possible so to record the figures that none is written more than once and the computations may be most readily performed. Further, by arranging the computations so that each figure has a definite space on the diagram, the chance of using incorrect figures is entirely eliminated.

As a first application of the scientific arrangement of computations let the changes of angles in a triangle as given by equation (2) be obtained on a truss diagram as shown in Fig. 21. First write down the values of P on the sides of the triangle and then insert these values within the triangle, each value being entered twice. Next form the differences, $P_{12} - P_{23} = a$, and $P_{12} - P_{13} = b$, and multiply each by the cotangent of the corresponding angle, which is written above the columns for stresses P. The sum of these products, c, and d, gives $E\delta C$. Next find $P_{23} - P_{13} = e$; also $e \cot C = f$. Add f to negative of d, $E\delta B$ is obtained. Similarly for $E\delta A$. If C is a right angle f and h will be zero, then $E\delta B = -d$, and $E\delta A = -c$. This is the case shown in Plate IV.

XIII. EXAMPLE DETAILS OF PROCEDURE

Let the truss analyzed in Section IX be used to illustrate this analytic method and also the scientific arrangement of computations. Refer to Plate IV.

On the upper portion of the drawing construct a small truss

diagram to record the given data, and solve for the values of K. Construct another truss diagram with a scale as large as the drawing will permit. This diagram will be used for the analytic solution of the stresses. On each of the web members construct two small tables one with seven rows and the other three rows, each having three vertical columns and placed one above the other. In the first table extend the third and fifth rows one space to the right, as shown. For each of the chord members construct a small table of three columns and six rows with the second row extended one space to the right. For each of the joints construct two tables with one adjacent to the truss. The size of the table adjacent to the truss is determined as follows:

Number of vertical columns = number of members meeting at the joint.

Number of horizontal rows = number of vertical columns. The other table for the joint has four vertical columns and as many horizontal rows as may be needed in the process.

The table drawn on the chord member and the large table on the web member will be known as F-tables. The two tables for the joint will be known as K- and M-tables, the former applying to the one adjacent to the truss. The small tables below the F-tables for the web members will be known as C-tables. Mark all the vertical columns by letters as shown in the first horizontal rows. For the F-tables these will be $x,y,$ and k and for the M-tables, $S,$ $x,y,$ and k. For K-tables the identification marks for all the members entering the joint should be noted, except those for the reference members.

Next fill the K-table with the values of K found in the small truss diagram. For example, at joint 5, the value of K_{B5} is put in the second row and second column; that of K_{C5} in the third row and second column; that of K_{D5} in the fourth row and third column; and, finally, that of K_{E5} in the fifth row and fourth column. There will then be one column vacant at the right of K_{E5}. At joint 2, the value of K_{A2} with its sign reversed, is put in the second row and second column; that of K_{B2}, sign reversed, in the third row and second column, and finally that of K_{C2}, with sign reversed, in the fourth row and third column. One column is left vacant at the right of K_{C2}. Similarly, all the K-tables may be filled with the values of K transferred from the truss diagram at the top of the drawing. Next there are formed the summations of the various values of K, as follows: Find the algebraic sum of the figures in the second column and write the result in the third column and third row. This figure is next added algebraically to the figure below, the result being put in the fourth column and fourth row. Similarly, this figure may again be added to the figure below if there is any, and, finally, the space at the lower right hand corner of the table is filled. Thus, at joint 5, the sum of $-$ 0.55 and -2.10 is -2.65, that of -2.65 and -0.75 is $-3.$ 40 and that of -3.40 and 0.94 is -2.46. Similarly, for the joints at the top chord. It will be seen that the first figure in the second column is the value of $\sum K$ for the first member met with in a clockwise direction around the joint from the reference member, if the joint is at the bottom chord, and in a counter clockwise direction if the joint is at the top chord. Similarly, the first figure in the third column is the value of $\sum K$ for the second member met with in the proper direction from the reference member, the first figure in the fourth column is that of the third member from the reference member, and so on. The proper members to which these sums apply are written down in the first row. Thus, at joint 5, the figure -0.55 is for member 52,

-2.65 is for member 54, and so on.

Above and below the tables constructed on the members marked as F and C there are recorded the values of $U = \dfrac{L}{y}$, where L is in feet and y in inches, and $S = \dfrac{I}{y}$. For web members the values of $3U$ are also recorded beside U.

If the various sums in the K-tables are divided by U, the constant term in equation (16) will be obtained for the respective members. As mentioned before, the value of $\dfrac{r}{3}$ is to be used for the web members and $\sum K$ should then be divided by $3U$ for the web members. These values of $\dfrac{\sum K}{U}$ and $\dfrac{\sum K}{3U}$ are recorded in the F-tables, in the extended portions provided at the right. Thus, for member 25 at joint 5, the value of $\dfrac{\sum K}{3U}$ from reference member 53 is $-\dfrac{0.55}{15.43} = -0.036$ and is put in the extended portion of the fifth row in the F-table for member 25. This is the increment for r_{52}. Similarly, for r_{25} the value of $\dfrac{\sum K}{3U}$ from reference member 21 is $-\dfrac{2.04}{15.43} = -0.133$ and is put in the extended portion at the right of the third row of the F-table for member 25. As another example, the value of $\dfrac{\Sigma K}{U}$ for chord member 57 at joint 5 from reference member 53 is $-\dfrac{2.56}{2.93} = -$ 0.84. By the same process all the extended portions of F-tables will be filled up with the values of $\dfrac{\Sigma K}{U}$.

Now, to solve the problem, let $f_{12} = x$ and $f_{21} = y$, and put the proper figures in the F-table for member 12. Subtracting the coefficients of x and y for f_{21} from those for f_{12}, the difference $f_{12} - f_{21}$ is obtained. Reverse the signs of the difference and put them above the original signs. By using these reversed signs the difference $f_{21} - f_{12}$ is obtained. When f_{12} is added to $f_{12} - f_{21}$, the result is r_{12}, and if f_{21} is added to $f_{21} - f_{12}$, the result is r_{21}. In other words, r_{12} is obtained by adding f_{12} to $f_{12} - f_{21}$ with the lower signs, and r_{21} is obtained by adding f_{21} to $f_{12} - f_{21}$ with the upper signs. With the value of r_{21} thus found, apply it to members at joint 2 by equation (16) , as follows: Multiply the coefficient of x, -1, for r_{21}, by $U = 4.38$ giving -4.38. Divide this by $3U$ of member $23 = 13.55$ and put the quotient, -0.324, in the colum for x and row for $\frac{1}{3}r_{23}$ in the F-table for member 23. Divide

-4.38 by $3U$ of member $25 = 15.43$ and put the quotient, -0.284, in the proper space in the F-table for member 25. Similarly divide -4.38 by U of member $24 = 3.03$ and put the quotient, -1.46, in the proper space for member 24. Since the rows for r have an extended portion at the right, there is no chance of recording the figures in the wrong places. By a similar process fill the columns for y in the F-tables for the members around joint 2. For values of constant k the same process is to be pursued except that the figures are put directly above the extended portions of the respective horizontal rows. Thus, k for

$25 = k$ for 21 multiplied by $\dfrac{4.38}{15.43} = 0$, since k for member 12 is

zero. This figure, 0, is then put above the extended portion of the third row. Now add the figures in the extended portion and the one above and write the algebraic sum in the same row in column k. Thus, $0 - 1.333 = -1.333$ for member 25. Similarly, the constant term k is obtained for all the members around the joint and the deformation contour for joint 2 is then completely determined by analytical expressions. By the same method, complete the contour for joint 1.

To find f_{13} from f_{12}, apply the equilibrium equation to joint 1. Multiply by S_{12} the coefficients of x, y, and k of f_{12} and enter the products (equal to the moment due to f_{12}) in the M-table at joint 1. This part is believed to be self evident. Reverse the sign of the moment and enter the result in the row for member 13, whose S is equal to 155. Dividing the moment of 13 by S_{13}, f_{13} is obtained. Enter the result in the third row of the F-table for member 13. There are already in this table figures in the second row representing r_{13}. Subtract f_{13} from r_{13} giving $2f_{13} - f_{31} - f_{13} = f_{13} - f_{31}$. Thus, for coefficient of $x, 2.995 - (-2.405) = 5.400$. Next, reverse the signs of $f_{13} - f_{31}$ and put the new signs below the old. Thus, for coefficient of x, ± 5.400. Next, subtract $f_{13} - f_{31}$ from f_{13} or, using the lower signs, add $f_{13} - f_{31}$ to f_{13}. Thus, the coefficient of x is $-2.405 - 5.400 = -7.805$. This gives the value of f_{31}. Again subtract $f_{13} - f_{31}$ from f_{31} giving $2f_{31} - f_{13} = r_{31}$, or, using the lower signs of $f_{13} - f_{31}$, add to it f_{31}. Thus the coefficient of x for r_{31} is $-5.400 - 7.805 = -13.205$. From the value of r_{31} thus obtained, a deformation contour around joint 3 may be obtained by analytic expressions as explained for joint 2, obtaining thereby $\frac{1}{3}$ r_{32} and r_{35}. Since for member 23 there are

already computed the values of $\frac{1}{3}$ r_{23} and $\frac{1}{3}$ r_{32} in the F-table,

the difference of these two quantities may next be obtained.
This, as may be shown by subtracting equation (13), is equal to
$f_{23} - f_{32}$. These figures are to be placed in the sixth row. Next

find the sum of $\frac{1}{3}$ r_{23} and $\frac{1}{3}$ r_{32} and put the result in the fourth

row. This quantity when added to that in the third row gives f_{23} in
the second row, and when added to that in the fifth row gives f_{32}
in the seventh row. All the stresses being completely known in
the first triangle, it is necessary to apply the test of equation
(17). Noting the direction of the contour by which the
differences of f are taken, it will be seen that the figures in the
sixth row of the F-table for 23, plus the figures in the fourth row
of the F-table for 13, using the lower signs, plus the figures in the
fourth row of the F-table for 12, using the lower signs, each
multiplied by the corresponding value of U, should be zero.
These figures are recorded in the C-table under the F-table for
member 23.

Now, around joint 3 the stresses are known in all members except that in member 35, which can then be found from the equilibrium equation. This summation is performed in the M-table for joint 3. From f_{35} thus obtained and r_{35} already found, the F-table for member 35 may be filled as for member 13. From r_{53} obtained, complete the deformation contour around joint 5.

In the F-table for member 25, since $\frac{1}{3} r_{25}$ and $\frac{1}{3} r_{52}$ are known, it can then be filled as for member 23. As before, the figures in the sixth row of the F-table for 25, and those in the fourth row of the F-table for 35, using the lower signs, each multiplied by the proper U are next put in the first two rows of the C-table. The last row in the table is copied from the first row of the C-table for member 23, with the signs reversed. The sum of the figures in each column of the table must be zero.

By the same process the stresses in all members are obtained in terms of x and y in linear equations. The two simultaneous equations derived from the fact that f_{67} and f_{76} should be zero furnish the means to solve x and y. From these values of x and y the stress f in any member may be found in the following manner: On the side of the member where the fiber stress is obtained; that is, on the side first met with in passing around the end in a clockwise direction from the outside, the values of the terms containing x and y are written one above the other after substitution. The constant k is next written on the side of these two figures toward the end of the member. Find the algebraic sum of the terms of x and y and write the result below k. Add this to k, algebraically, the result is the value of f, which is to be written near the end of the member.

XIV. CONSTRUCTION OF INFLUENCE LINES

Analogous to the relations between deformation contours there is also a linear law existing between secondary stresses f. Consider a truss composed of triangular elements. In each triangle there are then six values of f, one at each end of the member. But the triangles are connected together and for every pair of adjacent triangles there is one side in common. One of the triangles may be considered as added to the other by the introduction of two sides, and therefore has only four values of f. If the structure is built up of triangles by successive addition to a "base triangle" all of the added triangles will then have only four unknowns while the base triangle will have six. To find these unknowns there is required an equal number of equations which must be found from the nature of the problem. Let the number of triangles in the structure be n. If one of the triangles be considered as the base triangle and the other $(n-1)$ triangles be added to this triangle by successive introduction of two sides at a time, there will be $6+4(n-1)=4n+2$ unknowns. For each triangle there are three angles giving three deformation equations, (3). This furnishes $3n$ equations and only $n+2$ more are required. But this is equal to the number of joints in the n triangles and for each joint there is also an equilibrium equation, (1). Since both the deformation and equilibrium equations are linear in form it follows that all the secondary stresses are connected by equations linear in form.

Since the base triangle has six values of f and there are only four equations available, three from the deformation equations and one from the equilibrium equation for the joint that is free from other triangles, two of the values of f could not be found until two more equations are introduced. This base triangle therefore has two unknowns at the outset of the solution. As the stresses in the other triangles depend on those in the base triangle, all of them will be expressed in terms of only two unknown quantities. Hence if f be the stress in any member and f_x and f_y be the two unknowns in the base triangle then,

$$f = af_x + bf_y + c, \qquad (18)$$

where a, b, and c are constants. To determine f_x and f_y it is only necessary to notice that the very last triangle added to the structure has three joints but only one is used for the equilibrium equation. There are therefore two more equations expressing the moments at these two joints, which furnish the means to solve f_x and f_y.

An examination of the methods by which the above equation is derived (see Section XI) will show that the constants a and b are functions of U only while c is a function of both U and K. Hence a and b depend only on the truss dimensions and sections of members while c depends in addition on the loading, as K is a function of the change of angles which depends on the loading. This fact shows that in constructing influence lines the only change in the analytic method that is necessary is to extend the various tables on the truss diagram so that there will be room for each change of loading which means a change of k.

XV. COMPARISON OF RESULTS

The secondary stresses worked out in the examples of this paper are compared with those obtained by the ordinary methods in a table shown on next page. The figures in the sixth and seventh columns of this table are taken respectively from Plate Ⅲ and Ⅳ. The figures in the fourth column are obtained from the article in Engineering News, vol. 66, p. 398 and those in the fifth column are obtained from the article in Engineering News, vol. 68, p. 439. These two articles have been mentioned before as to contain the analyses of secondary stresses by the method of Mohr and the method of Manderla. The truss and loading for the four examples are precisely the same.

In order to compare the results some computations were made so that the stresses will be given for the same fibers. In the graphic method of deformation contour the stress is found for the fiber where the moment and the stress have the same sign. In the analytic method the stress is found for the fiber which is first met with in passing around the end of the member in a clockwise direction. These two conventions are identical in result but are different from those adopted in the above mentioned articles. For members having symmetrical sections this difference of conventions makes a difference in sign but for members having unsymmetrical sections, it also makes a difference in magnitude. To facilitate comparison only one convention should be used in the table—the one adopted agrees with this paper.

The top chord members of the truss analyzed in the above examples have an unsymmetrical section and the values of y are different for the top and bottom fibers, being 8. 8″ for the top and 11. 6″ for the bottom. Since the examples worked out in this paper used a value of y that is for the top fiber of the chord (equal to 8. 8″), the fiber stress is not correct if it applies to the bottom fiber. The ratio of (y for bottom fiber) to (y for top fiber)—equal to $\dfrac{11.6}{8.8} = 1.32$—should therefore be applied, in order to secure the correct results.

TABLE OF COMPARISON

Member	End	Fibre	Secondary Stresses in 1000 lbs. /in.² by			
			Mohr's Method	Manderla's Method	Graphic Method	Analytic Method
1 – 2	1	T	– 0. 12	– 0. 12	– 0. 12	– 0. 12
	2	B	– 0. 01	– 0. 04	– 0. 03	– 0. 03
2 – 4	2	T	0. 11	0. 11	0. 09	0. 11
	4	B	0. 43	0. 45	0. 44	0. 45
4 – 6	4	T	– 0. 33	– 0. 33	– 0. 33	– 0. 33
	6	B	0. 13	0. 13	0. 13	0. 12
1 – 3	1	T	0. 29	0. 29	0. 29	0. 29
	3	B	0. 36	0. 35	0. 34	0. 35
3 – 5	3	T	– 0. 29	– 0. 28	– 0. 29	– 0. 28
	5	B	0. 08	0. 10	0. 10	0. 09
5 – 7	5	T	0. 02	0. 03	0. 03	0. 03
	7	B	0. 41	0. 42	0. 42	0. 39
2 – 5	2	T	– 0. 16	– 0. 17	– 0. 18	– 0. 16
	5	B	0. 01	0. 01	0	0
5 – 6	5	T	– 0. 17	– 0. 16	– 0. 16	– 0. 17
	6	B	0. 06	0. 06	0. 06	0. 06
2 – 3	2	R	– 0. 41	– 0. 44	– 0. 44	– 0. 44
	3	L	– 0. 46	– 0. 50	– 0. 50	– 0. 51
4 – 5	4	R	– 0. 36	– 0. 34	– 0. 34	– 0. 33
	5	L	– 0. 36	– 0. 31	– 0. 31	– 0. 31
6 – 7	6	R	0	0	0	0
	7	L	0	0	0	0

XVI. SECONDARY STRESSES IN SPECIAL FORMS OF BRIDGE TRUSSES

Bridge trusses of unusual types require special treatment in the analysis of secondary stresses. Each is a problem of itself and can only be solved by judicious changes in the procedure. A thorough understanding of the principles involved is very necessary as no hard and fast rules can be laid for all cases. Suffice it to say that as long as the truss is composed of triangular elements and there are at least two joints where only two members meet, the solution can always be effected by the methods presented in this paper (see Section XIV), although in certain cases the graphical method may not be as expedient as for simple trusses.

To illustrate the wide application of the methods given in this paper let a Baltimore truss as shown in Fig. 22 be analyzed for secondary stresses by the analytic method with scientific arrangement of computations.

As for simple trusses the first step of procedure is to find the change of angles due to the action of primary stresses. For a truss with subdivided panels, as that shown in Fig. 22, difficulties are encountered in applying equation (2) to a figure like 4698, as the members 46 and 69 are no longer on a straight line when the bridge is under load. In such a case assume a member inserted within the four – sided figure so that it will divide the space into two triangles, as member 68 shown by dotted lines in the figure. The deformation of this imaginary member can be calculated by finding the relative deflection of the two joints which it connects, assuming the sectional area of the inserted member to be zero. From the deformation thus obtained the changes in angles of figure 4698 may be determined in the usual way by using equation (2). As this equation contains unit stresses in the three sides of the triangle it will be necessary to find the unit stress in the imaginary member due to its deformation. For a panel as shown in Fig. 23a or Fig. 23b, the unit stress (or strain) in the imaginary member is

$$P_8 = 2(P_1 + P_2)\frac{d^2}{c^2} + 2(P_3 + P_4)\frac{h^2}{c^2} - (P_5 + P_6 + P_7).$$

By the use of this equation the changes in angles in panel 59 of Fig. 22 may be found by substituting the values of P in equation (2). After these values are known the K-tables in the analytic method may be filled with proper values as described in Section XIII.

In the solution of the problem there is no difficulty in the process until member 45 is reached. The values of f_{45}, f_{54}, r_{45}, and r_{54} are all expressed in terms of x and y, the secondary stresses in member 12 at ends 1 and 2 respectively. To find the stresses in member 56 it is necessary to know r_{56} and r_{65}. The value of r_{56} is obtained by passing a deformation contour, analytically, around joint 5 with the known value of r_{53}, but to find r_{65} from r_{56} it is necessary to know f_{56}. Now, around joint 5 the stresses f are known only for three members—f_{53}, f_{52} and f_{54}— while there are five members entering the joint. The value of f_{56} cannot, therefore, be found from an equilibrium equation. To overcome this difficulty assume the value of r_{65} be known. If z be used to denote this value, then $z = ax + by + c$, where a, b, and c are constants. From the assumed value of r_{65} and r_{56} the values of f_{56} and f_{65} are obtained. An equilibrium equation around joint 5 then determines the value of f_{57} in terms of x, y and z. The value of r_{57} being known from r_{53}, the value of r_{75} may be determined from f_{57}. With this modulus, r_{75}, a deformation contour may be passed around joint 7, giving r_{76}, and r_{79}—all expressed in terms of x, y and z. A deformation contour passed around joint 6 with r_{65} equal to z furnishes the values of r_{64}, r_{67} and r_{69}. From r_{67} and r_{76} the values of f_{67} and f_{76} are obtained. An equilibrium equation around joint 7 then determines f_{79}. This value of f_{79} together with that of r_{79} gives r_{97}. From r_{97}, r_{96} is obtained, and from r_{96} and r_{69}, f_{69} is obtained. Around joint 4, r_{45} is known from r_{42} and when combined with r_{64} from the assumed r_{65} the stresses f_{46} and f_{64} are determined. Up to this point all the values in panel 59 are expressed in terms of x, y and z and all stresses around joint 6 being now known it is possible to find the value of z by an equilibrium equation. That is,

$$S_{65}f_{65} + S_{67}f_{67} + S_{69}f_{69} + S_{64}f_{64} = 0.$$

This will give the value of z in terms of x and y. By substituting this value of z, all the values of f in panel 59 are then expressed in terms of x and y. This may be done by writing the values of z (in terms of x and y) on a strip of paper and slide this strip under the values of f for addition or subtraction.

The stresses in member 48 and 89 are obtained in the usual way. The imaginary member 68 is not considered any more after the values of K are known.

To arrange the computations for this truss it is necessary to make some slight changes in the form as shown in Fig. 24. Being familiarized with the method of constructing Plate Ⅳ a detailed explanation for this change is not necessary.

Ⅻ. FEATURES OF THE NEW METHODS

1. Features common to Both Methods

(1) No simultaneous equations involving more than two unknowns are required.

(2) Both methods handle a quantity which is of the same dimensional degree as the secondary stress itself. This gives more accurate results than when using a quantity involving large multiples of f.

(3) Both methods take much less time than the ordinary methods.

(4) There are a number of checks available for each method, especially for the graphic method.

(5) Both methods can be easily applied to influence lines of secondary stresses.

(6) In both methods the complete information is contained only one sheet of paper, including the rough computations, figuring, data, and formula.

2. Features of the Graphic Method of Deformation Contour

(1) It is strictly graphical. From beginning to end no computations are required.

(2) It gives a complete representation of secondary stresses in every member of the truss.

(3) The method not only shows secondary stresses at the two ends of a member, but also the variation of the stress along its whole length.

(4) It gives the point of inflexion in a member and also its form of bending.

(5) With the aid of the force diagram for secondary stress lines this method shows how the secondary stresses are affected by changing the section moduli of the members.

(6) It is more accurate than the ordinary analytic methods if the latter be computed by the use of a slide rule. This is due to the fact that in analytic expressions the secondary stress is often found from the difference of two large numbers which must be

very accurate, since the difference is small.

(7) The errors made in locating the base lines and vertex lines are not cumulative, as the effect of one line upon the other is very small.

(8) It has a sufficient number of checks at every step of procedure.

3. Features of the Analytic Method with Scientific Arrangement of Computations

(1) In bridge offices where a large number of standardized trusses are to be analyzed for secondary stresses, the truss diagram together with the small tables could be blue printed.

(2) This method is semi-graphical in that it shows the stresses at the proper places and the effects of one upon the other.

(3) The entire process is mechanical. Every figure has a definite space; every procedure has a definite order.

(4) No figure is to be recorded twice and no necessary figures are omitted.

(5) All rough computations and figuring are shown on the diagram thus rendering possible a check at any future time.

(6)The figures on the diagram are so arranged that all the additions and subtractions do not involve more than two lines of figures and can be accomplished without the use of extra paper, as when three columns of figures are to be added, etc. All the other computations can be made on a slide rule.

(7)The method is capable of affording any degree of accuracy which may be desired.

(8)Practically all back references are eliminated.

(9)It has a unique check of secondary stresses as soon as they are found.

SECONDARY STRESS: A FACTOR IN DESIGN

XVIII. INTRODUCTION

While the importance of secondary stresses is still a moot question, it cannot be denied that great improvements have been made in the design of structures as a result of the studies and observations made on the subject during recent years. The achievement so far attained is very beneficial and convincing although it is not uncommon to hear it said that secondary stresses accompany good design, as it makes the structure rigid and stiff which is highly desirable. To be sure, a good design does aim at a rigid structure but certainly not at the expense of its strength. Structures may be of different types: Some are rigid, others are strong, and still others are both rigid and strong. A bridge may be designed and detailed in different ways: Some make the structure rigid, others make it strong, and still others make it both rigid and strong. Would it be logical to consider only rigidity in a structure—in the selection of types, proportioning of members, and designing of details—and neglect its effects on the most vital factor in the design—the strength?

The consideration of secondary stresses in a design tends to make the structure both rigid and strong, not only in type and design but also in details. Eventually the design is also economical as the uncertain "factors of safety" may be greatly reduced. It is for this reason that the secondary stress is so important in design and has been so widely considered. It must be admitted, however, that the subject is a difficult one—not in the theoretical analysis of course, but in its practical application. Oftentimes a design made to satisfy secondary stresses violates other good principles which must be respected. In other cases, the gain in reduced secondary stresses does not balance the loss of rigidity. Here good judgement must be exercised to determine which course to pursue.

Before going into the details of design some general conclusions about the secondary stresses will be given.

(1) The secondary stresses are, in general, proportional to the primary stresses, and, therefore, are conveniently expressed in percentages of primary stresses.

(2) Other things being equal, the percentages of secondary stresses are proportional to the distance from the gravity axis to the outer fiber in the plane of bending, and inversely proportional to the length of the member.

(3) The secondary stresses in any member depend on the distortion of all the members of the truss, but primarily upon the distortion of members of the triangles of which this member is a part.

(4) A design in any individual member cannot be changed without affecting the secondary stresses in the other members.

In the following sections are given some principles of design which must be observed in reducing secondary stresses. These are largely derived from theoretical studies and practical observations and are very valuable on that account. During the past few years as increasing attention was paid to secondary stresses, considerable space in books and periodicals has been devoted to the design of structures from this point of view. This information, however, is not within easy reach as it is scattered in a vast amount of literature and the subject has never been systematically treated. As one of the purposes of the present paper is to present the facts about secondary stresses which must be understood by the designer, these principles of design will be listed under proper headings and discussed and digested in a logical order. For each principle stated the source of information is accredited by using a letter(explained in the following key) to indicate the title of the work. This is followed by a number referring to the page.

Key

a 4 = Proceedings of the American Railway Engineering Ass'n, 1914, p. 438.

a 6 = Proceedings of the American Railway Engineering Ass'n, 1916, p. 129.

e = Engineering News Record.

g = Grimm's Secondary Stresses in Bridge Trusses.

j = Merriman & Jacoby's Roofs and Bridges, Part Ⅲ.

k = Kunz's Design of Steel Bridges.

M = Thesis by T. Mao Presented to Cornell University.

m = Molitor's Kinetic Theory of Engineering Structures.

p = Secondary Stresses in Framed Structures by Pitman in Proceedings of Engineers' Society of Western Penn. , Vol. 25.

r = Notes on Design by Reichman in Journal of Western Society of Engineers, Vol. 17.

s = Stress Measurement on the Hell Gate Arch Bridge by Steinman in Transactions of the American Society of Civil Engineers, Vol. 82.

T II = Modern Framed Structures, Part II.

T III = Modern Framed Structures, Part III.

t = Thayer's Structural Design.

W = Waddell's Bridge Engineering.

w = Effects of Secondary Stresses on Design by Wilson in the Journal of Western Society of Engineers, Vol. 21.

XIX. SELECTION OF TRUSSES

(1) In Choosing between different styles of trusses, those of

the statically determinate class should always receive preference, other things being equal. The primary stresses will usually be less than in similar indeterminate systems, especially when the temperature stresses are included. Yet the deformation and secondary stresses may be less, and frequently the connections may be simpler for the indeterminate type. (m 268)

(2) Among the statically indeterminate structures the two-hinged arch is especially immune from secondary stresses. (s 1840 , M 497)

(3) The secondary stresses in continuous trusses are very large at the centre and in its neighborhood and for this reason pins at these points appear to be advisable in order to reduce the stresses. (g 129)

(4) The best modern practice in bridge engineering does not countenance the building of trusses having more than a single system of cancellation. (W 271) The secondary stresses in such a system are generally very high, often reaching 100 percent of the primary (T II 490) , as the distortion in truss members due to loads on one system only is very great. (t 235 , g 129)

(5) In double triangular trusses the secondary stresses can be greatly reduced by the insertion of verticals connecting every pair of upper and lower joints. These verticals effectively connect the two single systems. (T Ⅱ 491 , g 123)

(6) Trusses consisting of approximately equilateral triangles , and without hangers or vertical struts , present the most uniform condition and will have , in general , the lowest secondary stresses. (TⅢ 93 , a 4)

(7) A truss composed of right-angled triangles will show somewhat higher secondary stresses , and such stresses will be large if the ratio of height to panel length is large. (TⅢ 94 , a 4)

(8) The truss systems should be as simple as possible and all members which make the stress distribution uncertain should be avoided. (k 169)

(9) As far as the elimination of secondary stress is concerned , the Pratt truss , the Warren truss with verticals , and the K – truss are very desirable. (W 200)

(10) The amount of secondary stress in ordinary Pratt trusses varies from 15 to 35 percent of the primary stress. (a 6)

(11) The secondary stresses in the Warren system are often less than those in the Pratt system. (T Ⅱ 487)

(12) In single Warren trusses without verticals the secondary stresses in chords increase from the center of the span toward the end, while for the web members these stresses increase in the opposite direction. This circumstance is fortunate, as the web members in the region of the center of the span often have an excess in section and the end members of both the top and the bottom chords show this surplus still oftener. (g 105)

(13) In a riveted truss, the span should be made as deep as good designing will permit, for two reasons: (1) The connections of the diagonals to the chords may be made smaller, (2) the deformation of the truss under load may be minimized. Both of these effects tend to decrease the secondary stresses. (r 95, T II 492)

(14) In trusses with subdivided panels, the secondary members cause great secondary stresses in the main members of the truss on account of their distortion. (T II 492, t 235, w 201)

(15) The best arrangement, so far as the secondary stress is concerned, is where each web member forms an integral part of the entire truss so that its stress will gradually change as the load progresses.

(16) In a Baltimore truss, the use of a sub-strut is much more advisable than the use of a sub-tie. (T Ⅱ 494)

(17) The use of collision struts cannot be recommended, because they divide the end posts into two fragments, decreasing the ratio between length and width of these members, pulling the posts out of line and increasing the secondary stresses. The collision struts are, as a rule, rather weak members and it is a question whether it would not be better to use the material on the end posts instead of on these struts, thereby increasing their strength in the direction of the plane of the truss as well as at right angles to their plane. (g 130, T Ⅱ 458)

(18) The double triangular truss with sub-panels is satisfactory so far as the action of the secondary stress is concerned. (T Ⅱ 495)

(19) A truss with polygonal chord as the parabolic truss, or the Schwedler truss, is a good selection. (g 129)

(20) In a truss, aim to make the curve of deflection an approximate circle with vertical members radial. This means uniform stress in chords and low stresses elsewhere. (t 235)

(21) Curved members, in which the neutral axis is not

straight before the application of the load, should never be tolerated under any circumstance. (W 272 ,g 129 ,k 170)

XX. ARRANGEMENT OF SPAN

(22) Skew bridges and bridges on curves should never be built except in very rare cases where no other disposition is possible. These types should be regarded as measures of last resort. (m 267 ,W 271)

(23) Wherever possible the loads should be applied at the panel points only. (k 170)

(24) The plane of the lateral system should coincide with that of the chords; otherwise, the stresses in webs and chords will produce bending moments in the posts. (r 102 ,m 268 ,k 185)

(25) The plane of the floor should be as close as possible to that of the lateral system. (m 268)

(26) The gravity axes of all the main members of trusses and lateral systems coming together at any apex of a truss should intersect at a common point wherever such an arrangement is practicable. In other words, the panel points of the bracing

should coincide with the panel points of the main trusses. (W 273 , k 185 , g 130)

(27) If horizontal trusses are used in the floor system to resist traction stresses , these should be placed near the center of a span or midway between expansion joints. (T Ⅲ 105 , a 6)

(28) The double intersection Warren truss with verticals is very well adapted to lateral bracing. (T Ⅱ 492)

(29) Brackets on posts should be avoided wherever possible. (g 130)

(30) The pedestal pins in riveted trusses should not be placed below the bottom chords but on the center-lines. This eliminates the bending moments that may be developed in the end posts and the end panels of the bottom chords by the train thrust. (W 284)

XXI. DESIGN OF FLOOR SYSTEM

(31) The floor system should be so designed that it is rigid in itself with respect to its main duty , but independent , so far as practicable , of the action of the chord system of the main trusses.

(T Ⅱ 505)

(32) Floor-beams should be made deep to reduce secondary stresses in the cross frames. (m 268 , r 101)

(33) The flanges of the floor – beams should be made relatively narrow. (T Ⅱ 505)

(34) The floor-beams should be centrally connected to the post and not too far below the plane of chords. (r 101)

(35) Flexible connections are advisable between floor-beams and truss members. These should be so arranged that the end connections are rigid in a vertical plane but flexible in a horizontal plane. (m 268 , g 130)

(36) Floor-beams should be provided with brackets or any other suitable construction to prevent the bending of the horizontal flanges. (g 132)

(37) The stringers should be made heavy and continuous and should be designed to transmit the tractive forces to the panel points of the loaded chord instead of to the floor-beams by inserting proper tie members between the stringers. (W 400 , m 268)

(38) When the floor-beams extend considerably below the

bottom of the stringers they should have stiffeners at or near the stringer connection to transmit to the flanges of the floor-beam the bending caused by the deflection of the stringers.

(39) In long-span bridges, stringers should be provided with expansion joints at intervals of a few panels. (T Ⅱ 505 , r 101)

(40) The connection of lower laterals to stringers is of doubtful value. Owing to the relative movement between the chords and the floor system, such a connection would cause considerable lateral bending in the stringers. The laterals are therefore better merely supported on the stringers. (T Ⅲ 104 , a 6)

XⅦ. PROPORTIONING OF MEMBERS

(41) The more uniform the proportions of a truss, the less, in general, will be the secondary stresses. Sudden changes in length, in width, or in moment of inertia, care likely to result in relatively large secondary stresses. (T Ⅲ 93)

(42) A reduction of moment of inertia of any member tends to increase the deflection of that member and reduce that of

others. To maintain a balance in the fiber stresses themselves, it is necessary to make the width of a member correspond in a measure with its moment of inertia wide members of small moment of inertia are likely to have high secondary stresses; narrow and compact members will have low secondary stresses. This statement does not take account of the long column action in compression members, the effects of which tend in the opposite direction. (T Ⅱ 487)

(43) It is advisable to keep the moments of inertia of web members as small as possible. (t 235)

(44) The axes of all members of a truss should be in the same plane and should intersect at a common point at all connections. (m 267, W 273)

(45) In any truss, no torsion on any member should be allowed if it can possibly be avoided; otherwise, the greatest care must be taken to provide ample strength and rigidity for every portion of the structure affected by such torsion. (W 272)

(46) If the axes of members do not intersect at one point, the secondary stresses developed due to eccentric connections may be as high as 1 to $1\frac{1}{2}$ times the primary stresses. (p)

(47) Excessively deep members should be avoided (W 495) , especially in tension chords and in chords of trusses with sub-divided panels (a 6). However, the members should not be so slender as to impair economy of design (k 177) or to reduce the effectiveness of the riveted connections. (r 95)

(48) To secure a comparatively narrow width, and at the same time preserve the stiffness against buckling, a compression member may be tapered from the center toward both ends. (m 267)

(49) To avoid excessive bending stresses in posts to which floor-beams are riveted, the posts should be made only of moderate width in a transverse direction. In the best design these bending stresses are likely to be as much as 25 percent of the primary stresses. (T Ⅲ 105 , g 129)

(50) If the floor-beams connected to the posts are shallow, the latter should be given increased sectional areas to provide for the secondary stresses due to the deflection of the floor-beams. (j 401)

(51) Truss members and portions of truss members should always be arranged in pairs symmetrical about the plane of the truss (W 273). This will tend to equalize the stresses carried by the component parts of the member.

(52) The cross-sections of members should be so chosen that the material is concentrated as far from the neutral axis as possible, thus securing the largest moments of inertia for the smallest over all dimensions (g 129). The cross form is thus the least advantageous while a square box form is the most desirable. However, consideration must be given to the fact that when the secondary stresses occur simultaneously in the plane of the truss and in a cross frame, the stresses are additive in members of the box type but not in those with cross forms. (m 267)

(53) Since the effect of the hip verticals and suspenders in a truss is to pull the chord members out of line, they should be provided with liberal cross-sections. (TII 487 , a 6)

(54) Where the suspenders are of considerable length and attached to the lower chord, it is desirable to make them slightly shorter than the calculated length. (a 6)

(55) Where a top chord in a subdivided truss is supported by means of secondary members, these should be made slightly longer than the calculated length. (a 6)

(56) The secondary stresses in chord members may be reduced by the proper selection of cross-sections for the diagonals and their attachments to the chords. (r 101)

(57) In case a member is made up of a single part, like an angle, the gage lines for rivets should be as close to the gravity axis as possible. (p)

(58) For a truss with built-up continuous chords and built-up diagonals, pin-connected throughout, the secondary stresses will be very low if the chords are made sufficiently shallow and the diagonals sufficiently wide in the plane of the truss, so as to overcome the pin friction. (r 96)

(59) In eye-bar trusses, the eye-bars should be made as wide as permissible in the plane of the truss. (r 97)

XIII. DESIGNING OF DETAILS

(60) Riveted connections must be made concentric by so grouping the rivets that they will balance about center-lines and center planes to as great an extent as possible. (W 274 , g 129)

(61) The end lateral connections should be as concentric as practicable. This is very important in the case of the end panel of the lower chord, as eccentric connections at this point result in heavy secondary stresses in end posts and lower chords. (T Ⅲ 105)

(62) The use of thick gusset plates and large diameter rivets is advisable as this practice will materially reduce the number of rivets and the size of the plates. (m 267 , W 491)

(63) In case two gusset plates are used at each joint these should be firmly connected together by the use of diaphragm plates.

(64) In a pin-connected truss, the diameters of the pins should not be so large that the friction developed around the joints will virtually make the connections rigid. They should therefore be made as small as is consistent with the design. (m 268 , g 132)

(65) If the diameter of the pin be made three quarters the width of the eye-bar, as is usually done, the secondary stresses developed will be about 45 percent of the primary stresses with the coefficient of friction equal to 0. 2. (g 77)

(66) A double pin arrangement has been evolved for

bettering the pin bearings in cantilever spans, whereby a second pin is placed side by side with the bearing in the bottom chord. The object is to relieve the bending stresses in the bottom chord, which would result from the simultaneous deflections of the two adjacent spans. (W 1067)

(67) For lacing of compression members, the arrangement of double lacing with transverse cross-bars is advisable. (T Ⅱ 498)

(68) Long diagonals which are subject to the bending due to deflection of floor-beams should preferably be provided with lacings. (j 401)

(69) For heavy members composed of two parts, one on each side of the centre-line, the use of cross diaphragms spaced at 6 to 10 feet is necessary. (W 506) This not only holds the members true to shape and line but also tends to equalize the stresses in the two component parts.

(70) The use of lug angles for connections is advisable for members having single angle sections.

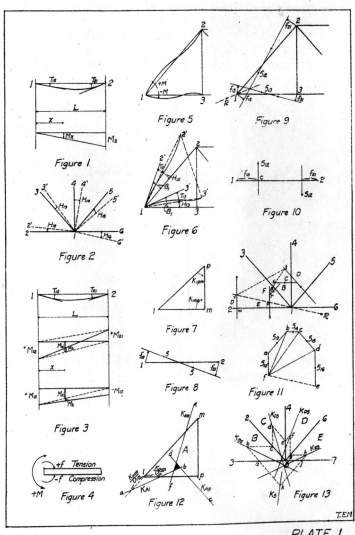

Figure 5

Figure 9

Figure 1.

Figure 2

Figure 6

Figure 10

Figure 3

Figure 7

Figure 11

Figure 8

+f Tension
-f Compression

+M   Figure 4

Figure 12

Figure 13

PLATE 1

T.E.M.

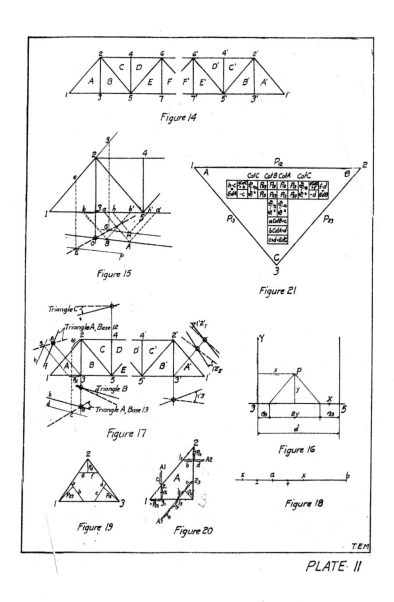

Figure 14

Figure 15

Figure 21

Figure 17

Figure 16

Figure 19

Figure 20

Figure 18

T.E.M.

PLATE II

760

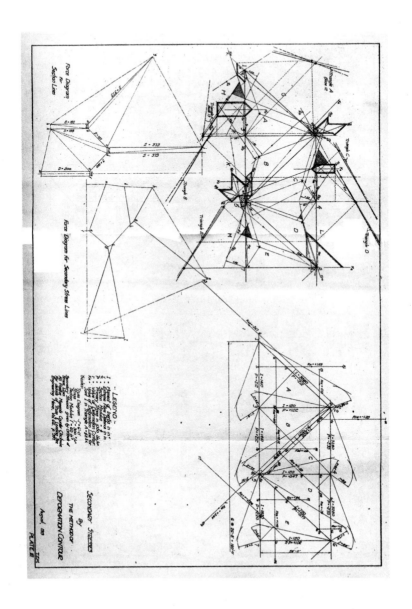

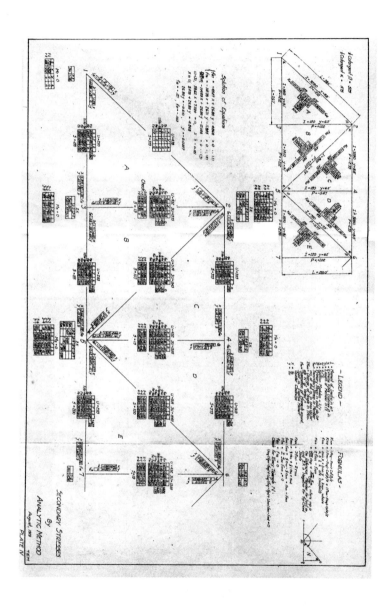

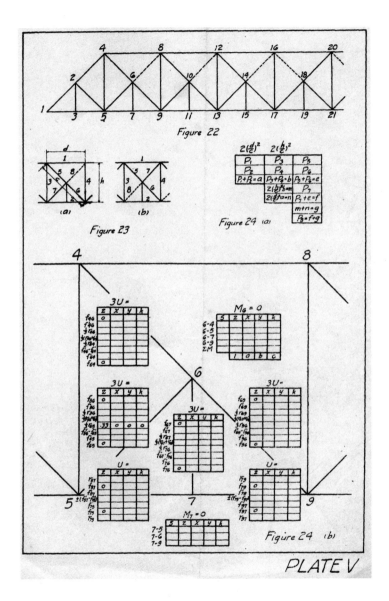

Figure 22

Figure 23

Figure 24 (a)

Figure 24 (b)

PLATE V